Factorization

Unique and Otherwise

CMS Treatises in Mathematics

Published by the Canadian Mathematical Society

Traités de mathématiques de la SMC

Publié par la Société mathématique du Canada

Factorization
Unique and Otherwise

Steven H. Weintraub
Lehigh University

Canadian Mathematical Society
Société mathématique du Canada
Ottawa, Ontario

A K Peters, Ltd.
Wellesley, Massachusetts

Sales and Customer Service

A K Peters, Ltd.
888 Worcester Street, Suite 230
Wellesley, MA 02482
www.akpeters.com

CMS Executive Office
Bureau administratif de la SMC

Canadian Mathematical Society
Société mathématique du Canada
577 King Edward
Ottawa, Ontario
Canada K1N 6N5
www.cms.math.ca/Publications

Library of Congress Cataloging-in-Publication Data

Weintraub, Steven H.
Factorization : unique and otherwise / Morgens Esrom Larsen.
p. cm. -- (CMS Treatises in mathematics)
Includes index.
ISBN 978-1-56881-241-0 (alk. paper)
1. Factorization (Mathematics). 2. Rings of integers. 3. Rings (Algebra). I. Title.

QA161.F3W45 2008
512.7′2--dc22

2007049328

Printed in Canada
12 11 10 09 08 10 9 8 7 6 5 4 3 2 1

To my nephews, nieces, and grandkids:

Wendy, Jenny, and William;
Erica, Jordan, and Allison;
Blake, Natalie, and Ethan

Contents

Preface ix

Introduction 1

1 Basic Notions 7
- 1.1 Integral Domains 7
- 1.2 Quadratic Fields 12
- 1.3 Exercises 16

2 Unique Factorization 19
- 2.1 Euclidean Domains 20
- 2.2 The GCD-L Property and Euclid's Algorithm 31
- 2.3 Ideals and Principal Ideal Domains 45
- 2.4 Unique Factorization Domains 51
- 2.5 Nonunique Factorization: The Case $D < 0$ 60
- 2.6 Nonunique Factorization: The Case $D > 0$ 67
- 2.7 Summing Up 78
- 2.8 Exercises 80

3 The Gaussian Integers 91
- 3.1 Fermat's Theorem 92
- 3.2 Factorization into Primes 101
- 3.3 Exercises 105

4 Pell's Equation 111
- 4.1 Representations and Their Composition 112
- 4.2 Solving Pell's Equation 118
- 4.3 Numerical Examples and Further Results 127
- 4.4 Units in $\mathcal{O}(\sqrt{D})$ 137
- 4.5 Exercises 139

5 Towards Algebraic Number Theory 143
5.1 Algebraic Numbers and Algebraic Integers 144
5.2 Ideal Theory 147
5.3 Dedekind Domains 150
5.4 Algebraic Number Fields and Dedekind Domains 154
5.5 Prime Ideals in $\mathcal{O}(\sqrt{D})$ 158
5.6 Examples of Ideals in $\mathcal{O}(\sqrt{D})$ 166
5.7 Behavior of Ideals in Algebraic Number Fields 178
5.8 Ideal Elements 180
5.9 Dirichlet's Unit Theorem 182
5.10 Exercises 186

A Mathematical Induction 191
A.1 Mathematical Induction and Its Equivalents 191
A.2 Consequences of Mathematical Induction 196
A.3 Exercises 199

B Congruences 205
B.1 The Notion of Congruence 205
B.2 Linear Congruences 211
B.3 Quadratic Congruences 223
B.4 Proof of the Law of Quadratic Reciprocity 236
B.5 Primitive Roots 241
B.6 Exercises 245

C Continuations from Chapter 2 251
C.1 Continuation of the Proof of Theorem 2.8 251
C.2 Continuation of Example 2.26 255
C.3 Exercises 257

Index 259

Preface

In this book, we introduce the reader to some beautiful and interesting mathematics, which is not only historically important but also still very much alive today. Indeed, it plays a central role in modern mathematics.

The mathematical content of this book is outlined in the introduction, but we shall preview it here. It is a basic property of the integers, known as the Fundamental Theorem of Arithmetic, that every integer can be factored into a product of primes in an essentially unique way. Our principal objective in this book is to investigate somewhat more general but still relatively concrete systems (known as rings of integers in quadratic fields) and see when this property does or does not hold for them. We accomplish this objective in Chapters 1 and 2. But this investigation naturally leads us into further investigations—mathematics is like that—and we consider related questions in Chapters 3 and 4, where we investigate the Gaussian integers and Pell's equation, respectively.

The questions we investigate here were at the roots of the development of algebraic number theory. In Chapter 5 we provide an overview of algebraic number theory with emphasis on how the results for quadratic fields generalize to arbitrary algebraic number fields.

We envision several ways in which this book can be used. One way is for a first course in number theory. In our investigations, we begin at the beginning, so this book is suitable for that purpose. Indeed, one of the themes of this book is that one can go a long ways with only elementary methods. To be sure, the topics covered here are not the traditional topics for a first course in number theory (though there is considerable overlap), but there is no reason that the traditional topics need be sacrosanct.

Another way to use this book is for a more advanced course in number theory, and there is plenty of appropriate material here for such a course. Indeed, there is far more than a semester's worth of material here, even for an advanced course.

In this regard, we call the reader's attention to Appendices A and B, on mathematical induction and congruences, respectively. If this book is used as a text for a first course, much of the material in these two appendices should be covered. If this book is used as a text for a more advanced course, these appendices will serve as background.

We have not tried to write a textbook on algebraic number theory in Chapter 5, but rather to provide an overview of the field. But we feel that this overview can serve as a valuable introduction to, and guide for, the student who wishes to study this field, and can also serve as a concrete reference for some of the general results that a student of this field will encounter.

Steven H. Weintraub
Bethlehem, PA, USA
August 2007

Introduction

We shall here be concerned with the circle of ideas that surrounds the *Fundamental Theorem of Arithmetic.*

First we recall the usual definition of a prime: a *prime number* is a positive integer, other than 1, that has no divisors except itself and 1. For example 2 and 3 are primes, but $6 = 2 \cdot 3$ and $10 = 2 \cdot 5$ are not.

Then the Fundamental Theorem of Arithmetic states that *every positive integer can be factored into primes in an essentially unique way.* For example,

$$\begin{aligned} 1 &= 1, \\ 2 &= 2, \\ 6 &= 2 \cdot 3, \\ 10 &= 2 \cdot 5, \\ 15 &= 3 \cdot 5, \\ 2499 &= 3 \cdot 7^2 \cdot 17. \end{aligned}$$

By "essentially unique," we mean unique up to the order of the factors, so that we consider $6 = 2 \cdot 3 = 3 \cdot 2$ to be the same factorization. (Note that 1 is a special case. We think of it as having an "empty" factorization, as it is not divisible by any prime.)

As its name implies, unique factorization is a fundamental property of the positive integers, a property that was known to the ancient Greeks. We will prove this property, and indeed our proof will follow that of Euclid. But we will be interested in examining this proof and seeing what makes it really "work," with an idea of seeing when we can extend it to more general situations.

For example, let us consider numbers of the form $a + b\sqrt{-1}$ with a and b integers. It turns out, and we shall prove, that numbers of this form also have unique factorization. For example, we have the following factorization into primes for numbers of this form:

$$\begin{aligned}
3 &= 3, \\
5 &= (2+\sqrt{-1})(2-\sqrt{-1}), \\
7 &= 7, \\
11 &= 11, \\
13 &= (3+2\sqrt{-1})(3-2\sqrt{-1}), \\
17 &= (4+\sqrt{-1})(4-\sqrt{-1}).
\end{aligned}$$

On the other hand, let us consider numbers of the form $a+b\sqrt{-5}$ with a and b integers. Numbers of this form do *not* have unique factorization. For example, we have the following two factorizations of 6 into irreducibles:

$$6 = (2)(3) = (1+\sqrt{-5})(1-\sqrt{-5}).$$

We can also consider numbers of the form $a+b\sqrt{10}$ with a and b integers. Numbers of this form also do *not* have unique factorization. For example, we have the following two factorizations of 10 into irreducibles:

$$14 = (2)(5) = (\sqrt{10})^2.$$

We have used the word "irreducible" rather than "prime" here as that turns out to be the correct mathematical language.

In fact, we will prove the Fundamental Theorem of Arithmetic in a way that enables us to establish it in many cases, including the two we have mentioned—the ordinary integers, and numbers of the form $a+b\sqrt{-1}$ with a and b integers—simultaneously.

On the other hand, we will also be able to systematically show that in many cases, including the two we have mentioned—numbers of the form $a+b\sqrt{-5}$ with a and b integers, and numbers of the form $a+b\sqrt{10}$ with a and b integers—unique factorization does not hold.

As we will see, instead of unique factorization being the norm and nonunique factorization the exception, the situation is reversed! It is really a very special property, though a crucially important one, of the ordinary integers that the Fundamental Theorem of Arithmetic holds for them.

Chapters 1 and 2 of this book are basically devoted to proving unique and nonunique factorization for ordinary integers and for numbers of the form $a+b\sqrt{D}$. (Here a and b are not always integers, but this is a technical point we will defer until later.)

In Chapter 3, we investigate numbers of the form $a+b\sqrt{-1}$ with a and b integers. Numbers of this form are called the *Gaussian integers*. As we have remarked, in the Gaussian integers we do have unique factorization into primes, but we would like to know what the primes are. Here we will show that the following is always true (compare the factorizations above): every ordinary prime that leaves a remainder of 3 when divided by 4 remains a prime in the Gaussian integers, but every prime that leaves a remainder of 1 when divided by 4 factors into a product of two "conjugate" primes in the Gaussian integers. In fact, this is closely related to a famous theorem of Fermat: *every prime that leaves a remainder of 1 when divided by 4 can be written as a sum of two squares.* (For example, $5 = 2^2 + 1^2$, $13 = 3^2 + 2^2$, $17 = 4^2 + 1^2$, but also $97 = 8^2 + 5^2$, $101 = 10^2 + 1^2$, and $99989 = 230^2 + 217^2$.)

Actually, we will give several proofs of this theorem. One, due to Euler, is believed to be along the lines of Fermat's original proof (which he never wrote down for posterity). It uses a technique known as *composition.* Another one uses unique factorization to prove Fermat's theorem quickly and easily. (It is a bit surprising that this abstract result gives such a concrete fact, but mathematics is full of beautiful surprises.)

To describe our next objective, we have to get a bit more technical. The ancient Greeks considered the positive integers, but when we wish to generalize our investigations, we no longer have the idea of positivity. (We cannot make any sense out of saying that $a + b\sqrt{-1}$ is positive.) So we have to consider all integers. But when we do, we see that we have typical factorizations:

$$6 = (2)(3) = (-1)(-2)(3) = (-2)(-3).$$

These are different factorizations, but we do not want to consider these to be essentially different. How do they differ? The answer is that 1 can be factored as $1 = (-1)(-1)$ and we are simply distributing the factors of 1 differently. We give a name to this situation. Factors of 1 are called *units*, and factorizations that differ merely because we have redistributed the units are essentially the same.

We can show that the units in the Gaussian integers are precisely 1, -1, $\sqrt{-1}$, and $-\sqrt{-1}$. Then we also have the prime factorization in the Gaussian integers:

$$2 = (-\sqrt{-1})(1 + \sqrt{-1})^2.$$

Here the first factor is a unit, so what we see is that, up to a unit factor, 2 is a square in the Gaussian integers.

In factoring numbers, we do not really care about units, but still it is an interesting question—indeed, a very interesting question—to ask what the units are. We have given the answer for the Gaussian integers, but we can ask the same question in other cases as well. Here we ask the question for numbers of the form $a + b\sqrt{D}$ for D positive and not a perfect square. If $D = 2$ we have units

$$\begin{aligned} 1 &= (3 + 2\sqrt{2})(3 - 2\sqrt{2}), \\ 1 &= (17 + 12\sqrt{2})(17 - 12\sqrt{2}), \\ 1 &= (99 + 70\sqrt{2})(99 - 70\sqrt{2}), \\ 1 &= (577 + 408\sqrt{2})(577 - 408\sqrt{2}). \end{aligned}$$

Note that a factorization $1 = (a + b\sqrt{D})(a - b\sqrt{D})$ gives a solution of the equation $a^2 - b^2D = 1$, and vice versa. Thus the search for units is intimately related to the search for solutions of the equation $a^2 - b^2D = 1$. The units above correspond to solutions for $D = 2$: $1 = 3^2 - 2^2 \cdot 2 = 17^2 - 12^2 \cdot 2 = 99^2 - 70^2 \cdot 2 = 577^2 - 408^2 \cdot 2$. But we can consider this equation for other values of D as well. For example, for $D = 61$ we have the solution

$$1 = (1766319049)^2 - (226153980)^2 \cdot 61$$

and for $D = 109$ we have the solution

$$1 = (158070671986249)^2 - (15140424455100)^2 \cdot 109.$$

In fact, for any such D there are infinitely many solutions (and hence infinitely many units). We shall prove this in Chapter 4 where we investigate the equation $a^2 - b^2D = 1$, known as *Pell's equation*. Our proof is a variant of the *cakravala* method experimentally developed by Indian mathematicians between the ninth and twelfth centuries. This is also a result known to Fermat, and his proof of this result may well have been along the lines of ours, as our proof uses a method of "composition" very closely related to our method in Chapter 3. Also, our proof is constructive, enabling us to find solutions by hand for values of D that are not too large. The above solutions for $D = 61$ and for $D = 109$ were known to have been found by Fermat (by hand, obviously, since computers did not exist in the seventeenth century).

Our investigations in Chapters 1 through 4 can be considerably generalized. To use the appropriate technical language, in these chapters we are considering *quadratic fields*, and we can consider analogous problems for *algebraic number fields*. Indeed, our treatment here parallels the historical development of the subject. Quadratic fields were investigated first, and the phenomena that arose there motivated the development of the general theory. This subject is known as *algebraic number theory.* In Chapter 5 we survey some of the highlights of this subject. As we have seen, unique factorization of elements holds in the integers $\mathbb{Z}$, but it does not always hold. While unique factorization of elements is the most straightforward generalization of the situation in the integers, it is not the right generalization. The right generalization is unique factorization of ideals, which does hold. Therefore in Chapter 5 we will focus (though not exclusively) on ideals in general. But we will also provide a wealth of examples, interesting in themselves, that show how quadratic fields fit into the general case. For a precise description of the scope of our investigations in this chapter, we simply refer the reader to the table of contents.

We have three appendices. Appendix A is a careful treatment of mathematical induction, an essential proof technique. Appendix B is a treatment of congruences. Here we begin with the definition, and proceed through linear congruences (including the Chinese Remainder Theorem) and quadratic congruences (including the Law of Quadratic Reciprocity). Appendix C is a technical one, dealing with some of the more complicated cases of results in Chapter 2.

Chapter 1

Basic Notions

In this chapter we introduce the objects we will be studying. First we introduce the general class of objects, and then we focus on the particular examples that will concern us.

1.1 Integral Domains

We begin by carefully defining the class of objects we shall be studying.

Definition 1.1. An *integral domain* is a set R equipped with two operations, addition and multiplication, that satisfy the following properties:

(1) R is closed under addition, i.e., if a and b are any two elements of R, then $a + b$ is an element of R.

(2) Addition is commutative, i.e., if a and b are any two of R, then $a + b = b + a$.

(3) Addition is associative, i.e., if a, b, and c are any elements of R, then $(a + b) + c = a + (b + c)$.

(4) There is an additive identity 0 in R, i.e., there is an element 0 of R with the property that for any element a of $0 + a = a + 0 = a$.

(5) Every element of R has an additive inverse, i.e., if a is any element of R, there is an element $-a$ of R with the $a + (-a) = (-a) + a = 0$.

(6) R is closed under multiplication, i.e., if a and b are any two elements of R, then ab is an element of R.

(7) Multiplication is commutative, i.e., if a and b are any two of R, then $ab = ba$.

(8) Multiplication is associative, i.e., if a, b, and c are three elements of R, then $(ab)c = a(bc)$.

(9) There is a multiplicative identity 1 in R, i.e., there is an element 1 of R with the property that for any element of R, $1a = a1 = a$.

(10) Multiplication distributes over addition, i.e., if a, b, and c are any three elements of R, then $a(b + c) = ab + ac$ and $(b + c)a = ba + ca$.

(11) R has no zero divisors, i.e., if a and b are any two non-zero elements of R, then their product ab is also non-zero. Note that, by taking the contrapositive, this condition may be equivalently rephrased as follows: if a and b are any two elements of R with $ab = 0$, then $a = 0$ or $b = 0$.

An important property of integral domains is the cancellation property, which holds for both addition and multiplication.

Lemma 1.2. *Let R be an integral domain and let a, b, and c be elements of R.*

(1) If $a + b = a + c$ then $b = c$.

(2) If $a \neq 0$ and $ab = ac$ then $b = c$.

Proof:

(1)

$$\begin{aligned} a + b &= a + c \\ -a + (a + b) &= -a + (a + c) \\ (-a + a) + b &= (-a + a) + c \\ 0 + b &= 0 + c \\ b &= c \end{aligned}$$

(2)

$$\begin{aligned} ab &= ac \\ ab + a(-c) &= ac + a(-c) \\ a(b - c) &= a(c - c) \\ a(b - c) &= a0 \\ a(b - c) &= 0 \end{aligned}$$

But by property (11), $a(b - c) = 0$ implies $a = 0$ or $b - c = 0$. Since we are assuming $a \neq 0$, we must have $b - c = 0$, so $b = c$. $\square$

Here are some examples of integral domains.

Example 1.3.

(1) The ordinary integers $\mathbb{Z}$ form an integral domain. (Indeed, the term "integral domain" has its origin in this fact.)

(2) The rational numbers $\mathbf{Q}$ form an integral domain. $\mathbf{Q}$ is just the set of fractions $\{a/b\}$ with a and b integers, with the usual addition and multiplication of fractions. (Note that $\mathbf{Q}$ includes $\mathbb{Z}$, as the integer a is equal to the fraction $a/1$.)

(3) Fix an integer D that is not a perfect square, and let

$$R = \{a + b\sqrt{D} \mid a \text{ and } b \text{ are integers } \}.$$

Then R is an integral domain. Let us examine R a little more carefully. First, if $a = a + b\sqrt{D}$ and $\beta = c + d\sqrt{D}$ are in R, then

$$\alpha + \beta = (a + b\sqrt{D}) + (c + d\sqrt{D}) = (a + b) + (c + d)\sqrt{D}$$

is in R, and

$$\alpha\beta = (a + b\sqrt{D})(c + d\sqrt{D}) = (ac + bdD) + (ad + bc)\sqrt{D}$$

is in R. The remaining properties of R follow directly from the corresponding properties of $\mathbb{Z}$, except for the last one, the absence of zero divisors. We leave this for the exercises.

(4) Fix an integer $D \equiv 1 \pmod 4$ that is not a perfect square, and let

$$R = \{(a + b\sqrt{D})/2 \mid a \text{ and } b \text{ are integers, and either they are both even or they are both odd}\}.$$

We shall abbreviate this condition by saying that a and b have the same parity. Then R is also an integral domain. This requires more care.

First, if $\alpha = \frac{a+b\sqrt{D}}{2}$ with a and b having the same parity, and $\beta = \frac{c+d\sqrt{D}}{2}$ with c and d having the same parity, then

$$\alpha + \beta = \frac{a + b\sqrt{D}}{2} = \frac{c + d\sqrt{D}}{2} = \frac{(a + b) + (c + d)\sqrt{D}}{2}$$

and it is easy to check that $a+b$ and $c+d$ have the same parity, so $\alpha+\beta$ is in R. Also,

$$\alpha\beta = \left(\frac{a+b\sqrt{D}}{2}\right)\left(\frac{c+d\sqrt{D}}{2}\right) = \frac{(ac+bdD)+(ad+bc)\sqrt{D}}{4}$$

and now it is a little more work to check that, in all cases, since $D \equiv 1 \pmod 4$, $ac+bdD = 2e$, with e an integer, and $ad+bc = 2f$, with f an integer, and e and f have the same parity, so $\alpha\beta = \frac{e+f\sqrt{D}}{2}$ is in R. Again, we leave the proof that R has no zero divisors to the exercises.

(5) Let D be a fixed integer that is not a perfect square, and let

$$R = \{a+b\sqrt{D} \mid a \text{ and } b \text{ are rational numbers}\}.$$

Then R is an integral domain. Once again, properties (1)–(9) of an integral domain are easy to check and we defer the proof of property (11) to Example 1.8(5).

The integral domains in Example 1.3(3) look pretty natural, but the integral domains in Example 1.3(4) look rather artificial. It turns out to be the case that, depending on the value of D, we sometimes want to consider the former and sometimes the latter. See the exercises for why this is the case.

We now make a further definition.

Definition 1.4. $\mathbf{F}$ is a *field* if it satisfies properties (1)–(10) and the following additional property:

(12) Every nonzero element of $\mathbf{F}$ has a multiplicative inverse, i.e., if a is any nonzero element of $\mathbf{F}$ there is an element a^{-1} with $aa^{-1} = a^{-1}a = 1$.

Lemma 1.5. *Every field is an integral domain.*

Proof: Suppose $\mathbf{F}$ satisfies properties (1)–(10) and (12). We need to show it satisfies properties (1)–(10) and (11). The fact that it satisfies properties (1)–(10) is immediate, as that is part of our hypothesis. So we need only show that it satisfies property (11). That is, we must show that a field has no zero divisors.

So let a and b be two elements of $\mathbf{F}$ with $ab = 0$. We wish to show that $a = 0$ or $b = 0$. If $a = 0$, we are done, so suppose $a \neq 0$. Then a has an inverse a^{-1}, and we see

$$\begin{aligned} ab &= 0 \\ a^{-1}(ab) &= a^{-1}(0) \\ (a^{-1}a)b &= 0 \\ 1b &= 0 \\ b &= 0, \end{aligned}$$

so $b = 0$ as required. □

Let us make one more definition.

Definition 1.6. An element a of an integral domain R is a *unit* if a has a multiplicative inverse. We let

$$R^* = \{\text{units of } R\}.$$

Remark 1.7. Note that an integral domain R is a field if and only if every nonzero element of R is a unit.

Example 1.8. Let us consider the integral domains in Example 1.3.

(1) $\mathbb{Z}$ is not a field. The only elements of $\mathbb{Z}$ that have inverses are 1 (where $1^{-1} = 1$) and -1 (where $(-1)^{-1} = -1$).

(2) $\mathbf{Q}$ is a field. The inverse of the element a/b is $(a/b)^{-1} = b/a$.

(3) For any fixed integer D that is not a perfect square, $R = \{a + b\sqrt{D} \mid a \text{ and } b \text{ integers}\}$ is not a field.

(4) For fixed integer $D \equiv 1 \pmod 4$ that is not a perfect square, $R = \{(a+b\sqrt{D})/2 \mid a \text{ and } b \text{ integers having the same parity}\}$ is not a field.

(5) For any fixed integer D that is not a perfect square, $R = \{a + b\sqrt{D} \mid a \text{ and } b \text{ rational numbers}\}$ is a field.

To show this, we explicitly find the inverse of any nonzero element α of R. Let $\alpha = a + b\sqrt{D}$. We define $\overline{\alpha}$ to be $\overline{\alpha} = a - b\sqrt{D}$. Then $\alpha\overline{\alpha} = (a + b\sqrt{D})(a - b\sqrt{D}) = a^2 - b^2D$, so

$$\alpha^{-1} = \frac{\overline{\alpha}}{a^2 - b^2D} = \frac{a}{a^2 - b^2D} + \frac{-b}{a^2 - b^2D}\sqrt{D} = e + f\sqrt{D}.$$

In particular, α^{-1} is of the form $e + f\sqrt{D}$ where e and f are rational numbers (to be precise, $e = (a/(a^2 - b^2D))$ and $f = (-b/(a^2 - b^2D))$), so

α^{-1} is an element of R. Thus we see that α indeed has an inverse in R, as claimed.

For this to make sense we need to know that the denominator is nonzero. But $a^2 - b^2 D = 0$ gives $D = (a/b)^2$, contradicting our choice of D not a perfect square.

(Actually, to be totally honest, a perfect square is usually defined to be the square of an integer, so, with this definition, in order to conclude that $D \neq (a/b)^2$, we need to know that if D is not the square of an integer, it is not the square of a rational number, and this fact already uses unique factorization of the integers.)

Also, since R is a field, we conclude from Lemma 1.5 that it is also an integral domain.

Remark 1.9. Let a and b be elements of an integral domain R with $b \neq 0$, and consider the equation $bx = a$. This equation may or may not have a solution, but if it has a solution, that solution x is unique. In this case we say that b *divides* a and we write $x = a/b$. With this definition, the "usual" rules of fractions hold—see the exercises. (Note that b divides 1 if and only if b is a unit, and then $1/b = b^{-1}$. In particular, note that in a field we can divide any element by any nonzero element.)

1.2 Quadratic Fields

We denote the field R of Example 1.3(5) by $\mathbf{Q}(\sqrt{D})$, i.e.,

$$\mathbf{Q}(\sqrt{D}) = \{a + b\sqrt{D} \mid a \text{ and } b \text{ are rational numbers}\}.$$

We have imposed the restriction that D not be a perfect square, but now we want to impose a further restriction: we want D also to be *square-free*, i.e., not divisible by any perfect square, except for 1. This is purely to avoid duplication. For suppose D were not square-free, i.e., that $D = n^2 D'$ for some integers n and D'. Then we would have (as you can check in the exercises) $\mathbf{Q}(\sqrt{D}) = \mathbf{Q}(\sqrt{D'})$, and we would just be repeating ourselves.

With this restriction, we let

$$\mathcal{O}(\sqrt{D}) = \{a + b\sqrt{D} \mid a \text{ and } b \text{ integers}\}$$

if $D \equiv 2$ or $3 \pmod 4$, i.e., if D leaves a remainder of 2 or 3 when divided by 4, (i.e., $D = \ldots, -10$, (not -9), $-6, -5, -2, -1, 2, 3, 6, 7, 10, 11, 14, 15$, (not 18), $19, \ldots$), and

$$\mathcal{O}(\sqrt{D}) = \{\frac{a + b\sqrt{D}}{2} \mid a \text{ and } b \text{ integers having the same parity}\}$$

if $D \equiv 1 \pmod 4$, i.e., if D leaves a remainder of 1 when divided by 4 (i.e., $D = \ldots, -31$, (not -27), -23, -19, -15, -11, -7, -3, (not 1), 5, (not 9), 13, 17, $\ldots$).

(Note that D cannot be divisible by $4 = 2^2$, as we are assuming that D is square-free.)

Definition 1.10. Let $D \neq 1$ be a square-free integer. $\mathbf{Q}(\sqrt{D})$ is called a *quadratic field* and $\mathcal{O}(\sqrt{D})$ is called the *ring of integers* of $\mathbf{Q}(\sqrt{D})$. More precisely, $\mathbf{Q}(\sqrt{D})$ is called a *real quadratic field* if $D > 0$ and an *imaginary quadratic field* if $D < 0$.

In computing with quadratic fields, there are some quantities that are extremely useful.

Definition 1.11. Let $\alpha = a + b\sqrt{D}$ be an element of $\mathbf{Q}(\sqrt{D})$. Then its *conjugate* $\overline{\alpha}$ is defined by

$$\overline{\alpha} = a - b\sqrt{D},$$

its *norm* $\mathrm{N}(\alpha)$ is defined by

$$\mathrm{N}(\alpha) = \alpha\overline{\alpha} = a^2 - b^2 D,$$

and its *trace* $\mathrm{Tr}(\alpha)$ is defined by

$$\mathrm{Tr}(\alpha) = \alpha + \overline{\alpha} = 2a.$$

The following properties are crucial.

Lemma 1.12. *Let α and β be any two elements of $\mathbf{Q}(\sqrt{D})$. Then,*

(1) $\overline{\alpha + \beta} = \overline{\alpha} + \overline{\beta}$ *and* $\overline{\alpha\beta} = \overline{\alpha}\overline{\beta}$;

(2) $\mathrm{Tr}(\alpha) = \mathrm{Tr}(\overline{\alpha})$ *and* $\mathrm{N}(\alpha) = \mathrm{N}(\overline{\alpha})$;

(3) $\mathrm{Tr}(\alpha + \beta) = \mathrm{Tr}(\alpha) + \mathrm{Tr}(\beta)$ *and* $\mathrm{N}(\alpha\beta) = \mathrm{N}(\alpha)\,\mathrm{N}(\beta)$;

(4) If $\mathrm{N}(\alpha) = 0$ *then* $\alpha = 0$.

Proof: (1), (2), and (3) are easy to check by direct computation, and we leave them as exercises. We prove (4).

Let $\alpha = a + b\sqrt{D}$ and suppose $\mathrm{N}(\alpha) = 0$. Then

$$0 = \mathrm{N}(\alpha) = a^2 - b^2 D$$

so $a^2 = b^2 D$ and, if $b \neq 0$, then $(a/b)^2 = D$. But we assumed D was not a perfect square, so this is impossible. Hence $b = 0$ and then $a = 0$, so $\alpha = 0$. □

Lemma 1.13. *For any element x of $\mathcal{O}(\sqrt{D})$, $\mathrm{N}(x)$ is an integer.*

Proof: If $x = a+b\sqrt{D}$, then $\mathrm{N}(x) = a^2-b^2D$, so if a and b are integers, $\mathrm{N}(x)$ is certainly an integer. Thus the only case we need to check is that of $x = (a+b\sqrt{D})/2$ where a and b are both odd and $D \equiv 1 \pmod 4$. In this case we write $a = 2m+1, b = 2n+1, D = 4E+1$. Then $\mathrm{N}(x) = (a^2-b^2D)/4 = ((2m+1)^2-(2n+1)^2(4E+1))/4 = m^2+m-4n^2E-4nE-E-n^2-n$ is an integer. □

Lemma 1.14. *Let $R = \mathcal{O}(\sqrt{D})$. Then the units of R are precisely those elements x of R with $|\mathrm{N}(x)| = 1$.*

Proof: First suppose $|\mathrm{N}(x)| = 1$. Then $\mathrm{N}(x) = \pm 1$. But $\mathrm{N}(x) = x\overline{x}$. Thus either $x\overline{x} = 1$, in which case x has inverse $x^{-1} = \overline{x}$, or $x\overline{x} = -1$, in which case x has inverse $x^{-1} = -\overline{x}$, so in either case x is a unit.

Conversely, suppose that x is a unit. Then there is an element y of R with $xy = 1$. Then on the one hand $\overline{xy} = \overline{1} = 1$, and on the other hand $\overline{xy} = \overline{x}\overline{y}$, by Lemma 1.12. Thus $\overline{x}\overline{y} = 1$. Multiplying, we see that $x\overline{x}y\overline{y} = 1$, i.e., that $\mathrm{N}(x)\,\mathrm{N}(y) = 1$. However, by Lemma 1.13, $\mathrm{N}(x)$ and $\mathrm{N}(y)$ are both integers. Therefore, we must have either $\mathrm{N}(x) = \mathrm{N}(y) = 1$ or $\mathrm{N}(x) = \mathrm{N}(y) = -1$. But in either case, we conclude $|\mathrm{N}(x)| = 1$. □

Let us use Lemma 1.14 to try to find the units in $\mathcal{O}(\sqrt{D})$.

Corollary 1.15.

(1) The units in $\mathcal{O}(\sqrt{-1})$ are $\{\pm 1, \pm i\}$.

(2) The units in $\mathcal{O}(\sqrt{-3})$ are $\{\pm 1, \pm(1+\sqrt{-3})/2, \pm(1-\sqrt{-3})/2\}$.

(3) For any other negative value of D, the units in $\mathcal{O}(\sqrt{D})$ are $\{\pm 1\}$.

Proof: We leave this proof for the exercises. □

Thus, we have completely found the units of $\mathcal{O}(\sqrt{D})$ when D is negative. But when D is positive the situation is much more involved.

So let D be positive. Then, by Lemma 1.14, $x = a+b\sqrt{D}$ is a unit exactly when $\mathrm{N}(x) = a^2-b^2D = \pm 1$ and $x = (a+b\sqrt{D})/2$ is a unit exactly when $\mathrm{N}(x) = (a^2-b^2D)/4 = \pm 1$, i.e., when $a^2-b^2D = \pm 4$, where in the first case a and b are both integers, and in the second case a and b are both integers having the same parity. (The first case happens for all values of D, regardless of $D \pmod 4$, while the second case happens only for $D \equiv 1 \pmod 4$.) Here it is certainly not a priori clear that $|\mathrm{N}(x)| = 1$ has a solution, other than $x = \pm 1$, and, even if we know there are solutions, it is completely unclear how to find them.

Nevertheless, let us experiment a bit, by taking small values of D.

Example 1.16.

(1) Let $D = 2$. Then $\mathrm{N}(1+\sqrt{2}) = 1^2 - 1^2 \cdot 2 = -1$ so $x = 1+\sqrt{2}$ is a unit and its inverse is $-\overline{x} = -(1-\sqrt{2}) = -1+\sqrt{2}$. Since $xx^{-1} = 1, x^{-k} = (xx^{-1})^k = 1^k = 1$ so x^k is also a unit for any k. So, for example, $x^2 = (1+\sqrt{2})^2 = 3+2\sqrt{2}$ is a unit, as is $x^3 = (1+\sqrt{2})^3 = 7+5\sqrt{2}$, etc. Note that $x > 1$ so $\{1, x, x^2, x^3, \ldots\}$ is a steadily increasing sequence of numbers, so in particular they are all distinct. Moreover, we see that $\{\ldots, \pm x^{-3}, \pm x^{-2}, \pm x^{-1}, \pm 1, \pm x, \pm x^2, \pm x^3, \ldots\}$ are all distinct as well, giving an infinite set of distinct units in this case.

(2) Let $D = 3$. Then $\mathrm{N}(2+\sqrt{3}) = 2^2 - 1^2 \cdot 3 = 1$ so $x = 2+\sqrt{3}$ is a unit and its inverse is $\overline{x} = 2-\sqrt{3}$. Again, x^k is a unit for any k, so for example, $x^2 = (2+\sqrt{3})^2 = 7+4\sqrt{3}$ and $x^3 = (2+\sqrt{3})^3 = 26+15\sqrt{3}$ are units. Again, $\{\ldots, \pm x^{-3}, \pm x^{-2}, \pm x^{-1}, \pm 1, \pm x, \pm x^2, \pm x^3, \ldots\}$ is an infinite set of distinct units.

(3) Let $D = 5$. Then $\mathrm{N}((1+\sqrt{5})/2) = (1^2 - 1^2 \cdot 5)/4 = -1$ so $x = (1+\sqrt{5})/2$ is a unit and its inverse is $-\overline{x} = -(-1-\sqrt{5})/2$. Again, x^k is a unit for any k, so for example, $x^2 = (3+\sqrt{5})/2$ and $x^3 = 2+\sqrt{5}$ are units, and once again $\{\ldots, \pm x^{-3}, \pm x^{-2}, \pm x^{-1}, \pm 1, \pm x, \pm x^2, \pm x^3, \ldots\}$ is an infinite set of distinct units.

Remark 1.17. Suppose we have integers a and b satisfying the equation $a^2 - b^2 D = 1$. We have the identity $a^2 - b^2 D = (a + b\sqrt{D})(a - b\sqrt{D})$ and so these values of a and b give a factorization $1 = (a+b\sqrt{D})(a-b\sqrt{D})$. Thus we see that in this case $a + b\sqrt{D}$ is a unit of $\mathcal{O}(\sqrt{D})$ (as is $a - b\sqrt{D}$). This equation, $a^2 - b^2 D = 1$, is known as *Pell's equation*, and has a long history. A priori it is unclear that Pell's equation *has* a solution other than $a = \pm 1$, $b = 0$ (which just gives the units ± 1) but indeed it *does*. We devote Chapter 4 to studying Pell's equation, where we will show that for any positive integer D that is not a perfect square, Pell's equation always has a solution. That is the hard part, but once we have that we will also show that it always has infinitely many solutions, and furthermore that the pattern of units in $\mathcal{O}(\sqrt{D})$ is always as in Example 1.16.

1.3 Exercises

Exercise 1.1. Let R be an integral domain and let S be a subset of R that satisfies the following four conditions:

(1) 1 is in S;

(2) if a is in S, then $-a$ is in S;

(3) if a and b are in S, then $a+b$ is in S;

(4) if a and b are in S, then ab is in S.

(a) Show that S is an integral domain.

(b) Give examples to show that if S satisfies any three of these four conditions then S may not be an integral domain. (Thus you will need four examples, one for each omitted condition.)

Exercise 1.2. Show that the usual rules of signs hold in any integral domain R:

(a) $-(-a) = a$;

(b) $-(a+b) = (-a) + (-b)$;

(c) $(-1)a = -a$;

(d) $(-a)b = a(-b) = -(ab)$;

(e) $(-a)(-b) = ab$.

Exercise 1.3.

(a) Let R be an integral domain and consider the equation $bx = a$ in R. Show that if this equation has a solution, that solution is unique.

(b) Suppose that b is a unit. Show that for any a, $bx = a$ has the solution $x = ab^{-1}$.

Exercise 1.4. Recall from Remark 1.9 that if a and b are elements of an integral domain R, we say that b divides a if there is an element x satisfying the equation $bx = a$, in which case we write $x = a/b$. Show that with this definition, the usual rules of fractions hold:

(a) $b(a/b) = a$;

(b) $(a/b)(b/a) = 1$;

(c) $a(b/c) = (ab)/c$;

(d) $(ab)/(ac) = b/c$;

(e) $(a/b)(c/d) = (ac)/(bd)$;

(f) $(a/c) + (b/c) = (a+b)/c$;

(g) $(a/b) + (c/d) = (ad + bc)/(bd)$; (h) $(a/b) = (c/d) \Leftrightarrow ad = bc$.

(Note that in some cases the right-hand side of the above equalities may be defined when the left-hand side is not. We mean these equalities to hold when both sides are defined.)

Exercise 1.5. Let R be an arbitrary integral domain. Show that if α is a unit of R then α^k is a unit of R for any integer k.

Exercise 1.6. Show that R as in Example 1.3(3) and as in Example 1.3(4) has no zero divisors (and hence is an integral domain).

Exercise 1.7. If $D = n^2 D'$ for some integer (or more generally some rational number) n, show that $\mathbf{Q}(\sqrt{D}) = \mathbf{Q}(\sqrt{D'})$.

Exercise 1.8. Prove Lemma 1.12(1), (2), and (3).

Exercise 1.9. Let $R = \mathcal{O}(\sqrt{D})$ and suppose that $\alpha = a + b\sqrt{D}$ is a unit of R. Show that $\overline{\alpha} = a - b\sqrt{D}$ is also a unit of R.

Exercise 1.10. Let $R = \mathcal{O}(\sqrt{D})$ with $D > 0$ and suppose that $\alpha = a + b\sqrt{D}$ is a unit of R, $\alpha \neq \pm 1$. Show that $\{\ldots, \pm\alpha^{-3}, \pm\alpha^{-2}, \pm\alpha^{-1}, \pm 1, \pm\alpha, \pm\alpha^2, \pm\alpha^3, \ldots\}$ are all distinct (and hence that R has infinitely many units).

Exercise 1.11. Prove Corollary 1.15(3): for $D < 0$, $D \neq -1$, and $D \neq -3$, the only units in $\mathcal{O}(\sqrt{D})$ are $\{\pm 1\}$.

Exercise 1.12. Let $\alpha = a + b\sqrt{D}$ be an element of $\mathbf{Q}(\sqrt{D})$. Show that α is a root of a monic quadratic polynomial (i.e., a quadratic whose x^2 coefficent is 1). Furthermore, express the coefficients of this quadratic in terms of $\mathrm{N}(\alpha)$ (the norm of α) and $\mathrm{Tr}(\alpha)$ (the trace of α).

Exercise 1.13. Recall Definition 1.10:

$$\mathcal{O}(\sqrt{D}) = \{a + b\sqrt{D} \mid a \text{ and } b \text{ integers}\}$$

if $D \equiv 2$ or $3 \pmod 4$, while

$$\mathcal{O}(\sqrt{D}) = \{\frac{a + b\sqrt{D}}{2} \mid a \text{ and } b \text{ integers having the same parity}\}$$

if $D \equiv 1 \pmod 4$. You may wonder why we made a distinction between these two cases. The answer is that we want the ring of integers to be naturally defined in terms of some property that it satisfies. Here is the property:

$\mathcal{O}(\sqrt{D})$ is the set of elements of $\mathbf{Q}(\sqrt{D})$ that are roots of a monic quadratic with integer coefficients (i.e., roots of a quadratic polynomial $f(x) = x^2 + mx + n$ with m and n integers).

(a) Verify that this is true for the following elements α of $\mathcal{O}(\sqrt{D})$:

(i) $\alpha = 3 + 8\sqrt{6}$;

(ii) $\alpha = 7 - 10\sqrt{11}$;

(iii) $\alpha = 2 + 9\sqrt{5}$;

(iv) $\alpha = 4 + 5\sqrt{-2}$;

(v) $\alpha = -6 + 11\sqrt{-5}$;

(vi) $\alpha = \frac{3}{2} + \frac{7}{2}\sqrt{-3}$.

(b) Show that this is the case in general. That is, show that

(i) if $D \equiv 2$ or $3 \pmod 4$, then $\alpha = c + d\sqrt{D}$ is a root of a monic quadratic with integer coefficients if and only if c and d are both integers;

(ii) if $D \equiv 1 \pmod 4$, then $\alpha = c+d\sqrt{D}$ is a root of a monic quadratic with integer coefficients if and only if either c and d are both integers or $c = a/2$ and $d = b/2$ with a and b both odd integers.

In the text of this book, we treat integral domains of the form $\mathcal{O}(\sqrt{D})$. But many of the statements we make have analogs for polynomials, and we leave the treatment of the polynomial situation to the exercises. Here is the first case: Let R be an integral domain. Then

$$\begin{aligned} R[X] &= \{\text{polynomials in the variable } X \text{ with coefficients in } R\} \\ &= \{a^n X^n + a_{n-1}X^{n-1} + \ldots + a_1 X + a_0 \mid a_i \text{ in } R \text{ for every } i\}. \end{aligned}$$

(In considering $R[X]$ you may assume the usual properties of polynomial arithmetic. The cases we will be most concerned with here are $R = \mathbf{Q}$ and $R = \mathbb{Z}$ and indeed for the purposes of this book you may confine your attention to these.)

Exercise 1.14. Show that $R[X]$ is an integral domain.

Exercise 1.15. Show that $R[X]^* = R^*$ (i.e., that the units of $R[X]$ are the constant polynomials $\{a\}$ for those values of a that are units of R). In particular, if R is a field, the units of $R[X]$ are the nonzero constant polynomials.

Chapter 2

Unique Factorization

We now embark on the proof that a number of the integral domains we are interested in satisfy unique factorization. We have written "proof" rather than "proofs" as it is our goal to establish a framework that will enable us to come up with one proof that handles all these cases simultaneously. To be precise, our strategy will be as follows:

Step 1a. Define "Euclidean domain."

Step 1b. Prove that certain integral domains are Euclidean domains.

Step 2a. Define "Principal ideal domain."

Step 2b. Prove that every Euclidean domain is a principal ideal domain.

Step 3a. Define "Unique factorization domain."

Step 3b. Prove that every principal ideal domain is a unique factorization domain.

Thus, putting all of these steps together, we see that certain integral domains are unique factorization domains.

The obvious question now is: "Which ones?" As we shall see, these include the integers $\mathbb{Z}$, and the integral domains $\mathcal{O}(\sqrt{D})$ for some (definitely not all!) values of D.

Indeed, the first part of this chapter will be devoted to the general argument we have just described, and to proving that $\mathcal{O}(\sqrt{D})$ *is* a unique factorization domain in many cases. However, once we accomplish that we will turn our attention to the opposite phenomenon, and will prove that $\mathcal{O}(\sqrt{D})$ *is not* a unique factorization domain in many other cases.

We will not be able to settle the issue in all cases, and in fact, in complete generality the answer is unknown. We will describe our (that is, mathematicians') present state of knowledge about this question.

2.1 Euclidean Domains

A Euclidean domain is, roughly speaking, an integral domain in which we can divide one number by another, obtaining a quotient and a remainder that is smaller than the divisor. In order to say what smaller is, we must have a notation of size. We first define this:

Definition 2.1. Let R be an integral domain. Then $\|\cdot\|$ is a *norm* on R if

(1) for every nonzero element a of R, $\|a\|$ is a nonnegative integer;

(2) for every two nonzero elements a and b of R,

$$\|a\| \leq \|ab\|.$$

Remark 2.2.

(1) Note that we do not require $\|0\|$ to be defined, though it may be.

(2) Note that under this definition it is possible that $\|a\| = 0$ even though $a \neq 0$.

Lemma 2.3. *The following are norms:*

(1) $R = \mathbb{Z}$ *and* $\|a\| = |a|$.

(2) $R = \mathcal{O}(\sqrt{D})$ *and* $\|\alpha\| = |\mathrm{N}(\alpha)|$ *(where* $\mathrm{N}(\alpha)$ *is defined in Definition 1.11).*

Proof:

(1) We need to check both properties of the norm:

Property 1: Certainly if a is an integer, $|a|$ is a nonnegative integer.

Property 2: For any nonzero integer b, $1 \leq |b|$. Then for any nonzero integer a, by the properties of the absolute value,

$$\|a\| = |a| \leq |a| \cdot |b| = |ab| = \|ab\|.$$

(2) This follows from earlier work we have done. Let us see this.

Property (1): We showed in Lemma 1.13 that $\mathrm{N}(\alpha)$ is an integer, so $\|\alpha\| = |\,\mathrm{N}(\alpha)|$ is a nonnegative integer.

Property (2): For any β, $|\,\mathrm{N}(\beta)|$ is a nonnegative integer, and if $\beta \neq 0$, $\mathrm{N}(\beta) \neq 0$. (This is the contrapositive of Lemma 1.12(4).) Thus for any $\beta \neq 0$, $\|\beta\| \geq 1$. For any α, $\mathrm{N}(\alpha\beta) = \mathrm{N}(\alpha)\,\mathrm{N}(\beta)$ by Lemma 1.12(3). Thus

$$\|\alpha\| = |\,\mathrm{N}(\alpha)| \leq |\,\mathrm{N}(\alpha)| \cdot |\,\mathrm{N}(\beta)| = |\,\mathrm{N}(\alpha\beta)| = \|\alpha\beta\|.$$

□

Remark 2.4. Unfortunately, the word "norm" is used to mean two slightly different things. We called $\mathrm{N}(\alpha)$ a norm in Chapter 1. Note that $\mathrm{N}(\alpha)$ is defined on $\mathbf{Q}(\sqrt{D})$, may be negative, and need not be an integer. In our definition here, the norm is required to be a nonnegative integer, and so we must consider $|\,\mathrm{N}(\alpha)|$, and only for α in $\mathcal{O}(\sqrt{D})$. In this chapter, we will always use a norm in the sense of Definition 2.1, and we will always denote such a norm by $\|\cdot\|$.

Now we can define what we mean by division with a small remainder.

Definition 2.5. An integral domain R with a norm $\|\cdot\|$ is a *Euclidean domain* if it has the following property: for any element a of R, and any nonzero element b of R, there is an element q of R (the quotient) and an element r of R (the remainder) with

$$a = bq + r \text{ and } r = 0 \text{ or } \|r\| < \|b\|.$$

Remark 2.6.

(1) This is a familiar property for the integers, which you probably learned in elementary school: $75 = 17 \cdot 4 + 7, 93 = 11 \cdot 8 + 5, 105 = 23 \cdot 5 + 0$. Nevertheless, it requires proof! We shall prove it momentarily.

(2) Note we are not claiming that the quotient and remainder are unique. For example, $100 = 3 \cdot 33 + 1 = 3 \cdot 34 + (-2)$ both work.

(3) Strictly speaking, the definition of a Euclidean domain includes the integral domain R and the norm $\|\cdot\|$. We usually say, however, "R is an integral domain" when the norm is understood.

Lemma 2.7. *The integers $\mathbb{Z}$ are a Euclidean domain.*

Proof: We are claiming that for any integer a, and any integer $b \neq 0$, there is an integer q and an integer r with $a = bq + r$ and $r = 0$ or $\|r\| < \|b\|$.

For each fixed value of b, we prove this claim by complete induction on $\|a\|$. We shall prove this claim in the case $a \geq 0$ and $b > 0$ here.

So suppose $a \geq 0$ and $b > 0$. Note then that $\|a\| = |a| = a$ and $\|b\| = |b| = b$.

If $\|a\| = 0$, then $a = 0$, and this claim is certainly true: $a = b \cdot 0 + 0$ so $q = 0$ and $r = 0$.

Also, if $0 < \|a\| < \|b\|$, then $0 < a < b$ and this claim is also true: $a = b \cdot 0 + a$ so $q = 0$ and $r = a$ satisfies $\|r\| < \|b\|$.

Now assume that this claim is true for all integers a' with $\|a'\| < \|a\|$. Consider a. We have just proved this claim if $a = 0$ or $0 < \|a\| < \|b\|$, so we may restrict our attention to the case that $\|b\| \leq \|a\|$. But in this case $0 < b \leq a$ so $0 \leq a - b < a$. Set $a' = a - b$. Then we can apply the inductive hypothesis to a' to conclude that $a' = bq' + r'$ for some r' with $r' = 0$ or $\|r'\| < \|b\|$. Substituting, we see that $a - b = bq' + r'$ and hence that $a = b(q' + 1) + r' = bq + r$ with $q = q' + 1$ and $r = r'$. But then also $r = 0$ or $\|r\| < \|b\|$, as required.

Hence our claim is true for a, and so by complete induction we may conclude that our claim is true for every $a \geq 0$.

Thus, we have proved the lemma in this case. We leave the remaining cases as exercises. □

The point of introducing the notion of a Euclidean domain is that it applies in many cases other than that of the ordinary integers. In particular, we have the following theorem.

Theorem 2.8. *Let $R = \mathcal{O}(\sqrt{D})$. Then for the following values of D, R is a Euclidean domain:*

$$D = -11,\ -7,\ -3,\ -2,\ -1,\ 2,\ 3,\ 5,\ 6,\ 7,\ 11,\ 13,\ 17,\ 21,\ \textit{and}\ 29.$$

Proof: This is a very long proof, so let us begin by describing our strategy. We are trying to investigate an algebraic question—when $\mathcal{O}(\sqrt{D})$ is Euclidean—but we will convert this question to a geometric question. Then we will solve this question, only using basic analytic geometry. The geometric idea is simple, but we will work very hard at it and obtain our results. Thus, this proof is an illustration of the fact that often in mathematics one may start with a simple idea and by pushing it hard enough go a long way with it.

As we go further, the details—though not the basic idea—get more complicated. The easiest cases are $D = -1$, -2, and -3, and the second easiest cases are $D = 2$ and 3. We do these cases here. The other cases are considerably more involved, and we defer them to Appendix C.

Let α and β be elements of R with $\beta \neq 0$. We wish to show that we may always write $\alpha = \beta\gamma + \rho$ where γ is some element of R and ρ is an element of R with $\rho = 0$ or $\|\rho\| < \|\beta\|$. To be concrete, let us write $\alpha = a + b\sqrt{D}$ and $\beta = c + d\sqrt{D}$. Then we may form the quotient $\alpha/\beta = (a + b\sqrt{D})/(c + d\sqrt{D})$. In general this quotient will *not* be an element of $\mathcal{O}(\sqrt{D})$ but *will* be an element of $\mathbf{Q}(\sqrt{D})$. In fact, as we saw in Chapter 1,

$$\begin{aligned}\frac{a + b\sqrt{D}}{c + d\sqrt{D}} &= \frac{(a + b\sqrt{D}) \cdot (c - d\sqrt{D})}{(c + d\sqrt{D}) \cdot (c - d\sqrt{D})} \\ &= (a + b\sqrt{D})\Big(\frac{c}{c^2 - d^2 D} + \frac{-d}{c^2 - d^2 D}\sqrt{D}\Big) \\ &= \frac{ac - bdD}{c^2 - d^2 D} + \frac{-ad + bc}{c^2 - d^2 D}\sqrt{D} = (e + f\sqrt{D}),\end{aligned}$$

where $e = (ac - bdD)/(c^2 - d^2 D)$ and $f = (-ad + bc)/(c^2 - d^2 D)$. If it happens that $e + f\sqrt{D}$ is an element of $R = \mathcal{O}(\sqrt{D})$, then set $\gamma = e + f\sqrt{D}$. Then $\alpha = \beta\gamma$ with γ in R, so β divides α and $\alpha = \beta\gamma + 0$ so we simply set $\rho = 0$, and we are done.

Usually we will not be so lucky, however, so we turn to geometry in order to proceed further. We observe that e and f are certainly rational numbers. We now represent $\gamma_0 = e + f\sqrt{D}$ with e and f rational numbers, or equivalently with γ_0 in $\mathbf{Q}(\sqrt{D})$, by the point (e, f) in the plane. If $D = -1$ this is just the usual representation of the complex number $e + fi$ as the point (e, f) in the complex plane. If $D \neq -1$ it is a new representation of $e + f\sqrt{D}$, but an equally valid one. Note that the points in the plane that represent elements of $\mathbf{Q}(\sqrt{D})$ are precisely the points (e, f) where both e and f are rational numbers. We will call these points $\mathbf{Q}$-points.

Along with this new representation of points we introduce a new metric (i.e., measure of distance) in the plane. We let $\|(e, f)\|_D = |e^2 - f^2 D|$ and this measures the distance from the point (e, f) to the origin. Note that (e, f) corresponds to $e + f\sqrt{D}$. More generally, if (s, t) is another point of the plane, corresponding to $s + t\sqrt{D}$, then $(e, f) - (s, t) = (e - s, f - t)$ corresponds to the difference $(e + f\sqrt{D}) - (s + t\sqrt{D}) = (e - s) + (f - t)\sqrt{D}$, and then $\|(e, f) - (s, t)\|_D = \|(e - s, f - t)\|_D$ is the distance between these two points. (Remember that we are restricting ourselves to $\mathbf{Q}$-points as those are the points that correspond to elements of $\mathbf{Q}(\sqrt{D})$.)

Let us make some important observations about $\|\cdot\|_D$. First of all, if (e,f) corresponds to an element $e+f\sqrt{D}$ of $R=\mathcal{O}(\sqrt{D})$, then $\|(e,f)\|\|_D = \|e+f\sqrt{D}\|$. Thus, in our identification of $\mathbf{Q}(\sqrt{D})$ with $\mathbf{Q}$-points in the plane, $\|\cdot\|_D$ agrees with our norm $\|\cdot\|$ on $\mathcal{O}$-points, i.e., $\|\alpha_0\|_D = \|\alpha_0\|$ for any α_0 in $\mathcal{O}(\sqrt{D})$. Also, we know that $\|\alpha_1\alpha_2\| = \|\alpha_1\|\cdot\|\alpha_2\|$ for any two elements α_1 and α_2 of $\mathcal{O}(\sqrt{D})$. This is still true if α_1 and α_2 are any two elements of $\mathbf{Q}(\sqrt{D})$. For we see, from Definition 1.11 and Lemma 1.12, that, for any two elements α_1 and α_2 of $\mathbf{Q}(\sqrt{D})$, $\|\alpha_1\alpha_2\| = |\operatorname{N}(\alpha_1\alpha_2)| = |\operatorname{N}(\alpha_1)\operatorname{N}(\alpha_2)| = |\operatorname{N}(\alpha_1)|\cdot|\operatorname{N}(\alpha_2)| = \|\alpha_1\|\cdot\|\alpha_2\|$. But $\|\cdot\|_D$ is *not* a norm on $\mathbf{Q}$-points in the sense of Definition 2.1, as for a general $\mathbf{Q}$-point (e,f), $\|(e,f)\|\|_D$ need not be an integer. For example, $\|(1/2,0)\|_D = 1/4$. Similarly, it is not always the case that $\|\alpha_2| \leq \|\alpha_1\alpha_2\|_D$ for α_1 and α_2 elements of $\mathbf{Q}(\sqrt{D})$, as, for example, $\|(1/2)\alpha_2\|_D = (1/4)\|\alpha_2\|_D$ for any α_2.

Now let us return to our problem. We have $\alpha/\beta = e+f\sqrt{D}$. Write $\gamma_0 = e+f\sqrt{D}$, with γ_0 in $\mathbf{Q}(\sqrt{D})$. Our objective is to find $\gamma = s+t\sqrt{D}$ with γ in $R=\mathcal{O}(\sqrt{D})$ and $\|\gamma_0-\gamma\|_D < 1$.

Let us assume for the moment that we have succeeded in doing so. Then

$$\alpha = \beta\gamma_0 = \beta\gamma + \beta(\gamma_0-\gamma) = \beta\gamma+\rho$$

where we set $\rho = \beta(\gamma_0-\gamma)$. Then $\|\rho\|_D = \|\beta(\gamma_0-\gamma)\|_D = \|\beta\|\cdot\|\gamma_0-\gamma\|_D$ and, since $\|\gamma_0-\gamma\|_D<1$, we have $\|\rho\|_D < \|\beta\|_D$, so we have found values for γ and ρ that satisfy the conditions of a Euclidean domain. (We are assuming that γ is an element of R, and then we see that $\rho = \alpha-\beta\gamma$ is also an element of R, and furthermore we see that (and this is the crucial point!) $\|\rho\|_D < \|\beta\|_D$.)

Hence we have reduced our problem to the problem of showing that for any $e+f\sqrt{D}$ in $\mathbf{Q}(\sqrt{D})$, there is an $s+t\sqrt{D}$ in $\mathcal{O}(\sqrt{D})$ with $\|(e+f\sqrt{D})-(s+t\sqrt{D})\|_D < 1$. Translating this into our geometric language, we need to show that for any point (e,f) in the plane with rational coordinates, there is a point (s,t) in the plane corresponding to an element of R, with $\|(e,f)-(s,t)\|_D<1$.

We will need to know which points of the plane correspond to elements of $R=\mathcal{O}(\sqrt{D})$, so let us determine that now. The answer depends on D. If $D\equiv 2$ or $3 \pmod 4$, then $s+t\sqrt{D}$ is in R when both s and t are integers, so the points in the plane corresponding to elements of R are the points (s,t) with both coordinates integers. If $D\equiv 1 \pmod 4$, then $s+t\sqrt{D}$ is in R when both s and t are integers or when both s and t are half-integers, so the points in the plane corresponding to elements of R are the points (s,t)

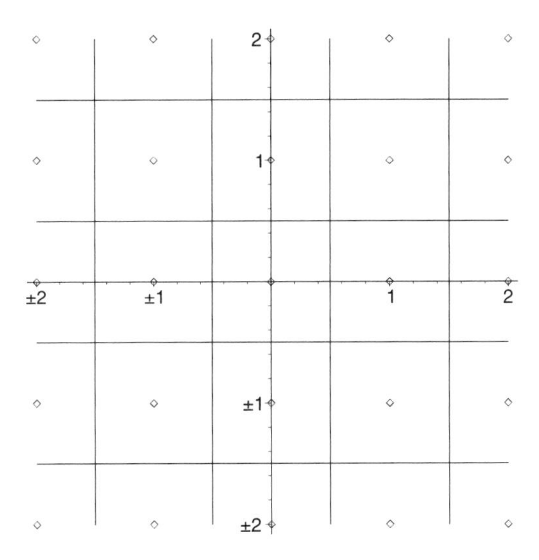

Figure 2.1. The case $D \equiv 2$ or $3 \pmod 4$.

with both coordinates integers or both coordinates half-integers. In any event, we will refer to the points corresponding to elements of $R = \mathcal{O}(\sqrt{D})$ as $\mathcal{O}$-points.

First, let us consider the case $D \equiv 2$ or $3 \pmod 4$. Here the $\mathcal{O}$-points are points with integer coordinates. Let us divide the plane into regions consisting of points that are nearest each of these points *in the usual metric* on the plane, *not in the new metric* $\| \cdot \|_D$. We shall call these points *apparently nearest*, since they "look" nearest when we look at the plane. The points apparently nearest the point (s, t) are the points in a square of side 1 centered at (s, t), as in Figure 2.1.

Next, we consider the case $D \equiv 1 \pmod 4$. Here the $\mathcal{O}$-points are points with both coordinates integers or half-integers, and the points apparently nearest the point (s, t) are the points in a diamond with diagonals of length 1 centered at (s, t), as in Figure 2.2.

These regions of apparently nearest points cover the plane, so for any **Q**-point $\gamma_0 = (e, f)$ there is an $\mathcal{O}$-point $\gamma_1 = (s, t)$ to which it is apparently nearest. (Usually there will be exactly one such point, but if γ_0 is on the border of one of these squares or diamonds there will be more than one such point. In that case, choose γ_1 to be any of them—it doesn't matter which we choose.)

Now the distance of γ_0 from γ_1, $\|\gamma_0 - \gamma_1\|_D$, is the same as the distance $\|(\gamma_0 - \gamma_1) - 0\|_D$ from $\gamma_0 - \gamma_1$ to the origin. (Note we are now using our

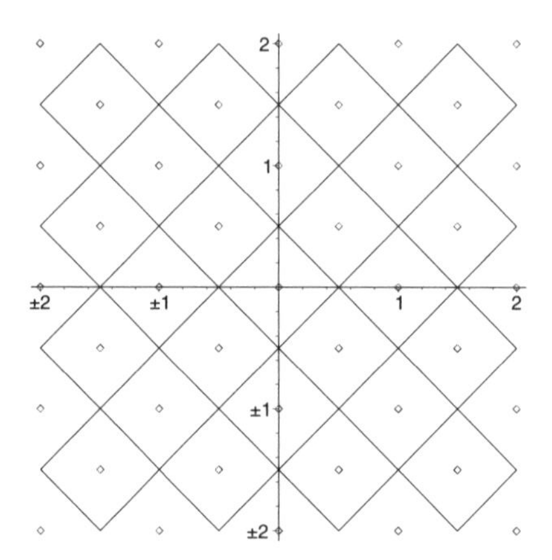

Figure 2.2. The case $D \equiv 1 \pmod 4$.

new metric, the one in which we are really interested.) A little thought shows that we have simply translated (i.e., shifted) the problem to points apparently nearest the origin, and we can always translate the problem in this way. Thus, if we can show every **Q**-point apparently nearest the origin is within a distance of 1 from some $\mathcal{O}$-point (s, t), this will be true of all **Q**-points in the plane, and again we will be done.

Thus, we have reduced our problem to considering the **Q**-points that are apparently nearest the origin. We shall denote this region by $\triangle_0$. Now the real work begins.

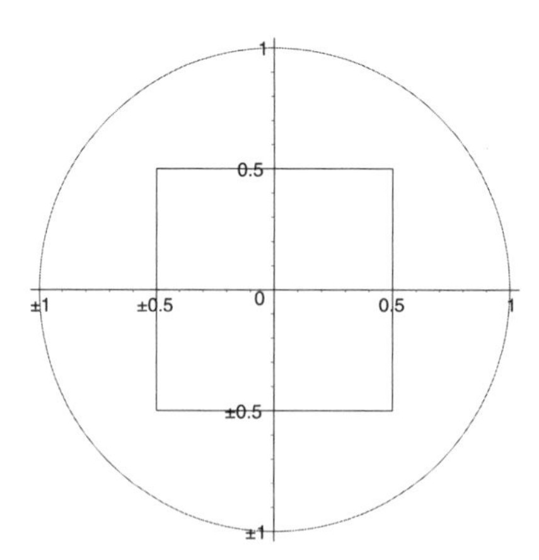

Figure 2.3. The case $D = -1$.

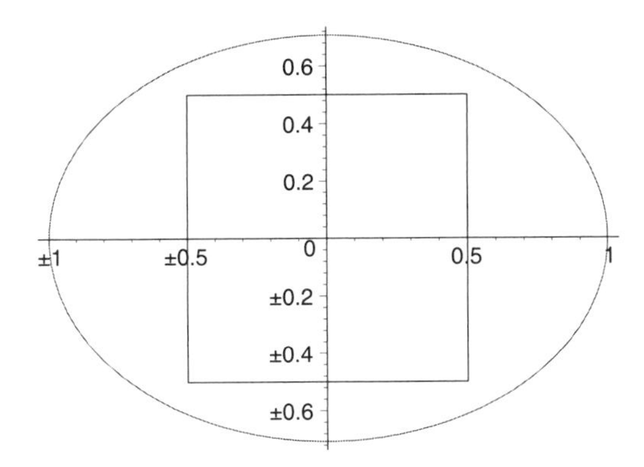

Figure 2.4. The case $D = -2$.

Now we bring in the analytic geometry. First, we consider the case when $D < 0$, and ask for what points (x, y) we have $\|(x, y)\|_D < 1$. Now $\|(x, y)\|_D = x^2 - Dy^2 = x^2 + (-D)y^2$ and since $D < 0$, $-D > 0$. Then $\|(x, y)\|_D = 1$ is the equation $x^2 + (-D)y^2 = 1$, which we recognize as an ellipse centered at the origin with semi-major axis of length 1 along the x-axis, and semi-minor axis of length $1/\sqrt{-D}$ along the y-axis. (Actually, there is one exception. When $D = -1$, the semi-minor axis also has length 1, and the curve is a circle.) Thus, the points with $\|(x, y)\|_D < 1$ are the points that are strictly inside (that is, inside and not on) this ellipse (or circle, when $D = -1$). But now, for $D = -1$, -2, or -3, every point apparently nearest the origin is in this region, as we see from Figures 2.3, 2.4, and 2.5.

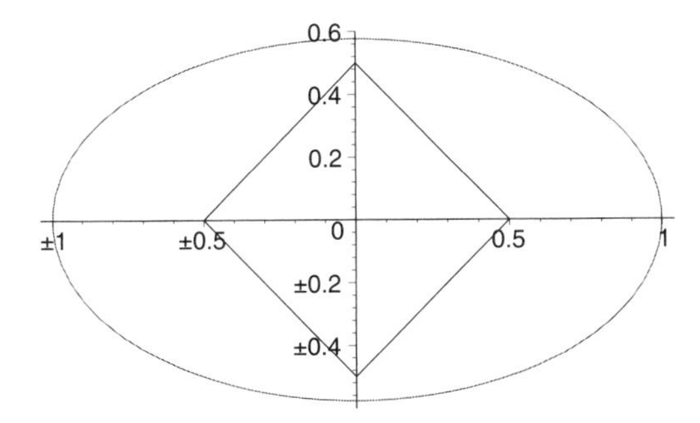

Figure 2.5. The case $D = -3$.

Tracing the argument back, we see that for the $\mathbf{Q}$-point $\gamma_0 = (e, f)$, corresponding to the element $e + f\sqrt{D}$ of $\mathbf{Q}(\sqrt{D})$, if $\gamma_1 = (s, t)$ is the $\mathcal{O}$-point, corresponding to the element $s + t\sqrt{D}$ of $R = \mathcal{O}(\sqrt{D})$, apparently nearest to γ_0, then choosing $\gamma = \gamma_1$ we have found an $\mathcal{O}$-point γ with $\|\gamma_0 - \gamma\|_D < 1$, as required, in the cases $D = -1$, -2, or -3, completing the proof in these cases.

Now we consider the case $D > 0$, and again ask for what points (x, y) we have $\|(x, y)\|_D < 1$. Here $\|(x, y)\|_D = x^2 - Dy^2$. We recognize $|x^2 - Dy^2| = 1$ as the equation of two pairs of hyperbolas. The equation $x^2 - Dy^2 = 1$ gives a pair of hyperbolas, one opening to the right and one opening to the left, having vertices 1 unit to the right and 1 unit to the left of the origin, respectively, and the equation $x^2 - Dy^2 = -1$ gives a pair of hyperbolas, one opening up and one opening down, having vertices $1/\sqrt{D}$ units above and $1/\sqrt{D}$ units below the origin, respectively. (We shall say these two pairs of hyperbolas are centered at the origin and have semi-major axis 1 and semi-minor axis $1/\sqrt{D}$, although those terms are usually just used for ellipses.) Now the points (x, y) with $\|(x, y)\|_D < 1$ are the points with $|x^2 - Dy^2| < 1$, i.e., with $-1 < |x^2 - Dy^2| < 1$, so they are the points "inside" of these hyperbolas. That is, they are the points in the region bounded by all four of these curves, consisting of a rectangular area in the center adjoined by four tails that go apparently infinitely far out toward the northeast, northwest, southeast, and southwest. (Actually, the exact direction they go out depends on D. To be precise, they go out around the asymptotes $y = x/\sqrt{D}$ and $y = -x/\sqrt{D}$.) But for $D = 2$ or 3, every point

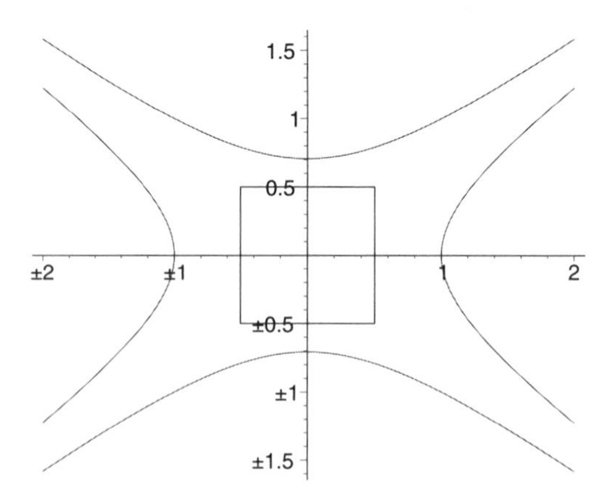

Figure 2.6. The case $D = 2$.

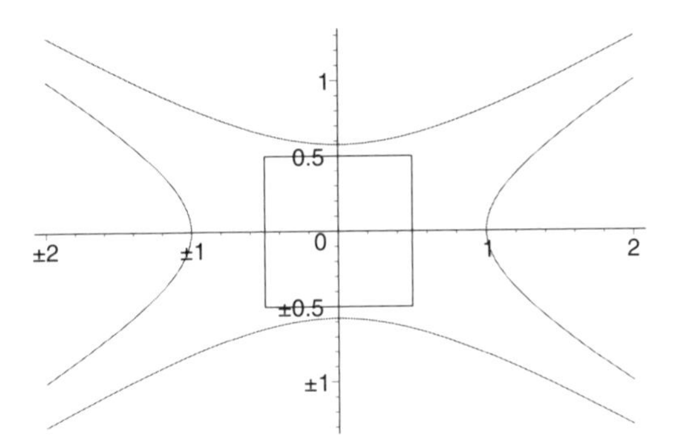

Figure 2.7. The case $D = 3$.

apparently nearest the origin is in this region, as we see from Figures 2.6 and 2.7.

Again, tracing the argument back, we see that for the $\mathbf{Q}$-point $\gamma_0 = (e, f)$, corresponding to the element $e + f\sqrt{D}$ of $\mathbf{Q}(\sqrt{D})$, if $\gamma_1 = (s, t)$ is the $\mathcal{O}$-point, corresponding to the element $s + t\sqrt{D}$ of $R = \mathcal{O}(\sqrt{D})$, apparently nearest to γ_0, then choosing $\gamma = \gamma_1$ we have found an $\mathcal{O}$-point γ with $\|\gamma_0 - \gamma\|_D < 1$, as required, in the cases $D = 2$ or 3, completing the proof in these cases as well.

This concludes the argument in the (relatively) easy cases. As we mentioned above, the other cases are handled in Appendix C, to which we refer the interested reader. □

We have just proved that, for some values of $D, \mathcal{O}(\sqrt{D})$ is a Euclidean domain. In fact, the *only* negative values of D for which $\mathcal{O}(\sqrt{D})$ is a Euclidean domain are the ones we have given, and we shall prove that now. Before we do, we will remark that our list for positive D is not complete. Also, it is much harder to prove that $\mathcal{O}(\sqrt{D})$ is not a Euclidean domain for a positive value of D. The problem here is that the "tails" of the hyperbolas apparently go infinitely far out and so it is possible that some point γ of $\mathcal{O}(\sqrt{D})$ apparently very far away from γ_0 will really be within a distance of 1 from it (or perhaps even closer). We saw some examples of this in the part of the proof of Theorem 2.8 that appears in Appendix C, but with more work can come up with very dramatic examples. For example, taking $D = 41$, we have that $\gamma = 46 - 7\sqrt{41}$, which is apparently very far away from $\gamma_0 = (23/125)\sqrt{41}$, is actually very close to γ_0. Calculation shows $\|\gamma_0 - \gamma\|_{41} = 1/250 = 0.004!$

Lemma 2.9. *If $D < 0$ and $D \neq -1, -2, -3, -7$, or -11, then $R = \mathcal{O}(\sqrt{D})$ is not a Euclidean domain with respect to its norm $\|\cdot\|$.*

Proof: We shall continue to use the language and notation of the proof of Theorem 2.8.

To show that $R = \mathcal{O}(\sqrt{D})$ is *not* a Euclidean domain with respect to its norm, we need only find a point γ_0 of $\triangle_0$ that is not within a distance of 1, in the norm $\|\cdot\|_D$, of any point γ of R, i.e., which is not in the interior of an ellipse centered at any γ.

First, suppose $D \equiv 2$ or $3 \pmod 4$. We are excluding $D = -1$ or -2, so we have $D \leq -5$, i.e., $|D| \geq 5$. Now each ellipse has semiminor axis $1/\sqrt{-D} < 1/2$ centered at a point $\gamma = s + t\sqrt{D}$ where both s and t, and in particular t, are integers. Thus, in order for $\gamma_0 = e + f\sqrt{D}$ to be in such an ellipse, we must have f within a distance of $1/\sqrt{-D}$ of some integer. But $f = 1/2$ is a distance of $1/2 > 1/\sqrt{-D}$ from the nearest integer, and hence any integer, so $\gamma_0 = (1/2)\sqrt{-D}$ is a suitable point.

Now suppose $D \equiv 1 \pmod 4$. We are excluding $D = -3, -7$, or -11, so we have $D \leq -15$, i.e., $|D| \geq 15$. Suppose in fact that $D \neq -15$, so $|D| \geq 17$. The argument here is very similar to the previous case. Each ellipse has semiminor axis $1/\sqrt{-D} < 1/4$ centered at a point $\gamma = s + t\sqrt{D}$ where both s and t, and in particular t, are integers or half-integers. Thus, in order for $\gamma_0 = e + f\sqrt{D}$ to be in such an ellipse, we must have f within a distance of $1/\sqrt{-D}$ from the nearest integer or half-integer, so $\gamma_0 = (1/4)\sqrt{D}$ is a suitable point.

Thus, we have proved the lemma for every value of D except for $D = -15$, and our proof does not work in that case, for the point $\gamma_0 = (1/4)\sqrt{-15}$ is indeed within a distance of 1 from $\gamma = 0$, as $\|(0, 1/4)\|_{-15} = 15/16 < 1$. But here we make a different choice of γ_0. Here we choose $\gamma_0 = (3/11)\sqrt{-15}$. If $\gamma = 0$, then $\|\gamma_0 - \gamma\|_{-15} = \|\gamma_0\|_{-15} = \|(0, 3/11)\|_{-15} = 135/121 > 1$; if $\gamma = (\pm 1/2) + (1/2)\sqrt{-15}$ then $\|\gamma_0 - \gamma\|_{-15} = \|(\pm 1/2, -5/22)\|_{-15} = 496/484 > 1$; and for any other value of γ we see that $\|\gamma_0 - \gamma\|_{-15}$ is even larger (as the difference of the y-coordinates is larger), so γ_0 is a suitable point. □

Remark 2.10.

(1) Actually, the complete list of positive values of D for which $R = \mathcal{O}(\sqrt{D})$ is a Euclidean domain with respect to its norm $\|\cdot\|$ is known. We state this result without proof. These values of D are $D = 2, 3, 5, 6, 7, 11, 13, 17, 19, 21, 29, 33, 37, 41, 57$, and 73.

(2) To be precise, what we showed is that for the values of D in Lemma 2.9, $\mathcal{O}(\sqrt{D})$ is not a Euclidean domain with respect to the norm $\|\cdot\|$ that we have defined. This leaves open the possibility that $\mathcal{O}(\sqrt{D})$ is a Euclidean domain with respect to some different norm. We shall not investigate this question.

Remark 2.11. We are left with a final question: where does the name "Euclidean" come from? The answer is that in a Euclidean domain, we may perform Euclid's algorithm. We shall save this for the next section, when will learn not only how to do it, but also what it is good for.

We close this section by recording a lemma that will enable us to easily tell when an element of a Euclidean domain is a unit. (Recall from Definition 1.6 that an element β of R is a unit if there is an element β' of R with $\beta\beta' = 1$.) This will provide a generalization of Lemma 1.14. (Actually, using this lemma would enable us to simplify some of our earlier proofs slightly, but in the interest of directness we did not do so.)

Lemma 2.12. *Let R be a Euclidean domain and let β be any nonzero element of R. Then $\|\beta\| \geq \|1\|$ and $\|\beta\| = \|1\|$ if and only if β is a unit.*

Proof: For the first inequality, set $a = 1$ and $b = \beta$ in Definition 2.11(2) to conclude that, for any element β,

$$\|1\| \leq \|1 \cdot \beta\| = \|\beta\|.$$

Now suppose β is a unit, and let $\beta\beta' = 1$. Then, setting $a = \beta$ and $b = \beta'$ in Definition 2.1,

$$\|\beta\| \leq \|\beta\beta'\| = \|1\|$$

so combining these two inequalities shows $\|\beta\| = 1$.

On the other hand, suppose $\|\beta\| = 1$. By Definition 2.5, we can find elements q and r of R with

$$1 = \beta q + r \text{ and } r = 0 \text{ or } \|r\| < \|\beta\|.$$

But by assumption, $\|\beta\| = \|1\|$, and by what we have just proved, there are no nonzero elements r of R with $\|r\| < \|1\|$, so we must have $r = 0$ and then $1 = \beta\beta'$ with $\beta' = q$, so β is a unit in R. □

2.2 The GCD-L Property and Euclid's Algorithm

Let us begin by recalling a definition that may be familiar to you in the case of positive integers: Let a and b be positive integers. Their greatest common divisor $g = \gcd(a, b)$ is the unique positive integer with the property that (1) g divides both a and b; and (2) if d is any integer dividing both a and b, then d divides g.

We should point out that a priori the gcd may not exist. We are claiming that there is one and only one positive integer with a certain property, and a priori there may be no such integer or more than one such integer.

But for the positive integers the gcd does indeed exist and can be found by taking the common prime factors of a and b. For example, if $a = 360 = 2^3 \cdot 3^2 \cdot 5$ and $b = 2268 = 2^2 \cdot 3^4 \cdot 7$, then $g = 2^2 \cdot 3^2 = 36$. If $a = 37 = 37$ and $b = 143 = 11 \cdot 13$, then $g = 1$ (as they have no prime factors in common). If $a = 280067 = 229 \cdot 1223$ and $b = 227168 = 2^5 \cdot 31 \cdot 229$, then $g = 229$.

This is not really a satisfactory answer, however, because this *assumes* unique factorization, which we have not shown yet. In fact, we will *use* the gcd to *prove* unique factorization, not the other way around. (It is also not really satisfactory from a practical viewpoint either, since this method of finding the gcd requires us to factor a and b into a product of primes, and this is not so easy, unless a and b are small.) Moreover, we will see that we have a gcd in any Euclidean domain. Thus, since we have already shown that $\mathcal{O}(\sqrt{-1})$ is Euclidean, we can consider elements of that ring.

For example, if $a = 23 - i$ and $b = 24 + bi$, then $g = -1 + i$. This comes from the prime factorization $23 - i = (-1 + i)(2 - i)(7 + 2i)$ and $24+6i = -(1+i)(-1+i)(3)(4+i)$, and as difficult as it may be to find prime factorizations in $\mathbb{Z}$, it is more difficult to find them in $\mathcal{O}(\sqrt{-1})$. (Actually, I have exaggerated here to make a point. We will develop a lot more theory, which will tell us how to do prime factorization in $\mathcal{O}(\sqrt{-1})$, and we will see that it is not too much more difficult than in $\mathbb{Z}$.)

What we shall use is not only the property that elements α and β of R have a gcd, but in addition, that the gcd can be written as a *linear combination* of α and β. That is, if γ is the gcd of α and β, then there are elements δ and ε of R with $\gamma = \alpha\delta + \beta\varepsilon$. For example,

$$\begin{aligned}
36 &= 360 \cdot 19 + 2268 \cdot (-3), \\
1 &= 37 \cdot 58 + 143 \cdot (-15), \\
229 &= 280067 \cdot (-73) + 227168 \cdot 90, \\
-1 + i &= (23 - i)(-6 + 5i) + (24 + 6i)(4 - 6i).
\end{aligned}$$

Even given the prime factorizations of the numbers involved, it is no simple task to find a linear combination that yields their gcd, as the above examples show.

We will develop an algorithm, known as *Euclid's algorithm*, and we will see that Euclid's algorithm provides an effective way to find the gcd of α and β (without having to factor them first), and as a byproduct yields an expression $\gamma = \alpha\delta + \beta\varepsilon$ of their gcd γ as a linear combination of α and β.

For example, $\alpha = 1123456789$ and $\beta = 876543210$ have gcd $\gamma = 1$, and furthermore

$$1 = 1123456789(356396689) + 876543210(-456790122).$$

(I have no idea what the prime factorizations of 1123456789 and 876543210 are.)

But, far more important than the practical aspect, we will be able to prove using Euclid's algorithm that, in this situation, it is always possible to express γ as $\gamma = \alpha\delta + \beta\varepsilon$, and this theoretical result will be the key to proving unique factorization.

With these examples in mind, we get to work.

Definition 2.13. Let R be an integral domain and let $\{\alpha_i\}$ be a set of elements of R, not all of which are zero. Then an element γ of R is a greatest common divisor (gcd) of $\{\alpha_i\}$, $\gamma = \gcd(\{\alpha_i\})$ if

(1) γ divides each α_i,

(2) if ζ is any element of R that divides each α_i, then ζ divides γ.

In general, a gcd may or may not exist. We shall soon explore the question of when it does. But for the moment, let us assume that a gcd does exist and explore the consequences of that assumption.

Lemma 2.14.

(1) Let γ be a gcd *of $\{\alpha_i\}$ and let ε be any unit of R. Then $\gamma' = \gamma\varepsilon$ is also a* gcd *of $\{\alpha_i\}$.*

(2) If γ and γ' are any two gcd*'s of $\{\alpha_i\}$, then $\gamma' = \gamma\varepsilon$ for some unit ε.*

Proof:

(1) By the definition of a unit, there is an element ε' of R with $\varepsilon\varepsilon' = 1$. Then $\gamma = \gamma 1 = \gamma(\varepsilon\varepsilon') = (\gamma\varepsilon)\varepsilon' = \gamma'\varepsilon'$. Thus, γ divides γ' and also γ' divides γ. With that in mind, let us check that γ' satisfies both properties of a gcd.

Property (1): Since γ' divides γ and γ divides each α_i, γ' divides each α_i.

Property (2): Since ζ divides γ and γ divides γ', ζ divides γ'.

(2) By the definition of a gcd, γ divides γ', so $\gamma' = \gamma\varepsilon$ for some ε, and, again by the definition of a gcd, γ' divides γ, so $\gamma = \gamma'\varepsilon'$ for some ε'. Then $\gamma = \gamma'\varepsilon' = \gamma\varepsilon\varepsilon'$ so $1 = \varepsilon\varepsilon'$ and hence ε is a unit. □

Still assuming that a gcd exists, we have the following important definition.

Definition 2.15. If $\{\alpha_i\}$ has a gcd of 1, then $\{\alpha_i\}$ is *relatively prime.*

The next lemma shows we can "factor out" a gcd.

Lemma 2.16. *Let $\{\alpha_i\}$ have a* gcd *of γ, and for each i, write $\alpha_i = \gamma\alpha_i'$. Then $\{\alpha_i'\}$ is relatively prime.*

Proof: We must show that 1 has the two properties of a gcd of $\{\alpha_i'\}$. Now 1 has property (1) of a gcd of $\{\alpha_i'\}$ as 1 certainly divides each α_i'.

Suppose now that ζ is any element of R that divides each α_i'. Then $\gamma\zeta$ divides each $\gamma\alpha_i'$. But $\gamma\alpha_i' = \alpha_i$ so $\gamma\zeta$ divides each α_i. By property (2) of a gcd of $\{\alpha_i\}$, we have that $\gamma\zeta$ divides γ, and hence ζ divides 1. Thus we see that 1 also has property (2) of a gcd of $\{\alpha_i'\}$, so we conclude that 1 is a gcd of $\{\alpha_i'\}$. □

It is sometimes convenient to have a stronger condition than that in Definition 2.15. For further reference, we define that now.

Definition 2.17. If $\{\alpha_i\}$ is a set of elements of R such that any two distinct elements of this set have a gcd of 1, then $\{\alpha_i\}$ is *pairwise relatively prime.*

To see the distinction between these two definitions, let us consider the set of integers $\{6, 10, 15\}$. This set is relatively prime, as it has a gcd of 1, but is not pairwise relatively prime. Looking at pairs of elements, we see that 6 and 10 have a gcd of 2, that 6 and 15 have a gcd of 3, and that 10 and 15 have gcd of 5. Thus in this set, no two distinct elements are relatively prime.

We have been proceeding a bit hypothetically, assuming a gcd exists and exploring the consequences of that assumption. Now let us turn to the question of when a gcd actually does exist.

We shall formulate a stronger property than the mere existence of the gcd, and investigate that.

Definition 2.18. An integral domain R has the *GCD-L property* if the following is true:

(1) Every set of elements $\mathcal{A} = \{\alpha_i\}$ in R, not all zero, has a gcd γ, and

(2) it is possible to write the gcd γ as a linear combination of the elements of $\mathcal{A}$, i.e., there are elements $\{\beta_i\}$ of R such that

$$\gamma = \sum \alpha_i \beta_i.$$

This is a very important property, as we shall see, but GCD-L is not standard mathematical language.

Remark 2.19. In this definition, $\mathcal{A} = \{\alpha_i\}$ may be infinite. In that case, we (implicitly) require that only finitely many of $\{\beta_i\}$ be nonzero, as otherwise we would have an infinite sum, which would not make sense.

Here is our main theorem. First, we will give a very short and slick (but nonconstructive) proof of this theorem. Then we will give a constructive proof that will lead us to Euclid's algorithm.

Theorem 2.20. *Every Euclidean domain R has the GCD-L property.*

First Proof: Let R be a Euclidean domain with norm $\|\cdot\|$, and let $\mathcal{A} = \{\alpha_i\}$ be any set of elements of R not all of which are zero.

Let $\mathcal{S}$ be the set of all linear combination of the elements of $\mathcal{A}$,

$$\mathcal{S} = \left\{ \sum \alpha_i \beta_i \mid \text{ each } \beta_i \text{ is in } R, \text{ and only finitely many } \beta_i \neq 0 \right\}.$$

Observe that $\mathcal{S}$ contains each element of $\mathcal{A}$, as for any value of i we may write $\alpha_i = \alpha_i \cdot 1$. (That is, we write α_i as a linear combination of the elements of $\mathcal{A}$ by choosing $\beta_i = 1$ and $\beta_{i'} = 0$ for $i' \neq i$.) In particular, since not all of the $\{\alpha_i\}$ are zero, $\mathcal{S}$ contains at least one nonzero element. Let

$$\mathcal{S}' = \{\text{nonzero elements of } \mathcal{S}\}.$$

Now let $\mathcal{T}'$ be the set of norms of the element of $\mathcal{S}'$,

$$\mathcal{T}' = \{\|\alpha\| \mid \alpha \text{ is in } \mathcal{S}'\}.$$

The set $\mathcal{T}'$ is a nonempty set of nonnegative integers, so by the Well-Ordering Principle has a smallest element t. Let γ be an element of $\mathcal{S}'$ with $\|\gamma\| = t$. By the definition of $\mathcal{S}'$, γ is a linear combination $\gamma = \sum \alpha_i \beta_i^0$ for some elements $\{\beta_i^0\}$. We claim that γ is a gcd of $\{\alpha_i\}$. To see this, we must verify the properties of the gcd.

Property (1): γ divides each α_i. Actually, we shall prove that γ divides each element of $\mathcal{S}$. Since, as we have observed, each α_i is in $\mathcal{S}$, that shows what we need. To see this, let α be an arbitrary element of $\mathcal{S}$. Then, by definition, $\alpha = \sum \alpha_i\beta_i$ for some $\{\beta_i\}$. Since R is a Euclidean domain, we know we can write

$$\alpha = \gamma\delta + \rho$$

for some δ in R and some ρ in R with $\rho = 0$ or $\|\rho\| < \|\gamma\|$. Substituting, we see

$$\sum \alpha_i\beta_i = \Big(\sum \alpha_i\beta_i^0\Big)\delta + \rho,$$

and solving for ρ we find

$$\rho = \sum \alpha_i(\beta_i - \beta_i^0\delta),$$

which we recognize as a linear combination of $\{\alpha_i\}$, i.e., as an element of $\mathcal{S}$. On the one hand, by our choice of ρ, we have that ρ is an element of $\mathcal{S}$ with $\rho = 0$ or $\|\rho\| < \|\gamma\|$. On the other hand, by the definition of γ, $\|\gamma\| = t$ is the smallest norm of any nonzero element of $\mathcal{S}$, so there are no elements ρ of $\mathcal{S}$ with $\|\rho\| < \|\gamma\|$. Hence, the only possibility for ρ is $\rho = 0$. But, substituting, this gives $\alpha = \gamma\delta$, and so γ divides α.

Property (2): If ζ divides each α_i, then ζ divides γ. To see this, observe that since ζ divides each α_i, we may write each $\alpha_i = \zeta\theta_i$ for some element θ_i of R.

We know

$$\gamma = \sum \alpha_i\beta_i^0,$$

so, substituting, we find

$$\gamma = \sum \zeta\theta_i\beta_i^0 = \zeta\Big(\sum \theta_i\beta_i^0\Big),$$

and hence we see that ζ divides γ. □

We shall build up to our second proof gradually.

Definition 2.21. For $\mathcal{A} = \{\alpha_i\}$ a set of elements of an integral domain R, not all of which are zero, let

$$D(\mathcal{A}) = \{\zeta \text{ in } R \mid \zeta \text{ divides each } \alpha_i \text{ in } \mathcal{A}\}.$$

Remark 2.22. We observe that $D(\mathcal{A})$ is a nonempty set, as it contains the identity element of 1 of R. (Certainly 1 divides every α_i.) Referring to Definition 2.13, we also observe that $\mathcal{A}$ has a gcd if and only if there is some element γ of $D(\{\alpha_i\})$ divisible by every element of $D(\{\alpha_i\})$, in which case γ is the gcd.

Our next lemma shows that we may modify a set of elements $\mathcal{A} = \{\alpha_1, \alpha_2\}$ in a controlled way without changing $D(\mathcal{A})$.

Lemma 2.23. *Let α_1 and α_2 be any two elements of R, and let δ be any element of R. Set $\alpha_2' = \alpha_2 + \delta\alpha_1$. Then $D(\{\alpha_1, \alpha_2'\}) = D(\{\alpha_1, \alpha_2\})$.*

Proof: Let us set $\mathcal{D} = D(\{\alpha_1, \alpha_2\})$ and $\mathcal{D}' = D(\{\alpha_1, \alpha_2'\})$. We want to show that these two sets are the same, and we show this by showing that every element of $\mathcal{D}$ is an element of $\mathcal{D}'$ and vice-versa.

First, suppose that β is in $\mathcal{D} = D(\{\alpha_1, \alpha_2\})$. Then β divides α_1, so $\alpha_1 = \beta\varepsilon_1$, and β divides α_2, so $\alpha_2 = \beta\varepsilon_2$, for some elements ε_1 and of R. But then $\alpha_2' = \alpha_2 + \delta\alpha_1 = \beta\varepsilon_2 + \delta\beta\varepsilon_1 = \beta(\varepsilon_2 + \delta\varepsilon_1)$, so β divides α_2'. Hence β is in $D(\{\alpha_1, \alpha_2'\}) = \mathcal{D}'$.

The argument in the other direction uses the identical logic. Suppose that β' is in $\mathcal{D}' = D(\{\alpha_1, \alpha_2'\})$. Then β' divides α_1, so $\alpha_1 = \beta'\varepsilon_1$, and β' divides α_2', so $\alpha_2' = \beta'\varepsilon_2'$, for some elements ε_1 and of R. But then $\alpha_2 = \alpha_2' - \delta\alpha_1 = \beta'\varepsilon_2' - \delta\beta'\varepsilon_1 = \beta'(\varepsilon_2' - \delta\varepsilon_1)$, so β' divides α_2. Hence β' is in $D(\{\alpha_1, \alpha_2\}) = \mathcal{D}$. □

Here is another lemma about changing sets in a (different) controlled way. Note that at this point we do not want to assume that every set of elements of R, not all of which are zero, has a gcd. (To be sure, we proved that in Theorem 2.20, but we are building up to a second, independent proof of that theorem here.) So we include the assumption that the sets $\mathcal{A}$ and $\{\alpha', \gamma\}$ each have a gcd as part of our hypothesis.

Lemma 2.24. *Let $\mathcal{A} = \{\alpha_i\}$ be any set of elements of R, not all of which are zero, and suppose that $\mathcal{A}$ has a* gcd *γ. Let α' be any element of R. If $\{\alpha', \gamma\}$ has a* gcd *γ', then γ' is also a* gcd *of the set $\mathcal{A}' = \{\alpha'\} \cup \mathcal{A}$. (In particular, the set $\mathcal{A}'$ has a* gcd.*)*

Proof: We shall show that $D(\{\alpha', \gamma\}) = D(\{\alpha'\} \cup \mathcal{A})$. In light of Remark 2.11, this proves the lemma.

Again, we show that these two sets are equal by showing that every element of one of them is also an element of the other.

Suppose δ is in $D(\{\alpha', \gamma\})$. Then δ divides α' and δ divides γ. Since γ divides each α_i, we see that δ divides each α_i, so δ is in $D(\{\alpha'\} \cup \mathcal{A})$.

On the other hand, suppose δ' is in $D(\{\alpha'\} \cup \mathcal{A})$. Then δ' divides α', and δ' divides each α_i. But by the definition of the gcd of $\mathcal{A}$, δ' divides γ. Hence δ is in $D(\{\alpha', \gamma\})$. □

With these results in hand, we can now give our second proof of Theorem 2.20. This proof only applies, however, to the case that $\mathcal{A} = \{\alpha_i\}$ is a finite set.

Second Proof of Theorem 2.20: Suppose that $\mathcal{A} = \{\alpha_i\} = \{\alpha_1, \ldots, \alpha_n\}$ is a finite set consisting of n elements. We prove the theorem by induction on n.

First, suppose $n = 1$. Then the gcd of $\{\alpha_1\}$ is clearly $\gamma = \alpha_1$. (α_1 divides α_1 and any β that divides α_1 divides α_1.) So $\{\alpha_1\}$ has a gcd, and furthermore $\alpha_1 = \alpha_1 \cdot 1$ so we see that both conditions in Definition 2.18 are satisfied.

Next, suppose $n = 2$. This is the crucial case. To handle this case we employ a procedure known as *Euclid's algorithm*. Consider $\{\alpha_1, \alpha_2\}$. If $\alpha_2 = 0$ then (as every element of R divides 0), the gcd of $\{\alpha_1, \alpha_2\}$ is the gcd of $\{\alpha_1\}$, which we have just observed is α_1. Also, $\alpha_1 = \alpha_1 \cdot 1 + \alpha_2 \cdot 0$. So in this case we are done. Similarly, if $\alpha_1 = 0$ then by the same logic the gcd of $\{\alpha_1, \alpha_2\}$ is α_2, and $\alpha_2 = \alpha_1 \cdot 0 + \alpha_2 \cdot 1$, and we are again done. Now suppose α_1 and α_2 are both nonzero.

To avoid notational confusion, we shall set $\theta_1 = \alpha_1$ and $\theta_2 = \alpha_2$. We may use the division algorithm in the Euclidean domain R to write

$$\theta_1 = \theta_2\delta_1 + \theta_3 \quad \text{with } \theta_3 = 0 \text{ or } \|\theta_3\| < \|\theta_2\|.$$

If $\theta_3 = 0$, we stop. Otherwise we continue the process, and write

$$\theta_2 = \theta_3\delta_2 + \theta_4 \quad \text{with } \theta_4 = 0 \text{ or } \|\theta_4\| < \|\theta_3\|.$$

If $\theta_4 = 0$, we stop. Otherwise we continue the process, and write

$$\theta_3 = \theta_4\delta_3 + \theta_5 \quad \text{with } \theta_5 = 0 \text{ or } \|\theta_5\| < \|\theta_4\|.$$

Keep going.

We claim this process eventually stops. If it never did, we would have an infinite sequence $\theta_2, \theta_3, \theta_4, \theta_5, \ldots$, with $\|\theta_2\| > \|\theta_3\| > \|\theta_4\| > \|\theta_5\| > \cdots$. But each of $\|\theta_i\|$ is a nonnegative integer, and it is impossible to have an infinite sequence of strictly decreasing nonnegative integers. (This is a consequence of the Well-Ordering Principle.) So it stops at some stage. Let that be stage k.

Thus, we see we have a sequence of divisions:

$$\begin{array}{lllll} \theta_1 & = & \theta_2\delta_1 + \theta_3 & \|\theta_3\| & < & \|\theta_2\|, \\ \theta_2 & = & \theta_3\delta_2 + \theta_4 & \|\theta_4\| & < & \|\theta_3\|, \\ & & \vdots & & & \\ \theta_{k-3} & = & \theta_{k-2}\delta_{k-3} + \theta_{k-1} & \|\theta_{k-1}\| & < & \|\theta_{k-2}\|, \\ \theta_{k-2} & = & \theta_{k-1}\delta_{k-2} + \theta_k & \|\theta_k\| & < & \|\theta_{k-1}\|, \\ \theta_{k-1} & = & \theta_k\delta_{k-1}. & & & \end{array}$$

We claim $\gamma = \theta_k$ is the gcd of $\theta_1 = \alpha_1$ and $\theta_2 = \alpha_2$. To prove this we use Lemma 2.23 repeatedly:

$$\begin{array}{rlll} \theta_k & = \gcd(\theta_k, 0) & & \\ & = \gcd(\theta_k, 0 + \theta_k\delta_{k-1}) & = \gcd(\theta_k, \theta_{k-1}) & = \gcd(\theta_{k-1}, \theta_k) \\ & = \gcd(\theta_{k-1}, \theta_k + \theta_{k-1}\delta_{k-2}) & = \gcd(\theta_{k-1}, \theta_{k-2}) & = \gcd(\theta_{k-2}, \theta_{k-1}) \\ & = \gcd(\theta_{k-2}, \theta_{k-1} + \theta_{k-2}\delta_{k-3}) & = \gcd(\theta_{k-2}, \theta_{k-3}) & = \gcd(\theta_{k-3}, \theta_{k-2}) \\ & = \cdots & \cdots & = \gcd(\theta_2, \theta_3) \\ & = \gcd(\theta_2, \theta_3 + \theta_2\delta_1) & = \gcd(\theta_2, \theta_1) & = \gcd(\theta_1, \theta_2), \end{array}$$

as required.

This shows the first condition in Definition 2.18. Now we need to show the second condition, that γ can be written as a linear combination of α_1 and α_2. In practice, as we will see from the examples we do after finishing the proof, we work up from the bottom. But it is easier to prove this instead by induction. In fact, we claim, and we shall prove by complete induction, that it is possible to write each θ_i, and hence in particular $\gamma = \theta_k$, as a linear combination of $\theta_1 = \alpha_1$ and $\theta_2 = \alpha_2$.

This claim is certainly true for θ_1, as $\theta_1 = \theta_1 \cdot 1 + \theta_2 \cdot 0$, and also for θ_2, as $\theta_2 = \theta_1 \cdot 0 + \theta_2 \cdot 1$.

So suppose $i \geq 3$ and the claim is true for θ_j for all $j < i$. Then in particular

$$\theta_{i-2} = \theta_1\varepsilon_1 + \theta_2\varepsilon_2,$$
$$\theta_{i-1} = \theta_1\zeta_1 + \theta_2\zeta_2$$

for some ε_1, ε_2, ζ_1, and ζ_2. We also know that

$$\theta_{i-2} = \theta_{i-1}\delta_{i-2} + \theta_i,$$

so, solving for θ_i and substituting, we have

$$\begin{aligned}\theta_i &= \theta_{i-2} - \theta i - 1\delta_{i-2} \\ &= (\theta_1\varepsilon_1 + \theta_2\varepsilon_2) - (\theta_1\zeta_1 + \theta_2\zeta_2)\delta_{i-2} \\ &= \theta_1(\varepsilon_1 - \zeta_1\delta_{i-2}) + \theta_2(\varepsilon_2 - \zeta_2\delta_{i-2}),\end{aligned}$$

and thus we have that θ_i is written as a linear combination of θ_1 and θ_2, completing the inductive step, and hence proving the claim.

This concludes the proof of the $n = 2$ case.

Now for the inductive step. Suppose the theorem is true for $\{\alpha_1, \ldots, \alpha_{n-1}\}$ and consider $\{\alpha_1, \ldots, \alpha_n\}$. We have already considered the cases $n = 1$ and $n = 2$, so we may suppose $n \geq 3$. There are two easy cases: if $\alpha_n = 0$, then the situation for $\{\alpha_1, \ldots, \alpha_n\}$ reduces immediately to the situation for $\{\alpha_1, \ldots, \alpha_{n-1}\}$, and if $\alpha_1 = \ldots = \alpha_{n-1} = 0$, the gcd is α_n. So suppose that neither of these is the case. Then, by the inductive hypothesis, $\{\alpha_1, \ldots, \alpha_{n-1}\}$ has a gcd γ and then by the $n = 2$ case $\{\alpha_n, \gamma\}$ has a gcd γ'. But by Lemma 2.24, γ' is then the gcd of $\{\alpha_1, \ldots, \alpha_n\}$. This verifies the first condition of Definition 2.18.

Now for the second condition. By the inductive hypothesis we may assume that we have written

$$\gamma = \alpha_1\beta_1 + \alpha_2\beta_2 + \ldots + \alpha_{n-1}\beta_{n-1}$$

for some elements β_1, β_2, ..., β_{n-1} of R, and by the $n = 2$ case we may assume that we have written

$$\gamma' = \alpha_n\zeta_1 + \gamma\zeta_2.$$

Substituting, we see that

$$\begin{aligned}\gamma' &= \alpha_n\zeta_1 + (\alpha_1\beta_1 + \alpha_2\beta_2 + \ldots + \alpha_{n-1}\beta_{n-1})\zeta_2 \\ &= \alpha_1(\beta_1\zeta_2) + \alpha_2(\beta_2\zeta_2) + \ldots + \alpha_{n-1}(\beta_{n-1}\zeta_2) + \alpha_n\zeta_1\end{aligned}$$

and γ' is a linear combination of the elements of $\{\alpha_1, \ldots, \alpha_n\}$.

Thus, the truth of the theorem for $n - 1$ implies its truth for n, and by induction we are done. □

We now use Euclid's algorithm, as given in our second proof of Theorem 2.20, to find the gcd of elements of a Euclidean domain R, and to express the gcd as a linear combination of those elements.

Example 2.25. We begin by considering the case $R = \mathbb{Z}$.

(1) Let $\alpha_1 = 2268$ and $\alpha_2 = 360$. Then

$$\begin{aligned} 2268 &= 360 \cdot 6 + 108, \\ 360 &= 108 \cdot 3 + 36, \\ 108 &= 36 \cdot 3, \end{aligned}$$

so we see that $\gcd(2268, 360) = 36$. We find a linear combination that works by solving for the gcd in the next-to-the-last equation and working our way up, substituting at each step:

$$\begin{aligned} 36 &= 360 + (108)(-3) \\ &= \quad 360 + (2268 + 360(-6))(-3) \\ &= 2268(-3) + 360(19). \end{aligned}$$

(2) Let $\alpha_1 = 2268$, $\alpha_2 = 360$, and $\alpha_3 = 552$. We find the gcd of α_1, α_2, and α_3 by first finding the gcd γ' of α_1 and α_2, and then finding the gcd of γ' and α_3. We have just done the first step. Here is the second:

$$\begin{aligned} 552 &= 36 \cdot 15 + 12, \\ 36 &= 12 \cdot 3, \end{aligned}$$

so the gcd is 12, and furthermore, also using our work above,

$$\begin{aligned} 12 &= 552 + 36(-15) \\ &= \quad 552 + (2268(-3) + 360(19))(-5) \\ &= 2268(45) + 360(-285) + 552(1). \end{aligned}$$

(3) Let $\alpha_1 = 143$ and $\alpha_2 = 37$. Then

$$\begin{aligned} 143 &= 37 \cdot 3 + 32, \\ 37 &= 32 \cdot 1 + 5, \\ 32 &= 5 \cdot 6 + 2, \\ 5 &= 2 \cdot 2 + 1, \\ 2 &= 1 \cdot 2, \end{aligned}$$

so the gcd is 1, and then

$$\begin{aligned} 1 &= 5 + 2(-2) \\ &= \quad 5 + (32 + 5(-6))(-2) \\ &= 32(-2) + 5(13) \\ &= \quad 32(-2) + (37 + 32(-1))(13) \\ &= 37(13) + 32(-15) \\ &= \quad 37(13) + (143 + 37(-3))(-15) \\ &= 143(-15) + 37(58). \end{aligned}$$

In this example, we have followed the usual practice of using positive remainders. But there is another practice, namely, using remainders whose absolute value is as small as possible. Let us redo this example that way:

$$\begin{aligned} 143 &= 37 \cdot 4 + (-5), \\ 37 &= -5(-7) + 2, \\ -5 &= 2(-3) + 1, \\ 2 &= 1 \cdot 2, \end{aligned}$$

so again we find the gcd is 1 (of course), and then

$$\begin{aligned} 1 &= -5 + 2(3) \\ &= \quad -5 + (37 + (-5)7)(3) \\ &= 37(3) + (-5)22 \\ &= \quad 37(3) + (143 + 37(-4))22 \\ &= 143(22) + 37(-85). \end{aligned}$$

(4) Let $\alpha_1 = 227168$ and $\alpha_2 = 280067$. Then

$$\begin{aligned} 280067 &= 227168 \cdot 1 + 52899, \\ 227168 &= 52899 \cdot 4 + 15572, \\ 52899 &= 15572 \cdot 3 + 6183, \\ 15572 &= 6183 \cdot 2 + 3206, \\ 6183 &= 3206 \cdot 1 + 2977, \\ 3206 &= 2977 \cdot 1 + 229, \\ 2977 &= 229 \cdot 13, \end{aligned}$$

so the gcd is 229, and then

$$\begin{aligned} 229 &= 3206 + 2977(-1) \\ &= \quad 3206 + (6183 + 3206(-1))(-1) \\ &= 6183(-1) + 3206(2) \\ &= \quad 6183(-1) + (15572 + 6183(-2))(2) \\ &= 15572(2) + 6183(-5) \\ &= \quad 15572(2) + (52899 + 15572(-3))(-5) \\ &= 52899(-5) + 15572(17) \\ &= \quad 52899(-5) + (227168 + 52899(-4))(17) \\ &= 227168(17) + 52899(-73) \\ &= \quad 227168(17) + (280067 + 227168(-1))(-73) \\ &= 280067(-73) + 227168(90). \end{aligned}$$

(Note, as a practical matter, that in this example we easily found that $\gcd(227168, 280067) = 229$, whereas it would have been quite difficult to factor 227168 and 280067 into primes.)

(5) Let $\alpha_1 = 1123456789$ and $\alpha_2 = 876543210$. Then

$$\begin{aligned}
1123456789 &= 876543210 \cdot 1 + 246913579, \\
876543210 &= 246913578 \cdot 3 + 135802473, \\
246913579 &= 135802473 \cdot 1 + 111111106, \\
135802473 &= 111111106 \cdot 3 + 24691367, \\
111111106 &= 24691367 \cdot 4 + 12345638, \\
24691367 &= 12345638 \cdot 2 + 91, \\
12345638 &= 91 \cdot 135666 + 32, \\
91 &= 32 \cdot 2 + 27, \\
32 &= 27 \cdot 1 + 5, \\
27 &= 5 \cdot 5 + 2, \\
5 &= 2 \cdot 2 + 1, \\
2 &= 1 \cdot 2,
\end{aligned}$$

so the gcd is 1, and then

$$\begin{aligned}
1 &= 5 + 2(-2) \\
&= \quad 5 + (27 + 5(-5))(-2) \\
&= 27(-2) + 5(11) \\
&= \quad 27(-2) + (32 + 27(-1))(11) \\
&= 32(11) + 27(-13) \\
&= \quad 32(11) + (91 + 32(-2))(-13) \\
&= 91(-13) + 32(27) \\
&= \quad 91(-13) + (12345638 + 91(-135666))(37)
\end{aligned}$$

$$\begin{aligned}
&= 12345638(37) + 91(-5019655) \\
&= \quad 12345638(37) + (24691367 + 12345638(-2))(-5019655) \\
&= 24691367(-5019655) + 12345638(10039347). \\
&= \quad 24691367(-5019655) + (111111106 + 24691367(-4))(10039347) \\
&= 111111106(10039347) + 24691367(-45177043) \\
&= \quad 111111106(10039347) + (135802473 + 111111106(-1))(-45177043) \\
&= 135802473(-45177043) + 111111106(55216390) \\
&= \quad 135802473(-45177043) + (246913579 + 135802473(-1))(55216390) \\
&= 246913579(55216390) + 135802473(-100393433) \\
&= \quad 246913579(55216390) + (876543210 + 246913579(-3))(-100393433) \\
&= 876543210(-100393433) + 246913579(356396689) \\
&= \quad 876543210(-100393433) + (1123456789 + 876543210(-1))(356396689) \\
&= 1123456789(356396689) + 876543210(-456790122).
\end{aligned}$$

(Note, as a practical matter, that this computation can easily be performed with only the use of an arbitrary precision calculator. It is also clear that this method can be implemented on a computer to find quickly and easily the gcd of huge numbers and to represent the gcd as a linear combination of those numbers.)

Example 2.26.

(1) Now we do an example with $R = \mathcal{O}(\sqrt{-1})$. In finding the quotient (and remainder) at each step, we follow the strategy of the proof of Theorem 2.8: if $(a + b\sqrt{D})/(c + d\sqrt{D}) = e + f\sqrt{D}$ with e, f in $\mathbf{Q}$, we let the quotient be $s + t\sqrt{D}$ where s and t are integers closest to e and f, respectively (and then the remainder is forced).

Let $\alpha_1 = 24 + 6i$ and $\alpha_2 = 23 - i$. Then

$$\begin{aligned}
24 + 6i &= (23 - i)(1) + (1 + 7i) \\
&\qquad\qquad (\text{as } (25 + 6i)/(23 - i) = (546 + 162i)/530), \\
23 - i &= (1 + 7i)(-3i) + (2 + 2i) \\
&\qquad\qquad (\text{as } (23 - i)/(1 + 7i) = (16 - 162i)/50), \\
1 + 7i &= (2 + 2i)(2 + i) + (-1 + i) \\
&\qquad\qquad (\text{as } (1 + 7i)/(2 + 2i) = (16 + 5i)/8), \\
2 + 2i &= (-1 + i)(-2i),
\end{aligned}$$

so the gcd is $-1 + i$, and then

$$\begin{aligned}
-1 + i &= (1 + 7i) + (2 + 2i)(-2 - i) \\
&= \quad (1 + 7i) + ((23 - i) + (1 + 7i)(3i))(-2 - i) \\
&= (23 - i)(-2 - i) + (1 + 7i)(4 - 6i) \\
&= \quad (23 - i)(-2 - i) + ((24 + 6i) + (23 - i)(-1))(4 - 6i) \\
&= (24 + 6i)(4 - 6i) + (23 - i)(-6 + 5i).
\end{aligned}$$

(By way of further explanation, $(25 + 6i)/(23 - i) = (546 + 162i)/530 = (546/530) + (162/530)i$, which is nearest to 1; $(23 - i)/(1 + 7i) = (16 - 162i)/50 = (16/50) + (-162/50)i$, which is nearest to $-3i$; and $(1+7i)/(2+2i) = (16 + 5i)/8 = 2 + (5/8)i$, which is nearest to $2 + i$.)

(2) In our next example, $R = \mathcal{O}(\sqrt{-7})$. The logic of this example depends on the proof of Theorem 2.8 in the case $D = -7$. Since we deferred the proof of that case of Theorem 2.8 to Appendix C, we similarly defer this example to Appendix C.

Remark 2.27. Note that the gcd is only defined up to multiplication by a unit in R (i.e., if γ is a gcd of α_i and α_2, so is $\gamma\epsilon$ for any unit ϵ of R—compare Lemma 2.14), so it would be, strictly speaking, better to speak of "a" gcd rather than "the" gcd. In particular, if $R = \mathbb{Z}$, with units $\{\pm 1\}$, then, for example, -36 is also a gcd of 2268 and 360, and $-36 = 2258(3) + 360(-19)$.

If $R = \mathcal{O}(\sqrt{-1})$, the units are $\{\pm 1, \pm i\}$, so $1 - i = -(-1 + i), -1 - i = i(-1 + i)$, and $1 + i = -i(-1 - i)$ are also gcd's of $24 + 6i$ and $23 - i$, and, for example, $1 + i = (24 + 6i)(-6 - 4i) + (23 - i)(5 + 6i)$.

2.3 Ideals and Principal Ideal Domains

We are interested in investigating questions of factoring elements in an integral domain R. With a lot of mathematical hindsight, it turns out that instead of just looking at individual elements α of R, we should also look at subsets of R consisting of all multiples of α. If we consider multiples of α, we observe that

(1) the sum of any two multiples of α is a multiple of α,

(2) any multiple of a multiple of α is a multiple of α.

Again, with a lot of mathematical hindsight, it is precisely these two properties that we wish to consider in general. So we are led to the definition of an ideal.

Definition 2.28. An *ideal* I in an integral domain R is a nonempty subset of R with the following properties:

(1) if α_1 and α_2 are in I, then $\alpha_1 + \alpha_2$ is in I,

(2) if α_1 is in I and β is any element of R, then $\alpha_1\beta$ is in I.

In other words, an ideal I is a nonempty subset of R that is closed under addition and also is closed under multiplication by any element of R (not just by elements of I).

Our first examples of ideals consist precisely of the multiples of some element of R.

Definition 2.29. Let α be an element of R. The *principal ideal* generated by α is

$$I_\alpha = \{\alpha' \mid \alpha' = \alpha\beta \text{ for some } \beta \text{ in } R\}.$$

Let us check that principal ideals are indeed ideals.

Lemma 2.30. *Let α be an element of R. Then I_α is an ideal.*

Proof: We need to check that I_α satisfies the two properties of an ideal.

Property (1): Let α_1 and α_2 be in I_α. Then $\alpha_1 = \alpha\beta_1$ and $\alpha_2 = \alpha\beta_2$ for some elements β_1 and β_2 of R. Then $\alpha_1 + \alpha_2 = \alpha\beta_2 = \alpha(\beta_1 + \beta_2)$, so $\alpha_1 + \alpha_2$ is in I.

Property (2): Let α_1 be in I_α, and let β be any element of R. Then $\alpha_1 = \alpha\beta_1$ for some element β_1 of R. Then $\alpha_1\beta = (\alpha\beta_1)\beta = \alpha(\beta_1\beta)$, so $\alpha_1\beta$ is in I. □

Here are two extreme cases.

Example 2.31.

(1) $I = R$ is an ideal. (In this case, I is called an *improper ideal.* Otherwise, I is called a *proper ideal.*) Note that R is a principal ideal as $R = I_1$. (Every element of R is a multiple of 1.)

(2) $I = \{0\}$. (In this case, I is called the *zero ideal.* Otherwise, I is called a *nonzero ideal.*) Note that $\{0\}$ is a principal ideal as $\{0\} = I_0$. (The only multiple of 0 is 0.)

Let us return to Definition 2.29. On the one hand, we see that the ideal generated by α is simply the set of multiples of α. On the other hand, we see that the set of multiples of any element α of R is an ideal, and in fact a principal ideal. So you may ask: are *all* ideals of R of this form? Excellent question! But we shall defer the answer to this question for a while. Right now, we continue with a general development of properties of ideals.

Lemma 2.32. *Let I be an ideal of R. The following are equivalent:*

(1) 1 is in I.

(2) $I = R$.

Proof:

(1) implies (2): suppose 1 is in I. Then for any element β of R, by part (2) of Definition 2.28, $1\beta = \beta$ is in I.

(2) implies (1): if $I = R$, then certainly 1 is in I. □

Corollary 2.33. *Let R be an integral domain. Then R is a field if and only if the only ideals in R are $I = \{0\}$ and $I = R$.*

Proof: First suppose R is a field, and let I be an ideal in R. If $I \neq \{0\}$, then I contains a nonzero element α of R. But any $\alpha \neq 0$ is a unit in R, so there is an element β of R with $\alpha\beta = 1$. Then 1 is in I so, by Lemma 2.32, $I = R$.

Conversely, suppose the only ideals in R are $\{0\}$ and R. Let α be any nonzero element of R. Then I_α is a nonzero ideal (as it contains α), so $I_\alpha = R$, and so 1 is in I_α. But I_α is the set of multiples of α, so $1 = \alpha\beta$ for some element β of R, i.e., α is a unit. Thus, every nonzero element of R is a unit, so, by Remark 1.7, R is a field. □

In Definition 2.29, we considered multiples of a single element. Now a multiple of an element is the same as a linear combination of that element. So in the following generalization of Definition 2.29, it is natural to consider linear combinations.

Definition 2.34. Let $\mathcal{A} = \{\alpha_i\}$ be a nonempty set of elements of R.
The ideal generated by $\mathcal{A}$ is

$$\begin{aligned} I_{\mathcal{A}} &= \{ \text{ linear combinations of elements of } \mathcal{A}\} \\ &= \{\textstyle\sum \alpha_i\beta_i \mid \text{ each } \beta_i \text{ is in } R, \text{ and only finitely many } \beta_i \neq 0\}. \end{aligned}$$

Let us see that $I_{\mathcal{A}}$ is always an ideal, and in fact that we get every ideal this way.

Lemma 2.35.

(1) *For any nonempty subset $\mathcal{A}$, $I_{\mathcal{A}}$ is an ideal.*

(2) *Every ideal I is $I_{\mathcal{A}}$ for some $\mathcal{A}$.*

Proof:

(1) We have to verify that $I_{\mathcal{A}}$ satisfies both properties of an ideal. First, $I_{\mathcal{A}}$ is closed under addition: Let α^1 and α^2 be any two elements of $I_{\mathcal{A}}$. Then we may write $\alpha^1 = \sum \alpha_i\beta_i^1$ and $\alpha^2 = \sum \alpha_i\beta_i^2$. But then $\alpha^1 + \alpha^2 = \sum \alpha_i(\beta_i^1 + \beta_i^2)$, and so we see that $\alpha^1 + \alpha^2$ is in $I_{\mathcal{A}}$. Second, $I_{\mathcal{A}}$ is closed under multiplication by any element of R: Let α be any element of $I_{\mathcal{A}}$ and let β be any element of R. Then we may write $\alpha = \sum \alpha_i\beta_i$. But then $\alpha\beta = \sum \alpha_i(\beta_i\beta)$ is in $I_{\mathcal{A}}$.

(2) By the properties of an ideal, $I = I_I$. (That is, we may choose $\mathcal{A} = I$ itself.) □

In case $\mathcal{A} = \alpha$ (i.e., if this set consists of the single element α), then $I_{\mathcal{A}} = I_\alpha$ is nothing other than the principal ideal I_α that we have already considered. Suppose instead, for example, that $\mathcal{A} = \{\alpha_1, \alpha_2\}$. Then we have the ideal I_{α_1,α_2}, and this is not of the form I_α, so is not a priori a principal ideal. But it may turn out that in fact $I_{\alpha_1,\alpha_2} = I_{\alpha_0}$ for some α_0, i.e., that it is indeed a principal ideal. In fact, it may turn out that this is always the case. This is a very important situation to which we give a name.

Definition 2.36. An integral domain R is a *principal ideal domain* (PID) if every ideal I in R is principal.

Recall that we have earlier defined the GCD-L property (Definition 2.18).

Proposition 2.37. *Let R be an integral domain. Then R is a PID if and only if R has the GCD-L property.*

Proof: Suppose that R is a PID. Let $\mathcal{A} = \{\alpha_i\}$ be a set of elements of R, not all zero. Consider the ideal $I_{\mathcal{A}}$. Since R is a PID, this is a principal ideal, so $I_{\mathcal{A}} = I_\gamma$ for some element γ of R. We claim that γ is the gcd of $\mathcal{A}$, and furthermore that γ can be written as a linear combination of the elements of $\mathcal{A}$, and this is precisely what we need to prove in order to show that R has the GCD-L property.

We verify condition (2) of Definition 2.18 first: note that γ is in $I_\gamma = I_{\mathcal{A}}$, so by the definition of $I_{\mathcal{A}}$ (Definition 2.34), $\gamma = \sum \alpha_i\beta_i$ for some $\{\beta_i\}$, i.e., γ is a linear combination of the elements of $\mathcal{A}$.

Now for condition (1) of Definition 2.18: note that each α_i is in $I_A = I_\gamma$, so $\alpha_i = \gamma\beta_i^0$ for some β_i^0 (as every element of I_γ is a multiple of γ), i.e., γ divides each α_i. Suppose that ζ is any element of R that divides each α_i, and write $\alpha_i = \zeta\epsilon_i$ for each i. Then $\gamma = \sum \alpha_i\beta_i = \sum(\zeta\epsilon_i)\beta_i = \zeta(\sum \epsilon_i\beta_i)$ so ζ divides γ. Hence, by Definition 2.13, γ is a gcd of $\{\alpha_i\}$.

Conversely, suppose that R has the GCD-L property. Let I be any ideal. Then I is a set of elements of R, so by the GCD-L property (Definition 2.18) there is an element γ of R such that (1) γ is the gcd of the elements of I, and (2) γ is a linear combination of the elements of I. We claim that $I = I_\gamma$.

First, according to (1), γ divides every element of I, i.e., every element of I is a multiple of γ. Since, by definition, I_γ consists of all the multiples of γ, we see that $I \subseteq I_\gamma$.

Second, (2) states that γ is a linear combination of the elements of I, and then every multiple of γ is also a linear combination of the elements of I, so is also in I. Again, since I_γ consists of all the multiples of γ, we see that $I_\gamma \subseteq I$.

Thus, each of I and I_γ is a subset of the other, so these two sets must be equal.

Thus, every ideal I of R is of the form $I = I_\gamma$ for some element γ of R, i.e., every ideal I of R is principal, and thus (Definition 2.36) R is a PID.□

Corollary 2.38. *Every Euclidean domain R is a PID.*

Proof: This follows directly from Theorem 2.20 and Proposition 2.37. □

Let us observe that the proof of Proposition 2.37 actually showed a more precise result than we stated. Not only did it show that every ideal in a Euclidean ring is always a principal ideal; it in fact identified that ideal. So we actually have the following result:

Corollary 2.39. *Let R be a PID.*

(1) Let $\mathcal{A}$ be any set of elements of R, not all zero. Let $\gamma = \gcd(\mathcal{A})$. Then $I_{\mathcal{A}} = I_\gamma$.

(2) Let I be any nonzero ideal of R, and let $\gamma = \gcd(I)$. Then $I = I_\gamma$.

Proof: (1) is the claim in the first paragraph of the proof of Proposition 2.37, and (2) is the claim in the fourth paragraph of the proof of Proposition 2.37. □

Remark 2.40. One can ask whether, conversely, every PID is a Euclidean domain. The answer is no, but the examples are not particularly illuminating (or useful), so we shall not give any.

We have just shown that any set $\{\alpha_i\}$ of elements of a PID R has a gcd γ. Recall Definition 2.15: if $\{\alpha_i\}$ has a gcd of 1, then $\{\alpha_i\}$ is *relatively prime*.

The following is a classical and extremely useful result.

Lemma 2.41 (Euclid's Lemma). *Let R be a PID and let α be any nonzero element of R. Let β and γ be elements of R and suppose that α divides $\beta\gamma$. If α and β are relatively prime, then α divides γ.*

First Proof: Since α and β are relatively prime we can write

$$1 = \alpha\zeta + \beta\theta$$

for some elements ζ and θ of R. Multiplying by γ, we find

$$\gamma = \alpha\gamma\zeta + \beta\gamma\theta.$$

Now α divides $\beta\gamma$, so $\beta\gamma = \alpha\delta$ for some δ. Substituting, we find

$$\gamma = \alpha\gamma\zeta + \alpha\delta\theta = \alpha(\gamma\delta + \delta\theta)$$

so α divides γ. □

Second Proof: Consider the two elements $\alpha\gamma$ and $\beta\gamma$ of R. By hypothesis, these two elements have a gcd. Call it δ.

Clearly γ divides both $\alpha\gamma$ and $\beta\gamma$. Then, by the definition of a gcd, γ divides δ. Write $\delta = \zeta\gamma$. Now δ divides $\beta\gamma$, i.e., $\zeta\gamma$ divides $\beta\gamma$, so ζ divides β. Also, δ divides $\alpha\gamma$, i.e., $\zeta\gamma$ divides $\alpha\gamma$, so ζ divides α. Thus, we see that ζ divides both α and β. But α and β are assumed to be relatively prime, i.e., to have a gcd of 1. Then, by the definition of a gcd, ζ divides 1, i.e., ζ is a unit. Write $1 = \zeta'\zeta$.

Finally, α clearly divides $\alpha\gamma$, and by hypothesis α divides $\beta\gamma$, so, again by the definition of a gcd, α divides $\delta = \zeta\gamma$. But then α also divides $\zeta'\delta = \zeta'(\zeta\gamma) = (\zeta'\gamma)\gamma = 1\gamma = \gamma$, as claimed. □

Here are two important applications of Euclid's Lemma.

Corollary 2.42. *Let R be a PID and let α and β be relatively prime nonzero elements of R. Let γ be an element of R and suppose that α divides γ and β divides γ. Then $\alpha\beta$ divides γ.*

Proof: Since α divides γ, we may write $\gamma = \alpha\delta$ for some element δ of R. Then β divides $\alpha\delta$, and β and α are relatively prime, so by Euclid's Lemma, β divides δ. Write $\delta = \beta\zeta$ for some element ζ of R. Then

$$\gamma = \alpha\delta = \alpha(\beta\zeta) = (\alpha\beta)\zeta,$$

so $\alpha\beta$ divides γ. □

Corollary 2.43. *Let R be a PID and let α, β, and γ be elements of R. Suppose that α and β are relatively prime, and also that α and γ are relatively prime. Then α and $\beta\gamma$ are relatively prime.*

Proof: Let δ be a gcd of α and $\beta\gamma$. Then δ divides α, and so $\gcd(\delta, \beta)$ divides $\gcd(\alpha, \beta)$. But α and β are assumed to be relatively prime, so δ and β are also relatively prime.

Now δ divides $\beta\gamma$, and δ and β are relatively prime, so by Euclid's Lemma, we conclude that δ divides γ. Thus, δ is a common divisor of α and γ. But α and γ were assumed to be relatively prime, so δ is a unit.

Thus, we see that $\gcd(\alpha, \beta\gamma)$ is a unit, i.e., that α and $\beta\gamma$ are relatively prime. □

Remark 2.44. We presented two proofs of Euclid's Lemma (Lemma 2.41). The first proof was Euclid's original proof and was simpler than the second proof. Thus, the question arises as to why we bothered to provide the second proof. Here is the answer.

If you look at the proofs carefully, you will notice that the first proof required us to use the GCD-L property, Definition 2.18. This definition had two conditions, (1) and (2), and we used both of them. The second proof, on the other hand, only used condition (1) of Definition 2.18 (that the relevant elements of R had a gcd), but did *not* use condition (2) of Definition 2.18 (that the gcd could be expressed as a linear combination of those elements). Hence the second proof works more generally. Thus, we see that

> Euclid's Lemma (Lemma 2.41) and its consequences (Corollary 2.42 and Corollary 2.43) are true in any integral domain R in which any two elements (not both zero) have a gcd,

whether or not that gcd can be expressed as a linear combination of those elements.

2.4 Unique Factorization Domains

In this section we reach our first main goal, of showing that in certain integral domains we do have unique factorization.

We must begin, however, by carefully defining our terms.

Definition 2.45. Let R be an integral domain.

(1) An element of α of R is a *unit* if α divides 1 (i.e., if there is an element α' of R with $\alpha\alpha' = 1$).

(2) Two elements α_1 and α_2 of R are *associates* if $\alpha_2 = \alpha_1\beta$ where β is a unit of R.

(3) An element α of R is *irreducible* if α is not a unit and if $\alpha = \beta\gamma$ implies that β is a unit (and hence that α and γ are associates) or that γ is a unit (and hence that α and β are associates).

(4) An element α of R is *prime* if α divides a product $\beta\gamma$ implies that α divides β or α divides γ.

Remark 2.46. You may have been a bit surprised by parts (3) and (4) of this definition. The usual definition of a positive integer a being prime is that $a \neq 1$ and if $a = bc$ for some positive integers b and c, then $b = 1$ or $c = 1$. The generalization of that to an integral domain R is in part (3) of Definition 2.45, but we are not calling this generalization prime. Rather, we are calling it irreducible, and using the term prime to denote something else, defined in part (4). Although surprising, this turns out to be the right thing to do. As we will show (see Lemma 2.48), in the integers, or in any integral domain with unique factorization, these two notions turn out to be equivalent, so it does not matter whether we use notion (3) or (4) in that case, but in general (4) is the right notion to use.

We will observe that (3) may be more practical to check. In the case of the positive integers, to check (3) we only have to try divisors of a, and we know any such divisor must be less than or equal to a, so there are only finitely many numbers to check.

On the other hand, to check (4), we must look at *any* number d and see if d is divisible by a. If so, we must look at *any* factorization $d = bc$ of d, and see if a divides one of the factors b or c, and here there are infinitely many numbers to check. This difference, as well as millennia of mathematical tradition, are the reasons the usual definition is preferred for the positive integers.

Actually, a prime element is always irreducible, although in general, an irreducible element may not be prime. Let us see the first of these claims now. (We defer examples of the second.)

Lemma 2.47. *Let R be an integral domain. If an element α of R is prime, then α is irreducible.*

Proof: Suppose α is prime. Let $\alpha = \beta\gamma$. Then α certainly divides $\beta\gamma$, so α divides β or α divides γ. Suppose α divides β, i.e., $\beta = \alpha\delta$. Then $\alpha = \beta\gamma = (\alpha\delta)\gamma = \alpha(\delta\gamma)$, so $1 = \delta\gamma$, and hence γ divides 1, so γ is a unit. Similarly, if α divides γ then β is a unit. Thus α is irreducible. □

Lemma 2.48. *Let R be a PID. Then an element α of R is prime if and only if α is irreducible.*

Proof: In Lemma 2.47 we showed that, in any integral domain, a prime is irreducible. Thus, we must show that here (in the case of a PID), an irreducible is prime.

Thus, let α be an irreducible element of R and suppose α divides a product $\beta_1\beta_2$. We want to show that α divides one of the factors. Let γ be the gcd of α and β_1. (Since R is a PID, we know that $\gcd(\alpha, \beta_1)$ exists.) Then γ certainly divides α, i.e, $\alpha = \gamma\delta$. But α is irreducible, so that means γ or δ is a unit. Suppose δ is a unit. Then α and γ are associates. Now γ certainly divides β_1, so α divides β_1 as well. On the other hand, suppose γ is a unit. Then α and β_1 are relatively prime. But then we can apply Euclid's Lemma (Lemma 2.41, also true because R is a PID) to conclude that α divides β_2. Hence α is prime. □

Definition 2.49. An integral domain R is a *unique factorization domain* (UFD) if every nonzero element α of R can be written essentially uniquely as $\alpha = up_1 \cdots p_k$ with u a unit and each p_i irreducible, i.e.,

(1) every nonzero α can be written as $\alpha = up_1 \cdots p_k$ with u a unit and each p_i irreducible, and

(2) if also $\alpha = vq_1 \cdots q_\ell$ with v a unit and $q_1 \cdots q_\ell$ irreducible, then $\ell = k$ and, after possibly reordering, q_i and p_i are associates for each i.

Remark 2.50. Observe that essential uniqueness is the best we can hope for. For example, in the integers, we have $6 = (1)(2)(3) = (-1)(-2)(3) = (-1)(2)(-3) = (1)(-2)(-3) = (1)(3)(2) = (-1)(-3)(2) = (-1)(3)(-2) = (1)(-3)(-2)$.

We need the following technical lemma in our proof of Theorem 2.52.

Lemma 2.51. *Let $\alpha_1, \alpha_2, \alpha_3, \ldots,$ be a sequence of nonzero elements in a PID R such that α_i is divisible by α_{i+1} for each i. Then there is some integer k such that $\alpha_k, \alpha_{k+1}, \ldots,$ are all associates.*

Proof: Let $\mathcal{A} = \{\alpha_1, \alpha_2, \alpha_3, \ldots\}$. Since R is a PID, it has the GCD-L property by Proposition 2.37, and so $\mathcal{A}$ has a gcd γ and we can write γ as a finite sum of multiples of elements of $\mathcal{A}$. If k is the highest index that appears in this sum, we claim that the terms from α_k on are all associates.

Let us write $\gamma = \sum_{i=1}^{k} \alpha_i \beta_i$ for some elements β_i of R. Now by assumption α_2 divides α_1 and α_3 divides α_2, so α_3 divides α_1 as well. Continuing in this fashion, we see that α_k divides α_i for each $i = 1, \ldots, k$. Thus, α_k divides each term in the above sum for γ, so we see α_k divides γ as well.

Now consider any term α_ℓ with $\ell \geq k$. Then, just as above, α_ℓ divides α_k. We have seen that α_k divides γ. But γ is the gcd of $\mathcal{A}$, so it divides each α_i, and in particular it divides α_ℓ. Thus α_k and α_ℓ divide each other, so are associates. □

Here is a very important result.

Theorem 2.52. *Every PID R is a UFD.*

Proof: We must show that any nonzero element of α of R has an essentially unique factorization. We show this in two stages: first, that α has a factorization, and second, that this factorization is essentially unique.

In the first stage we claim that either α is a unit or α is divisible by some irreducible element of R.

We prove this claim: If α is a unit, we are done, so suppose α is not a unit. If α is irreducible, we are done. If not, then $\alpha = \alpha_1\beta_1$, where neither α_1 nor β_1 are units. If α_1 is irreducible, we are done. If not, $\alpha_1 = \alpha_2\beta_2$ where neither α_2 nor β_2 are units, and so $\alpha = \alpha_1\beta_1 = \alpha_2\beta_2\beta_1$. If α_2 is irreducible we are done. If not, $\alpha_2 = \alpha_3\beta_3$ where neither α_3 nor β_3 are units, and then $\alpha = \alpha_2\beta_2\beta_1 = \alpha_3\beta_3\beta_2\beta_1$. Continue this process. If it eventually stops, at step k, say, then $\alpha = \alpha_k\beta_k\ldots\beta_2\beta_1$, so α is divisible by the irreducible element α_k. Thus to complete the proof, we need only show the process eventually stops. Suppose not. Then we get a sequence $\alpha_1, \alpha_2, \alpha_3, \ldots$ with each α_i divisible by α_{i+1}. Then we can apply Lemma 2.51 to conclude that from some point α_k on, the elements are all associates. In particular, α_k and α_{k+1} are associates, so $\alpha_k = \alpha_{k+1}\beta_{k+1}$ *with* β_{k+1} *a unit.* But in our process we assumed that *neither* α_{k+1} *nor* β_{k+1} *were units.* Thus, we have a contradiction if the sequence goes on forever. This is impossible, and so we conclude that the sequence stops.

With this claim in hand, we can complete the proof of the first stage by a very analogous argument.

We know that α is divisible by an irreducible element α_1. Write $\alpha = \alpha_1\beta_1$. If β_1 is a unit, we are done, so suppose not. Then β_1 is divisible by an irreducible element α_2. Write $\beta_1 = \alpha_2\beta_2$ so $\alpha = \alpha_1\beta_1 = \alpha_1\alpha_2\beta_2$. If β_2 is a unit we are done. If not, $\beta_2 = \alpha_3\beta_3$ with α_3 irreducible and $\alpha = \alpha_1\alpha_2\beta_2 = \alpha_1\alpha_2\alpha_3\beta_3$. Continue this process. If it eventually stops, at step k, say, then $\alpha = \alpha_1\alpha_2\cdots\alpha_k\beta_k$ with $\alpha_1, \alpha_2, \ldots, \alpha_k$ irreducibles and β_k a unit, and we have our desired factorization. Thus to complete the proof, we need only show that the process eventually stops. Suppose not. Then we get a sequence $\beta_1, \beta_2, \beta_3$ with each β_i divisible by β_{i+1}. Then we can apply the preceding lemma to conclude that from some point β_k on, the elements are all associates. In particular, $\beta_k = \beta_{k+1}\alpha_{k+1}$ *with* α_{k+1} *a unit.* But in our process, we assumed that α_{k+1} *is irreducible.* Thus, we have a contradiction if the sequence goes on forever. This is impossible, and so we conclude that the sequence stops.

We therefore see that we have completed the proof of the first stage: every α has a factorization. Now we must prove the second stage: this factorization is essentially unique.

Suppose there is an α with two factorizations

$$\alpha = up_1\cdots p_k = vq_1\cdots q_\ell$$

with u and v units and $p_1, \ldots, p_k$ and $q_1, \ldots, q_\ell$ all irreducible. We now crucially use Euclid's Lemma (Lemma 2.41), or rather, its consequence

(Lemma 2.48), which tells us that in the PID R, every irreducible element is prime.

Consider p_1. It is a prime and divides the product $vq_1 \cdots q_\ell = (vq_1)(q_2 \cdots q_\ell)$ and hence divides one of the factors vq_1 or $(q_2 \cdots q_\ell)$. If it divides vq_1, and hence q_1 (as v is a unit), fine.

Otherwise, it divides $q_2 \cdots q_\ell = q_2(q_3 \cdots q_\ell)$. If it divides q_2, fine. Otherwise it divides $q_3 \cdots q_\ell = q_3(q_4 \cdots q_\ell)$. Continue in this fashion to conclude that p_1 divides some q_i. Reordering the q_i's, if necessary, we may assume that p_1 divides q_1. But q_1 is irreducible, so p_1 and q_1 must be associates, so $q_1 = u'p_1$ for some unit u'. Then

$$\alpha = up_1p_2 \cdots p_k = vq_1q_2 \cdots q_\ell = vu'p_1q_2 \cdots q_\ell,$$

so, setting $v' = vu'$,

$$\alpha' = up_2 \cdots p_k = v'q_2 \cdots q_\ell.$$

Now apply the same argument to p_2 and α' to conclude that p_2 divides, and hence is an associate of, some one of $q_2, \ldots, q_\ell$, which by renumbering we may assume is q_2, and then

$$\alpha'' = up_3 \cdots p_k = v''q_3 \cdots q_\ell.$$

Continuing in this way, we see that, until the process stops, and possibly after reordering the q_i's, p_1 is an associate of q_1, p_2 is an associate of q_2, ... We claim that when the process stops, we have used all the p_i's and all the q_i's, so $k = \ell$ and we are just left with a unit on each side, proving the theorem.

Otherwise, either we have used all the p_i's but not all the q_i's, so $\ell > k$, and we are left with

$$u = wq_{k+1} \cdots q_\ell$$

for some unit w, which is impossible, as the left-hand side u is a unit but the right-hand side $wq_{k+1} \cdots q_\ell$ is not, or vice versa, so $k > \ell$ and we are left with

$$up_{\ell+1} \cdots p_k = w,$$

which is similarly impossible. □

We now assemble several of our previous results.

Corollary 2.53. *The following integral domains are unique factorization domains:*

(1) $\mathbb{Z}$, the integers,

(2) $\mathcal{O}(\sqrt{D})$, *the ring of integers in the quadratic field* $\mathbf{Q}(\sqrt{D})$, *for* $D = -11, -7, -3, -2, -1, 2, 3, 5, 6, 7, 11, 13, 17, 21,$ *and* 29.

Proof: Each of these integral domains is a Euclidean domain (Lemma 2.7, and Theorem 2.8); every Euclidean domain is a PID (Corollary 2.39); and every PID is a UFD (Theorem 2.52). □

Remark 2.54. It is natural to ask whether the converse of Theorem 2.52 is true: is every UFD a PID? The answer to this question is no, for general integral domains R. But it turns out to be the case that the answer is yes for $R = \mathcal{O}(\sqrt{D})$, i.e., that the ring of integers of a quadratic field is a UFD if and only if it is a PID.

In the remainder of this section we will investigate UFDs more deeply.

As we saw in Lemma 2.48, in a PID, primes and irreducibles are the same thing. This is true in the more general situation of a UFD as well.

Lemma 2.55. *Let R be a UFD. Then an element α of R is prime if and only if α is irreducible.*

Proof: In Lemma 2.47 we showed that, in general, a prime is irreducible. Thus, we must show that here (in the case of a UFD) an irreducible is prime. Let α be irreducible and let β and γ be elements of R with α dividing $\beta\gamma$. Then β and γ have factorizations into irreducibles $\beta = up_1 \cdots p_k$ and $\gamma = vp'_1 \cdots p'_\ell$, so $\beta\gamma$ has the factorization $\beta\gamma = (uv)p_1 \cdots p_k p'_1 \cdots p'_\ell$ into irreducibles. Since α divides $\beta\gamma$, $\beta\gamma = \alpha\delta$ for some δ, and then δ has a factorization into irreducibles $\delta = wp''_1 \cdots p''_m$. Then

$$(uv)p_1 \cdots p_k p'_1 \cdots p'_\ell = \beta\gamma = \alpha\delta = w\alpha p''_1 \cdots p''_m.$$

But factorization into irreducibles is essentially unique, so α is an associate of some p_i, in which case α divides β, or α is an associate of some p'_i, in which case α divides γ. Hence α is a prime. □

This lemma justifies the following definition.

Definition 2.56. Let $\alpha = up_1 \cdots p_k$ be as in Definition 2.49. Then this is called the *prime factorization* of α and $p_1, \ldots, p_k$ are the *prime divisors* (or *prime factors*) of α.

Remark 2.57. Let $\alpha = up_1 \cdots p_k$ be as in Definition 2.56. Then we may gather associated prime divisors of α together and write $\alpha = vp_1^{e_1} \cdots p_j^{e_j}$ for some positive integers $e_1, \ldots, e_j$, with each p_i a prime and no two distinct p_i's associated, and with v a unit.

Recall that we introduced the *greatest common divisor* (gcd) in our discussion of PIDs (Definition 2.13). Now we will reconsider this concept in the more general situation of UFDs. Of course, since every PID is a UFD, our discussion here will apply to PIDs as well.

The key to our discussion is the following lemma, which is very useful in its own right.

Lemma 2.58. *Let R be a UFD and let α and β be nonzero elements of R. Then α divides β if and only if*

(1) every prime p that divides α also divides β, and

(2) for every such prime p, if p^e is the highest power of p dividing α, and if p^f is the highest power of p dividing β, then $f \geq e$.

Proof: First, let us suppose that conditions (1) and (2) are satisfied. Then, for some primes $p_1, \ldots, p_k, q_1, \ldots, q_\ell$ and some exponents, and some units u and v, we have

$$\alpha = up_1^{e_1} p_2^{e_2} \cdots p_k^{e_k},$$
$$\beta = vp_1^{f_1} p_2^{f_2} \cdots p_k^{f_k} q_1^{g_1} q_2^{g_2} \cdots q_\ell^{g_\ell}.$$

Then, setting

$$\gamma = wp_1^{f_1-e_1} p_2^{f_2-e_2} \cdots p_k^{f_k-e_k} q_1^{g_1} q_2^{g_2} \cdots q_\ell^{g_\ell}$$

with $w = uv^{-1}$, we have

$$\beta = \alpha\gamma,$$

so in this situation α divides β.

On the other hand, suppose α divides β, so $\beta = \alpha\gamma$.

Factor α and γ into primes, where we allow the exponents $d_1, \ldots, d_k$ to be zero:

$$\alpha = up_1^{e_1} \cdots p_k^{e_k},$$
$$\gamma = wp_1^{d_1} \cdots p_k^{d_k} q_1^{g_1} \cdots q_\ell^{g_\ell}.$$

Then, setting $v = uw$,

$$\beta = vp_1^{f_1} \cdots p_k^{f_k} q_1^{g_1} \cdots q_\ell^{g_\ell}$$

with $f_i = e_i + d_i \geq e_i$ for each $i = 1, \cdots, k$, and so we see that conditions (1) and (2) are satisfied. □

Proposition 2.59. *Let R be a UFD and let α and β be nonzero elements of R with prime factorizations*

$$\alpha = up_1^{e_1} \cdots p_k^{e_k} q_1^{g_1} \cdots q_\ell^{g_\ell},$$
$$\beta = vp_1^{f_1} \cdots p_k^{f_k} r_1^{h_1} \cdots r_m^{h_m}.$$

Then α and β have a gcd δ, *and, moreover,*

$$\delta = p_1^{d_1} \cdots p_k^{d_k}$$

where $d_i = min(e_i, f_i)$ for each $i = 1, \ldots, k$. In case α and β have no common prime factors, $\delta = 1$.

Proof: By Lemma 2.58, δ divides both α and β. Furthermore, also by Lemma 2.58, any ζ dividing both α and β must be of the form

$$\zeta = wp_1^{c_1} \cdots p_k^{c_k}$$

with $c_i \le e_i$ and $c_i \le f_i$, i.e., $c_i \le \min(e_i, f_i) = d_i$ for each i, so, once again by Lemma 2.58, ζ divides δ. Thus, δ satisfies the properties of $\gcd(\alpha, \beta)$ (Definition 2.13). □

Remark 2.60. Recall that the gcd is only defined up to multiplication by a unit, so for any unit w,

$$\delta' = w\delta = wp_1^{d_1} \cdots p_k^{d_k}$$

is also a gcd of α and β.

Recall that we defined two elements α and β of a PID to be *relatively prime* if their gcd is 1. We use the same language in the more general situation of a UFD here.

Note that Lemma 2.61 is the direct (word-for-word) generalization of Euclid's Lemma (Lemma 2.41) to the case of a UFD. But in this more general situation we need a new proof.

Lemma 2.61. *Let R be a UFD and let α be a nonzero element of R. Let β_1 and β_2 be elements of R and suppose that α divides $\beta_1\beta_2$. If α and β_1 are relatively prime, then α divides β_2.*

Proof: Since α and β_1 are relatively prime, they have no common prime factors. Thus, we have prime factorizations

$$\alpha = p_1^{e_1} \cdots p_k^{e_k},$$
$$\beta_1 = q_1^{g_1} \cdots q_\ell^{g_\ell}.$$

Now we are assuming that α divides $\beta_1\beta_2$, so by Lemma 2.58 we see that the prime factorization of $\beta_1\beta_2$ must include $p_1^{f_1}\cdots p_k^{f_k}$ with $f_i \geq e_i$, for each i. But the prime factorization of $\beta_1\beta_2$ is the product of the prime factorization of β_1 and the prime factorization of β_2. Since $p_1^{f_1}\cdots p_k^{f_k}$ does not appear in the prime factorization of β_1, it must appear in the prime factorization of β_2, and hence α divides β_2. □

The following corollaries are word-for-word generalizations of Corollary 2.42 and Corollary 2.43 to the case of a UFD.

Corollary 2.62. *Let R be a UFD and let α and β be relatively prime nonzero elements of R. Let γ be an element of R and suppose that α divides γ and β divides γ. Then $\alpha\beta$ divides γ.*

Proof: Once we have the generalization of Euclid's Lemma 2.41 to UFDs in Lemma 2.61, the *identical* proof of Corollary 2.42 works for UFDs, so we may just quote that proof.

But we will give a second (albeit longer) proof that uses prime factorization directly: Again, since α and β are relatively prime, they have no common prime factors, so we have prime factorizations

$$\alpha = p_1^{e_1}\cdots p_k^{e_k},$$
$$\beta = q_1^{g_1}\cdots q_\ell^{g_\ell}.$$

Then, by Lemma 2.58, applied first to α and γ and then to β and γ, we see that the prime factorizations of γ must contain every $p_i^{e_i}$ and every $q_j^{g_j}$, so it contains $p_i^{e_1}\cdots p_k^{e_k}q_1^{g_1}\cdots q_\ell^{g_\ell}$ and hence $\alpha\beta$ divides γ. □

Corollary 2.63. *Let R be a UFD and let α, β, and γ be elements of R. Suppose that α and β are relatively prime, and also that α and γ are relatively prime. Then α and $\beta\gamma$ are relatively prime.*

Proof: Once we have the generalization of Euclid's Lemma 2.41 to UFDs in Lemma 2.61, the *identical* proof of Corollary 2.43 works for UFDs, so we may just quote that proof.

But again we will give a proof that uses prime factorization directly: Let α have prime factorization

$$\alpha = p_1^{e_1}\cdots p_k^{e_k}.$$

Since α and β are relatively prime, no p_i appears in the prime factorization of β, and since α and γ are relatively prime, no p_i appears in the prime factorization of γ. Hence no p_i appears in the prime factorization of $\beta\gamma$, from which we conclude that α and $\beta\gamma$ are relatively prime. □

2.5 Nonunique Factorization: The Case $D < 0$

So far we have devoted our efforts to proving that $\mathcal{O}(\sqrt{D})$ *is* a UFD for various values of D. Now we will concentrate our efforts on showing that $\mathcal{O}(\sqrt{D})$ *is not* a UFD for other values of D. In fact, we will show that this happens for infinitely many values of D.

As the reader will see, our results are much more complete for negative values of D than they are for positive values of D. This reflects our present state of knowledge—much more is known when $D < 0$ than is known when $D > 0$.

We will be discussing primes and irreducible elements in both $\mathcal{O}(\sqrt{D})$ and $\mathbb{Z}$. To keep these straight, in Lemma 2.64 and Corollary 2.65 we will refer to primes in $\mathbb{Z}$ as *ordinary primes*. Note in Lemma 2.64 and Corollary 2.65 that D may be negative or positive.

The following result will help us recognize irreducible elements of $\mathcal{O}(\sqrt{D})$.

Lemma 2.64. *Let $R = \mathcal{O}(\sqrt{D})$ and let p be an ordinary prime.*

(1) If α is an element of R with $\|\alpha\| = |N(\alpha)| = p$, then α is irreducible.

(2) Suppose that R does not have an element β with $\|\beta\| = p$. If γ is an element of R with $\|\gamma\| = pp'$ with p' also an ordinary prime (perhaps $p' = p$), then γ is irreducible. In particular, if R does not have an element β with $\|\beta\| = p$, then p is irreducible.

Proof:

(1) Suppose $\alpha = \beta_1\beta_2$. Then

$$p = \|\alpha\| = \|\beta_1\beta_2\| = \|\beta_1\| \cdot \|\beta_2\|,$$

so either $\|\beta_1\| = 1$, and β_1 is a unit (Lemma 2.12), or $\|\beta_2\| = 1$, and β_2 is a unit. Thus α is irreducible.

(2) Suppose $\gamma = \beta_1\beta_2$. Then

$$pp' = \|\gamma\| = \|\beta_1\beta_2\| = \|\beta_1\| \cdot \|\beta_2\|,$$

and we cannot have $\|\beta_1\| = p, \|\beta_2\| = p'$, or vice versa, as R has no elements of norm p. Hence, as in part (1), either $\|\beta\| = 1$ and β_1 is a unit, or $\|\beta_2\| = 1$ and β_2 is a unit. Thus γ is irreducible. Finally, note that $\|p\| = p^2$, so by setting $\gamma = p$, we conclude that p is irreducible.□

The following corollary will be our basic tool in showing that $\mathcal{O}(\sqrt{D})$ is not a UFD. We will be using it or its consequences in all of the examples in this section and in the next section. So once again this provides an example of working hard enough with a single idea and being able to go a long way with it. We use it so often that we give it a name (although this name is not standard mathematical language).

Corollary 2.65 (The Non-UFD Test). *Let $R = \mathcal{O}(\sqrt{D})$.*

If there is some ordinary prime p such that

(1) R does not have an element β with $\|\beta\| = p$, and

(2) R has an element α that is not divisible by p, but with $\|\alpha\|$ divisible by p,

then R is not a UFD.

Proof: Write $\|\alpha\| = pq$, for some $q > 1$. By Lemma 2.3, $\|\alpha\| = |\alpha\overline{\alpha}|$, so $\|\alpha\| = pq$ gives

$$\alpha\overline{\alpha} = \pm pq.$$

By Lemma 2.64, p is irreducible. Now p does not divide α, by assumption, and then p does not divide $\overline{\alpha}$ either. Thus, p divides the product $\alpha\overline{\alpha}$ without dividing either factor, so p is not a prime. Now in a UFD, every irreducible is prime (Lemma 2.55), so R cannot be a UFD. □

Remark 2.66. It is easy to tell when p divides α. Suppose p is odd and $\alpha = a + b\sqrt{D}$ or $(a + b\sqrt{D})/2$. Then $\alpha/p = a/p + (b/p)\sqrt{D}$ or $((a/p) + (b/p)\sqrt{D})/2$, so p divides α if and only if p divides a and p divides b. The case $p = 2$ is a little more complicated. If $D \equiv 2$ or $3 \pmod 4$, then 2 divides $\alpha = a + b\sqrt{D}$ if and only if a and b are both even. If $D \equiv 1 \pmod 4$, then 2 divides $\alpha = a + b\sqrt{D}$ with a and b integers if and only if either a and b are both even or both odd integers. (This justifies the claim in the proof of Corollary 2.65 that if p does not divide $\alpha = a + b\sqrt{D}$, then p does not divide $\overline{\alpha} = a - b\sqrt{D}$ either.)

We begin by studying the rings of integers in imaginary quadratic fields, i.e., $\mathcal{O}(\sqrt{D})$ for $D < 0$.

In order to use Corollary 2.65, we have to come up with a suitable prime p. In fact, $p = 2$ often works.

Lemma 2.67. *If $D < 0$ and $D \neq -1$, -2, or -7, then $\mathcal{O}(\sqrt{D})$ does not have an element of norm 2.*

Proof: We divide the proof into two cases: (1) $D \equiv 2$ or $3 \pmod 4$, and (2) $D \equiv 1 \pmod 4$.

(1) Suppose $D \equiv 2$ or $3 \pmod 4$. Then any element α of $\mathcal{O}(\sqrt{D})$ is of the form $\beta = a + b\sqrt{D}$ with a and b integers. Then

$$\|\beta\| = |\,\mathrm{N}(\beta)| = |a^2 - b^2 D| = |a^2 + (-D)b^2| = a^2 + (-D)b^2.$$

So, in order for there to be an element β with $\|\beta\| = 2$, the equation $a^2 + (-D)b^2 = 2$ must have a solution with a and b integers. But if $D < -2$ (so $-D > 2$), it does not.

(2) Suppose $D \equiv 1 \pmod 4$. Then any element β of $\mathcal{O}(\sqrt{D})$ is of the form (a) $\beta = a + b\sqrt{D}$ with a and b integers or (b) $\beta = (a + b\sqrt{D})/2 = (a/2) + (b/2)\sqrt{D}$ with a and b odd integers. In case (a) the argument is exactly the same as above, and $\|\beta\| = 2$ has no solutions. In case (b)

$$\begin{aligned}\|\beta\| &= |\,\mathrm{N}(\beta)| = |(a/2)^2 - (b/2)^2 D| = |(a/2)^2 + (-D)(b/2)^2| \\ &= a^2/4 + (-D)b^2/4.\end{aligned}$$

So, in order for there to be an element β with $\|\beta\| = 2$, the equation $a^2/4 + (-D)b^2/4 = 2$, i.e., $a^2 + (-D)b^2 = 8$, must have a solution with a and b odd integers. But if $D \neq -7$, (so $-D = 3$ or $-D > 7$), it does not.□

Theorem 2.68. *Let $D < 0$, $D \equiv 1, 2, 3, 6,$ or $7 \pmod 8$, and $D \neq -1, -2,$ or -7. Then $R = \mathcal{O}(\sqrt{D})$ is not a UFD.*

Proof: We begin by noting that, by Lemma 2.67, in no case does R have an element β with $\|\beta\| = 2$. To proceed further, we divide the proof into three cases: (1) $D \equiv 2$ or $6 \pmod 8$, (2) $D \equiv 3$ or $7 \pmod 8$, and (3) $D \equiv 1 \pmod 8$.

(1) In this case $D \equiv 2 \pmod 4$. Let

$$\alpha = \sqrt{D}.$$

Then

$$\|\alpha\| = -D.$$

Thus α is not divisible by 2 but $\|\alpha\|$ is divisible by 2 (as in this case D is even). Hence, by Corollary 2.65 (the Non-UFD Test), R is not a UFD.

(2) In this case $D \equiv 3 \pmod 4$. Let

$$\alpha = 1 + \sqrt{D}.$$

Then

$$\|\alpha\| = 1 - D.$$

Thus α is not divisible by 2 but $\|\alpha\|$ is divisible by 2 (as in this case D is odd, so $1 - D$ is even). Hence, by Corollary 2.65 (the Non-UFD Test), R is not a UFD.

(3) In this case $D \equiv 1 \pmod 8$. Let

$$\alpha = \frac{1 + \sqrt{D}}{2}.$$

Then

$$\|\alpha\| = (1 - D)/4.$$

Thus α is not divisible by 2 but $\|\alpha\|$ is divisible by 2 (as in this case $D \equiv 1 \pmod 8$, so $1 - D$ is divisible by 8 and so $(1 - D)/4$ is even). Hence, by Corollary 2.65 (the Non-UFD Test), R is not a UFD. □

Thus, we see that we have complete information for $D < 0$ and $D \equiv 1$, 2, 3, 6, or 7 $\pmod 8$. We showed in Corollary 2.53 that $\mathcal{O}(\sqrt{D})$ is a UFD for $D = -1$, -2, or -7, and we just showed in Theorem 2.68 that $\mathcal{O}(\sqrt{D})$ is not a UFD for any other such value of D.

Now, we need to investigate the case $D < 0$ and $D \equiv 5 \pmod 8$. Here we will only be able to get partial (but very suggestive) information. Again, our key tool will be Corollary 2.65 (the Non-UFD Test), but we will not be able to choose $p = 2$ in order to apply it. In fact, the value of p will depend on D. Actually, our results will apply more generally than the case $D \equiv 5 \pmod 8$, so for many values of D we will have a second proof that $\mathcal{O}(\sqrt{D})$ is not a UFD.

First, we make an observation that will make it easy for us to tell when one of the hypotheses of Corollary 2.65 (the Non-UFD Test) is satisfied.

Lemma 2.69. *Let $D < 0$.*

(1) If $D \equiv 2$ or $3 \pmod 4$ and p is a prime with $p < |D|$, then $R = \mathcal{O}(\sqrt{D})$ does not have an element β with $\|\beta\| = p$.

(2) If $D \equiv 1 \pmod 4$ and p is a prime with $p < |D|/4$, then $R = \mathcal{O}(\sqrt{D})$ does not have an element β with $\|\beta\| = p$.

Proof:

(1) Let β be an element of R. Then $\beta = a + b\sqrt{D}$ where a and b are integers. Suppose $\|\beta\| = p$. Then

$$p = \|\alpha\| = a^2 - b^2 D = a^2 + b^2(-D)$$

and since $p < |D|$ we must have $b = 0$, giving $p = a^2$, which is impossible.

(2) Let β be an element of R. Then (a) $\beta = a + b\sqrt{D}$ where a and b are integers or (b) $\beta = (a + b\sqrt{D})/2$ where a and b are odd integers. Suppose $\|\beta\| = p$. The argument in case (a) is the same as above. In case (b),

$$p = \|\beta\| = (a^2 - b^2 D)/4 = a^2/4 + b^2(-D)/4,$$

which is similarly impossible if $p < |D|/4$. □

Using this lemma we may show that in many cases $\mathcal{O}(\sqrt{D})$ is not a UFD.

Proposition 2.70. *Let $D < 0$, and $D \equiv 1 \pmod 4$.*

(1) If $|D|$ is composite, then $R = \mathcal{O}(\sqrt{D})$ is not a UFD.

(2) If $m = (1 - D)/4$ is composite, then $R = \mathcal{O}(\sqrt{D})$ is not a UFD.

Proof:

(1) We claim that if $D < 0$, $D \equiv 1 \pmod 4$, and $|D|$ is composite, then D has a prime factor p with $p < |D|/4$.

(To see this, note that if $D < 0$, $D \equiv 1 \pmod 4$, and $|D|$ is composite, then $|D| \geq 15$. If $|D| = 15$, then D is divisible by 3 and $3 < 15/4$. Otherwise $|D| > 15$. Let p be the smallest prime dividing $|D|$. Then $|D|/p \neq 1$, so $|D|/p$ is divisible by some prime $p' \geq p$. We now argue by contradiction. Suppose $p \geq |D|/4$. Then $|D| = p(|D|/p) \geq pp' \geq p^2 \geq (|D|/4)^2 = |D|^2/16$, so $|D| < 16$, a contradiction.)

By Lemma 2.69 we see that hypothesis (1) of Corollary 2.65 holds. Let

$$\alpha = \sqrt{D}.$$

Then

$$\|\alpha\| = |D|,$$

so we see hypothesis (2) of Corollary 2.65 holds as well. Hence, by Corollary 2.65 (the Non-UFD Test), we conclude that $\mathcal{O}(\sqrt{D})$ is not a UFD.

(2) If m is composite then m is divisible by a prime $p < m$ and then certainly $p < |D|/4$, so, by Lemma 2.69, hypothesis (1) of Corollary 2.65 holds. Let

$$\alpha = \frac{1+\sqrt{D}}{2}.$$

Then

$$\|\alpha\| = |m|,$$

so hypothesis (2) of Corollary 2.65 holds as well, and again, by Corollary 2.65 (the Non-UFD Test), we conclude that $\mathcal{O}(\sqrt{D})$ is not a UFD. □

Corollary 2.71. *For $D \equiv 5 \pmod 8), D < 0$, and $|D| < 1000$, $\mathcal{O}(\sqrt{D})$ is not a UFD except (possibly) for the cases $D = -3, -11, -19, -43, -67$, and -163 and the cases $D = -211, -283, -331, -547, -691, -787$, and -907.*

Proof: For every value of D with $D \equiv 5 \pmod 8, D < 0$, and $|D| < 1000$, except for those listed in the statement of the corollary, $|D|$ is composite or $m = (1 - D)/4$ is composite, so by Proposition 2.70, $\mathcal{O}(\sqrt{D})$ is not a UFD. □

Actually, using the same idea, we can get a stronger test than that in Proposition 2.70, and use it to strengthen Corollary 2.71.

Proposition 2.72. *Let $D < 0$ and $D \equiv 1 \pmod 4$.*

(1) If there is an nonnegative even integer $a < (1/4)\sqrt{D^2 + 16D}$ with $a^2 - D$ composite, then $\mathcal{O}(\sqrt{D})$ is not a UFD.

(2) If there is a nonnegative odd integer $a < (1/2)\sqrt{D^2 + 4D}$ *with* $(a^2 - D)/4$ *composite, then* $\mathcal{O}(\sqrt{D})$ *is not a UFD.*

Proof:

(1) Let
$$\alpha = a + \sqrt{D}.$$

Then
$$\|\alpha\| = a^2 - D.$$

Now $a < (1/4)\sqrt{D^2 + 16D}$ implies, by simple algebra, that $a^2 - D < (D/4)^2$, so $a^2 - D$ must have a prime factor p less than $|D|/4$. Thus, by Lemma 2.69, we see that hypothesis (1) of Corollary 2.65 holds. Since a is even, $a^2 - D$ is odd, so p is odd and hence α is not divisible by p. Thus, we see that hypothesis (2) of Corollary 2.65 holds as well. Hence we conclude that, by Corollary 2.65 (the Non-UFD Test), $\mathcal{O}(\sqrt{D})$ is not a UFD.

(2) Let
$$\alpha = (a + \sqrt{D})/2,$$

and note that α is in $\mathcal{O}(\sqrt{D})$ as a is odd. Then
$$\|\alpha\| = (a^2 - D)/4.$$

Then, similarly, $a < (1/2)\sqrt{D^2 + 4D}$ implies that $(a^2 - D)/4 < (D/4)^2$, so $(a^2 - D)/4$ must have a prime factor p less than $|D|/4$. Certainly α is not divisible by p. Thus, we see that hypothesis (2) of Corollary 2.65 holds as well. Hence we conclude that, by Corollary 2.65 (the Non-UFD Test), $\mathcal{O}(\sqrt{D})$ is not a UFD. □

Remark 2.73. Proposition 2.72 is a direct generalization of Proposition 2.70. Setting $a = 0$ in part (1) of Proposition 2.72 recovers part (1) of Proposition 2.70, and setting $a = 1$ in part (2) of Proposition 2.72 recovers part (2) of Proposition 2.70.

Corollary 2.74. *For* $D \equiv 5 \pmod 8$, $D < 0$, *and* $|D| < 1000$, $\mathcal{O}(\sqrt{D})$ *is not a UFD except (possibly) for the cases* $D = -3, -11, -19, -43, -67$, *and* -163.

Proof: We use Proposition 2.72 to decide some of the cases of D we could not handle with Proposition 2.70, as follows: for $D = -211$, -283, -331, -547, -691, or -787, we choose $a = 2$, and for $D = -907$, we choose $a = 4$.

Here the value of a given in the above list is a value of a that can be used in Proposition 2.72 to show that $\mathcal{O}(\sqrt{D})$ is not a UFD. (For example, if $D = -211$, $\alpha = (2 + \sqrt{-211})$ has $\|\alpha\| = 215$, a composite number divisible by the prime $p = 5$, and if $D = -907$, $\alpha = (4 + \sqrt{-907})/2$ has $\|\alpha\| = 923$, a composite number divisible by the prime $p = 13$.) $\square$

In fact, Proposition 2.72 is a very effective way of showing that $\mathcal{O}(\sqrt{D})$ is not a UFD.

Corollary 2.75. *For $D \equiv 5 \pmod 8$, $D < 0$, and $|D| < 1{,}000{,}000{,}000$, $\mathcal{O}(\sqrt{D})$ is not a UFD except (possibly) for the cases $D = -3$, -11, -19, -43, -67, and -163.*

Proof: A computer computation using Proposition 2.72 rules out all values of D except for those in the statement of the corollary. (As a matter of curiosity, the largest value of a we have to consider for D in this range is $a = 11$, which occurs for $D = -543{,}764{,}323$.) $\square$

Remark 2.76. What about the exceptions in Corollary 2.75? It turns out that $\mathcal{O}(\sqrt{D})$ is a UFD for the cases $D = -3$, -11, -19, -43, -67, and -163. Given the large range of values of D, the reader may (strongly) suspect that these are the only such values of D for which $\mathcal{O}(\sqrt{D})$ is a UFD. This turns out to be true, and is a very deep (and famous) theorem.

2.6 Nonunique Factorization: The Case $D > 0$

Now we turn to studying rings of integers in real quadratic fields, i.e., $\mathcal{O}(\sqrt{D})$ for $D > 0$. Actually, the methods we derive here will also apply to values of D with $D < 0$, giving a second proof of the results in those cases. Again our strategy is the same, but the details are harder to carry out, and our results are much less general. But once again it is Corollary 2.65 (the Non-UFD Test) that is our basic tool, which we use in deriving the criteria in Theorems 2.78, 2.82, and 2.87. Once again, in order to use it we begin by proving the nonexistence of elements of norm 2.

Lemma 2.77. *Suppose that*

(1) *D is divisible by a prime p congruent to* $5 \pmod 8$, *or*

(2) *D is divisible by a prime p' congruent to* $3 \pmod 8$ *and D is also divisible by a prime p'' congruent to* $7 \pmod 8$.

Then $\mathcal{O}(\sqrt{D})$ does not have an element β with $\|\beta\| = 2$.

Proof: Set $q = p$ if condition (1) holds and set $q = p'p''$ if condition (2) holds.

Once again, we divide the proof into two cases: (1) $D \equiv 2$ or $3 \pmod 4$, and (2) $D \equiv 1 \pmod 4$.

(1) Suppose $D \equiv 2$ or $3 \pmod 4$. Then any element β of $\mathcal{O}(\sqrt{D})$ is of the form $\beta = a + b\sqrt{D}$ with a and b integers. Suppose $\beta = a + b\sqrt{D}$ with $\|\beta\| = 2$. Then
$$a^2 - Db^2 = \pm 2,$$
so
$$a^2 \equiv \pm 2 \pmod D,$$
and hence
$$a^2 \equiv \pm 2 \pmod q.$$
But it is a result from number theory (see Corollary B.37(2) in the situation of condition (1) and Corollary B.38 in the situation of condition (2)) that the congruence $a^2 \equiv \pm 2 \pmod q$ does not have a solution.

(2) Suppose $D \equiv 1 \pmod 4$. Then any element β of $\mathcal{O}(\sqrt{D})$ is of the form (2a) $\beta = a + b\sqrt{D}$ with a and b integers or (2b) $\beta = (a + b\sqrt{D})/2 = (a/2) + (b/2)\sqrt{D}$ with a and b odd integers. In case (2a) the argument is exactly the same as above. We consider case (2b). Suppose $\beta = (a + b\sqrt{D})/2$ with $\|\beta\| = 2$. Then
$$(a^2 - Db^2)/4 = \pm 2,$$
giving the equation
$$a^2 - Db^2 = \pm 8,$$
so
$$a^2 \equiv \pm 8 \pmod D$$
and hence
$$a^2 \equiv \pm 8 \pmod q$$

and again this congruence has no solution. (Since $8 = 2^2 \cdot 2$, the congruence $a^2 \equiv \pm 8 \pmod{p}$ has a solution if and only if the congruence $a^2 \equiv \pm 2 \pmod{q}$ has a solution, and it does not.) □

Theorem 2.78. *Let $D \equiv 1, 2, 3, 6,$ or $7 \pmod{8}$, and suppose that D is divisible by a prime $p \equiv 5 \pmod{8}$ or that D is divisible by a prime $p' \equiv 3 \pmod{8}$ and a prime $p'' \equiv 7 \pmod{8}$. Then $R = \mathcal{O}(\sqrt{D})$ is not a UFD.*

Proof: We begin by noting that, by Lemma 2.77, in no case does R have an element β with $\|\beta\| = 2$. The remainder of the proof is *identical* to the proof of Theorem 2.68, so instead of repeating it we simply refer the reader back to that proof. □

Now let us give some examples of Theorem 2.78. Before we do so, we want to quote the Chinese Remainder Theorem (Lemma B.17 or Theorem B.18), which states that a pair of congruences

$$\begin{aligned} x &\equiv a_1 \pmod{b_1}, \\ x &\equiv a_2 \pmod{b_2}, \end{aligned}$$

with b_1 and b_2 relatively prime, always has a solution, and that this solution is unique $\pmod{b_1 b_2}$, or in other words, that this pair of congruences is equivalent to a single congruence

$$x \equiv a_3 \pmod{b_1 b_2}$$

for some a_3. (There are various methods for finding a_3. If b_1 and b_2 are small, trial and error is as good as any.) We will use this in the statement of our results.

Example 2.79. We wish to apply Theorem 2.78. First, we consider condition (1) of that theorem. To apply this condition, we need to find primes p with $p \equiv 5 \pmod{8}$. The first few of these primes are $p = 5$, $p = 13$, and $p = 29$.

(1) Consider $p = 5$. We want D to be divisible by $p = 5$, i.e., $D \equiv 0 \pmod{5}$. We first apply this to the case $D \equiv 1 \pmod{8}$, so we have the pair of simultaneous congruences

$$D \equiv 0 \pmod{5} \qquad \text{and} \qquad D \equiv 1 \pmod{8},$$

which is equivalent to the single congruence

$$D \equiv 25 \pmod{40}.$$

We next apply this to the case $D \equiv 2 \pmod 8$, so we have the pair of simultaneous congruences

$$D \equiv 0 \pmod 5 \qquad \text{and} \qquad D \equiv 2 \pmod 8,$$

which is equivalent to the single congruence

$$D \equiv 10 \pmod{40}.$$

Proceeding in this way, we find that the pair of congruences

$$D \equiv 0 \pmod 5 \qquad \text{and} \qquad D \equiv 1,\ 2,\ 3,\ 6,\ \text{or } 7 \pmod 8$$

is equivalent to the single congruence

$$D \equiv 10,\ 15,\ 25,\ 30,\ \text{or } 35 \pmod{40}.$$

So for these values of D, by Theorem 2.78, we have that $\mathcal{O}(\sqrt{D})$ is not a UFD. Remember that D must be square-free, and so the first few such positive values of D are $D = 10$, 15, 30, 35, 55, 65, 70, 95, 105, 110, 115, 130, 145, 155, 170, 185, 190, and 195.

(2) Consider $p = 13$. Then we have the pair of simultaneous congruences

$$D \equiv 0 \pmod{13} \qquad \text{and} \qquad D \equiv 1,\ 2,\ 3,\ 6,\ \text{or } 7 \pmod 8,$$

which is equivalent to the single congruence

$$D \equiv 26,\ 39,\ 65,\ 78,\ \text{or } 91 \pmod{104}.$$

So for these values of D, by Theorem 2.78, we have that $\mathcal{O}(\sqrt{D})$ is not a UFD. Remembering that D must be square-free, the first few such positive values of D are $D = 26$, 39, 65, 78, 130, 143, 182, and 195.

(3) Consider $p = 29$. Then we have the pair of simultaneous congruences

$$D \equiv 0 \pmod{29} \qquad \text{and} \qquad D \equiv 1,\ 2,\ 3,\ 6,\ \text{or } 7 \pmod 8,$$

which is equivalent to the single congruence

$$D \equiv 58,\ 87,\ 145,\ 174,\ \text{or } 203 \pmod{232}.$$

So for these values of D, by Theorem 2.78, we have that $\mathcal{O}(\sqrt{D})$ is not a UFD.

(4) Next, we consider condition (2) of Theorem 2.78. To apply this condition we need to find a pair of primes p' and p'' with $p' \equiv 3 \pmod 8$ and $p'' \equiv 7 \pmod 8$. The first such pair is $p' = 3$ and $p'' = 7$. Then we have the pair of simultaneous congruences

$$D \equiv 0 \pmod{21} \qquad \text{and} \qquad D \equiv 1,\ 2,\ 3,\ 6, \text{ or } 7 \pmod 8,$$

which is equivalent to the single congruence

$$D \equiv 42,\ 63,\ 105,\ 126, \text{ or } 147 \pmod{168}.$$

So for these values of D, by Theorem 2.78, we have that $\mathcal{O}(\sqrt{D})$ is not a UFD.

Just as in the $D < 0$ case, we applied Corollary 2.65 (the Non-UFD Test) with $p = 2$ to prove Theorem 2.78. Again, as in the $D < 0$ case, in order to proceed further we wish to apply Corollary 2.65 (the Non-UFD Test) more generally. Again, we will not just be able to choose $p = 2$ in order to apply it, but the value of p will depend on D. Note the similarity of the following lemma (Lemma 2.80) to Lemma 2.77, and note also the similarity of the proofs.

Lemma 2.80. *Let D be divisible by an odd prime p_2, and suppose furthermore that there is an odd prime p_1 such that the congruence $x^2 \equiv \pm p_1 \pmod{p_2}$ does not have a solution. Then $\mathcal{O}(\sqrt{D})$ does not have an element of norm p_1.*

Proof: Once again we divide the proof into two cases: (1) $D \equiv 2$ or $3 \pmod 4$, and (2) $D \equiv 1 \pmod 4$.

(1) Suppose $D \equiv 2$ or $3 \pmod 4$. Then any element β of $\mathcal{O}(\sqrt{D})$ is of the form $\beta = a + b\sqrt{D}$ with a and b integers. Suppose $\beta = a + b\sqrt{D}$ with $\|\beta\| = p_1$. Then

$$a^2 - Db^2 = \pm p_1,$$

so

$$a^2 \equiv \pm p_1 \pmod D,$$

and hence

$$a^2 \equiv \pm p_1 \pmod{p_2}.$$

But by hypothesis this congruence has no solution.

(2) Suppose $D \equiv 1 \pmod 4$. Then any element β of $\mathcal{O}(\sqrt{D})$ is of the form (a) $\beta = a+b\sqrt{D}$ with a and b integers or (b) $\beta = (a+b\sqrt{D})/2 = (a/2)+(b/2)\sqrt{D}$ with a and b odd integers. In case (a), the argument is exactly the same as above. We consider case (b). Suppose $\beta = (a+b\sqrt{D})/2$ with $\|\beta\| = p_1$. Then

$$(a^2 - Db^2)/4 = \pm p_1,$$

giving the equation

$$a^2 - Db^2 = \pm 4p_1,$$

so

$$a^2 \equiv \pm 4p_1 \pmod D,$$

and hence

$$a^2 \equiv \pm 4p_1 \pmod{p_2},$$

and again this congruence has no solution. (As in the proof of Lemma 2.77, since $4p_1 = 2^2 p_1$, $a^2 \equiv \pm 4p_1 \pmod{p_2}$ has a solution if and only if $a^2 \equiv \pm p_1 \pmod{p_2}$ has a solution, and by hypothesis it does not.) □

Remark 2.81. Let p_2 be an odd prime and let q be any integer relatively prime to p. If $p_2 \equiv 3 \pmod 4$, then, by Corollary B.34, exactly one of the congruences $x^2 \equiv q \pmod{p_2}$ and $x^2 \equiv -q \pmod{p_2}$ has a solution, so the hypothesis of Lemma 2.80 is never satisfied. If $p \equiv 1 \pmod 4$, then, by Corollary B.34, either both of the congruences $x^2 \equiv q \pmod{p_2}$ and $x^2 \equiv -q \pmod{p_2}$ have a solution, or neither does, so in this case the hypothesis in Lemma 2.80 that $x^2 \equiv \pm p_1 \pmod{p_2}$ does not have a solution is equivalent to the simpler hypothesis that $x^2 \equiv p_1 \pmod{p_2}$ does not have a solution.

Theorem 2.82. *Let D be divisible by a prime $p_2 \equiv 1 \pmod 4$, and suppose furthermore that there is an odd prime p_1 such that*

(1) the congruence $x^2 \equiv p_1 \pmod{p_2}$ does not have a solution, and

(2) the congruence $x^2 \equiv D \pmod{p_1}$ has a solution.

Then $\mathcal{O}(\sqrt{D})$ is not a UFD.

Proof: By Lemma 2.80 and Remark 2.81, hypothesis (1) implies that $\mathcal{O}(\sqrt{D})$ does not have an element of norm p_1.

We also claim that D has an element α not divisible by p_1 but with $\|\alpha\|$ divisible by p_1. By hypothesis (2), there is an a with $a^2 \equiv D \pmod{p_1}$. Let

$$\alpha = a + \sqrt{D}.$$

Clearly α is not divisible by p_1 as $\alpha/p_1 = (a + \sqrt{D})/p_1$ is not in $\mathcal{O}(\sqrt{D})$. (Note that p_1 is odd.) But then, by hypothesis (2),

$$\|\alpha\| = |a^2 - D|$$

is divisible by p_1. Thus, by Corollary 2.65 (the Non-UFD Test), $\mathcal{O}(\sqrt{D})$ is not a UFD. □

Now let us give some examples of the use of Theorem 2.82. Once again we will formulate our results with the help of the Chinese Remainder Theorem.

Example 2.83. We wish to apply Theorem 2.82. For this we need to find primes p_2 with $p_2 \equiv 1 \pmod 4$. The first such primes are $p_2 = 5$ and $p_2 = 13$.

(1) Consider $p_2 = 5$. We want to find odd primes p_1 with $x^2 \equiv p_1 \pmod 5$ not having a solution. Since the squares $\pmod 5$ are 0, 1, and 4, this means that $p_1 \equiv 2$ or $3 \pmod 5$. But also $p_1 \equiv 1 \pmod 2$ (as p_1 is odd), so by the Chinese Remainder Theorem, the condition on p_1 in this case is $p_1 \equiv 3$ or $7 \pmod{10}$. The first such primes are $p_1 = 3$ and $p_1 = 7$.

(a) Consider $p_1 = 3$. We want to find values of D with $x^2 \equiv D \pmod{p_1}$ having a solution. Since the squares $\pmod 3$ are 0 and 1, this means that $D \equiv 0$ or $1 \pmod 3$. Also, D is divisible by $p_2 = 5$, i.e., $D \equiv 0 \pmod 5$. So we have the pair of simultaneous congruences

$$D \equiv 0 \pmod 5 \qquad \text{and} \qquad D \equiv 0 \text{ or } 1 \pmod 3,$$

which is equivalent to the single congruence

$$D \equiv 0 \text{ or } 10 \pmod{15}.$$

So for these values of D, by Theorem 2.82, we have that $\mathcal{O}(\sqrt{D})$ is not a UFD. Remembering that D must be square-free, the first few such positive values of D are $D = 10$, 15, 30, 55, 70, 85, 105, 115, 130, 145, 165, and 195.

(b) Consider $p_1 = 7$. We want to find values of D with $x^2 \equiv D \pmod{p_1}$ having a solution. Since the squares (mod 7) are 0, 1, 2, and 4, this means that $D \equiv 0,\ 1,\ 2,$ or $4 \pmod 7$. So we have the pair of simultaneous congruences

$$D \equiv 0 \pmod 5 \qquad \text{and} \qquad D \equiv 0,\ 1,\ 2,\ \text{or } 4 \pmod 7,$$

which is equivalent to the single congruence

$$D \equiv 0,\ 15,\ 25,\ \text{or } 30 \pmod{35}.$$

So for these values of D, by Theorem 2.82, we have that $\mathcal{O}(\sqrt{D})$ is not a UFD. Remembering that D must be square-free, the first few such positive values of D are $D = 15, 30, 35, 65, 70, 85, 95, 105, 115, 130, 155, 165, 170,$ and 190.

(2) Consider $p_2 = 13$. We want to find odd primes p_1 with $x^2 \equiv p_1 \pmod{13}$ not having a solution. Since the squares (mod 13) are 0, 1, 3, 4, 9, 10, and 12, this means that $p_1 \equiv 2,\ 5,\ 6,\ 7,\ 8,$ or $11 \pmod{13}$. But also $p_1 \equiv 1 \pmod 2$, so by the Chinese Remainder Theorem, the condition on p_1 in this case is $p_1 \equiv 5,\ 7,\ 11,\ 15,\ 19,$ or $21 \pmod{26}$. The first such primes are $p_1 = 5$ and $p_1 = 7$.

(a) Consider $p_1 = 5$. We want to find values of D with $x^2 \equiv D \pmod{p_1}$ having a solution. Since the squares (mod 5) are 0, 1, and 4, this means that $D \equiv 0,\ 1,$ or $4 \pmod 5$. Also, $D \equiv 0 \pmod 5$. So we have the pair of simultaneous congruences

$$D \equiv 0 \pmod{13} \qquad \text{and} \qquad D \equiv 0, 1,\ \text{or } 4 \pmod 5,$$

which is equivalent to the single congruence

$$D \equiv 0,\ 26,\ \text{or } 39 \pmod{65}.$$

So for these values of D, by Theorem 2.82, we have that $\mathcal{O}(\sqrt{D})$ is not a UFD. Remembering that D must be square-free, the first few such positive values of D are $D = 26, 39, 65, 91, 130,$ and 195.

(b) Consider $p_1 = 7$. We want to find values of D with $x^2 \equiv D \pmod{p_2}$ having a solution. Since the squares (mod 7) are 0, 1, 2, and 4, this means that $D \equiv 0,\ 1,\ 2,$ or $4 \pmod 7$. So we have the pair of simultaneous congruences

$$D \equiv 0 \pmod{13} \qquad \text{and} \qquad D \equiv 0,\ 1,\ 2,\ \text{or } 4 \pmod 7,$$

which is equivalent to the single congruence

$$D \equiv 0,\ 39,\ 65,\ \text{or}\ 78 \pmod{91}.$$

So for these values of D, by Theorem 2.82, we have that $\mathcal{O}(\sqrt{D})$ is not a UFD. Remembering that D must be square-free, the first few such positive values of D are $D = 39, 65, 78, 91, 130$, and 182.

Remark 2.84. In Example 2.83, we used the conditions in Theorem 2.82 in the order they naturally arose: first we chose p_2, and then, for a given choice of p_2, we chose p_1. But, with the help of the Law of Quadratic Reciprocity (Theorem B.40), we can make the choice in the other order. The Law of Quadratic Reciprocity applies to the system of simultaneous congruences

$$x^2 \equiv p_1 \pmod{p_2},$$
$$x^2 \equiv p_2 \pmod{p_1},$$

where p_1 and p_2 are distinct odd primes. It states that if at least one of p_1 and p_2 is congruent to $1 \pmod 4$, then either both of these congruences have a solution or neither does, while if both p_1 and p_2 are congruent to $3 \pmod 4$, then exactly one of these congruences has a solution. Since in Theorem 2.82 we require that $p_2 \equiv 1 \pmod 4$, we are in the first of these cases, and so we may replace hypothesis (1) in Theorem 2.82 by hypothesis (1′):

(1′) The congruence $x^2 \equiv p_2 \pmod{p_1}$ does not have a solution.

We see how to use this in the next example.

Example 2.85. We wish to apply Theorem 2.82 with hypothesis (1) replaced by hypothesis (1′). For this we need to find an odd prime p_1. The first such prime is $p_1 = 3$.

(1) Consider $p_1 = 3$. In order to satisfy hypothesis (2) of Theorem 6.6 we need to find values of D such that $x^2 \equiv D \pmod{p_1}$ has a solution. Since the squares $\pmod 3$ are 0, and 1, this means that $D \equiv 0$ or $1 \pmod 3$. In order to satisfy hypothesis (1′) of Theorem 6.6 we need to find values of p_1 such that $x^2 \equiv p_2 \pmod{p_1}$ does not have a solution. Since the squares $\pmod 3$ are 0, and 1, this means that $p_2 \equiv 2 \pmod 3$. Since we are also requiring $p_2 \equiv 1 \pmod 4$, we apply the Chinese Remainder Theorem to conclude that we must have $p_2 \equiv 5 \pmod{12}$. The first such primes are $p_2 = 5$ and $p_2 = 17$.

(a) Consider $p_2 = 5$. This is the same as case (1a) of Example 2.83, so yields nothing new.

(b) Consider $p_2 = 17$. We proceed by exactly the same logic as before. We want to find values of D with

$$D \equiv 0 \text{ or } 1 \pmod 3 \qquad \text{and} \qquad D \equiv 0 \pmod{17},$$

and, by the Chinese Remainder Theorem, this pair of simultaneous congruences is equivalent to the single congruence

$$D \equiv 0 \text{ or } 34 \pmod{51}.$$

So for these values of D, by Theorem 2.82 we have that $\mathcal{O}(\sqrt{D})$ is not a UFD. Remembering that D must be square-free, the first few such positive values of D are $D = 34$, 51, 85, 102, and 187.

Remark 2.86. The techniques in Example 6.7 and Example 6.9 produce the same pairs (p_1, p_2) and hence the same values of D, but they produce them in a different order. Depending on circumstances, either one may be more convenient to use.

We now give an example of a rather trickier application of Corollary 2.65 (the Non-UFD Test) that enables us to show that $\mathcal{O}(\sqrt{D})$ is not a UFD in some additional cases.

Theorem 2.87. *Let D be congruent to $2 \pmod 8$ and suppose that D is divisible by a prime $p \equiv 3 \pmod 8$. Then $\mathcal{O}(\sqrt{D})$ is not a UFD.*

Proof: We shall show that $\mathcal{O}(\sqrt{D})$ does not have an element of norm p. We prove this by contradiction.

Suppose $\beta = a + b\sqrt{D}$ with $\|\beta\| = p$. Then $|a^2 - Db^2| = p$, so $a^2 - Db^2 = \pm p$ and hence $a^2 = \pm p + Db^2$. Since the right-hand side of this equation is divisible by p, the left-hand side must be divisible by p as well, and so $a = pc$ for some c. Substituting into this equation and dividing each term by p, we obtain the equation $pc^2 = \pm 1 + db^2$, where $d = D/p$. Since $D \equiv 2 \pmod 8$ and $p \equiv 3 \pmod 8$, we see that $d \equiv 6 \pmod 8$. Thus, this equation yields the congruence

$$3c^2 \equiv \pm 1 + 6b^2 \pmod 8.$$

But, as you can easily check, for any integer x, $x^2 \equiv 0,\ 1,$ or $4 \pmod 8$. Substituting these possibilities for c^2 and b^2, we see that this congruence has no solution. Let $\alpha = \sqrt{D}$. Then α is not divisible by p, but $\|\alpha\| = |D|$ is divisible by p. Hence, by Corollary 2.65 (the Non-UFD Test), $\mathcal{O}(\sqrt{D})$ is not a UFD. □

D	(q_1, q_2)	D	(q_1, q_2)	D	(q_1, q_2)	D	(q_1, q_2)
10	$(2,5)$	155	$(2,5)$	274	$(3,137)$	394	$(2,197)$
15	$(2,5)$	159	$(2,53)$	282	$(2,141)$	395	$(2,5)$
26	$(2,13)$	165	$(3,5)$	285	$(3,5)$	399	$(2,21)$
30	$(2,5)$	170	$(2,5)$	286	$(2,13)$	402	$(3,402)$
34	$(3,17)$	174	$(2,29)$	287	$(7,41)$	403	$(2,13)$
35	$(2,5)$	178	$(3,89)$	290	$(2,5)$	406	$(2,29)$
39	$(2,13)$	182	$(2,13)$	291	$(5,97)$	407	$(2,37)$
42	$(2,21)$	183	$(2,61)$	295	$(2,5)$	410	$(2,5)$
51	$(3,17)$	185	$(2,5)$	298	$(2,149)$	411	$(3,137)$
55	$(2,5)$	186	$(2,93)$	299	$(2,13)$	415	$(2,5)$
58	$(2,29)$	187	$(3,17)$	303	$(2,101)$	418	$(11,418)$
65	$(2,5)$	190	$(2,5)$	305	$(2,5)$	426	$(2,213)$
66	$(3,66)$	194	$(5,97)$	310	$(2,5)$	427	$(2,61)$
70	$(2,5)$	195	$(2,5)$	314	$(2,157)$	429	$(5,13)$
74	$(2,37)$	202	$(2,101)$	318	$(2,53)$	430	$(2,5)$
78	$(2,13)$	203	$(2,29)$	319	$(2,29)$	435	$(2,5)$
82	$(3,41)$	205	$(3,5)$	323	$(7,17)$	438	$(7,73)$
85	$(3,5)$	210	$(2,5)$	327	$(2,109)$	442	$(2,13)$
87	$(2,29)$	215	$(2,5)$	330	$(2,5)$	445	$(2,5)$
91	$(2,13)$	218	$(2,109)$	335	$(2,5)$	447	$(2,149)$
95	$(2,5)$	219	$(5,73)$	339	$(3,113)$	451	$(3,41)$
102	$(3,17)$	221	$(2,13)$	345	$(2,5)$	455	$(2,5)$
105	$(2,5)$	222	$(2,37)$	346	$(2,173)$	458	$(2,229)$
106	$(2,53)$	226	$(3,113)$	354	$(3,354)$	462	$(2,21)$
110	$(2,5)$	230	$(2,5)$	355	$(2,5)$	465	$(2,5)$
111	$(2,37)$	231	$(2,21)$	357	$(3,17)$	466	$(3,233)$
114	$(3,114)$	235	$(2,5)$	362	$(2,181)$	470	$(2,5)$
115	$(2,5)$	238	$(3,17)$	365	$(5,73)$	471	$(2,157)$
119	$(5,17)$	246	$(3,41)$	366	$(2,61)$	474	$(2,237)$
122	$(2,61)$	247	$(2,13)$	370	$(2,5)$	481	$(2,13)$
123	$(3,41)$	255	$(2,5)$	371	$(2,53)$	482	$(11,241)$
130	$(2,5)$	258	$(3,258)$	374	$(5,17)$	483	$(2,21)$
138	$(2,69)$	259	$(2,37)$	377	$(2,29)$	485	$(5,97)$
143	$(2,13)$	265	$(2,5)$	385	$(2,5)$	493	$(3,17)$
145	$(2,5)$	266	$(19,266)$	386	$(5,193)$	494	$(2,13)$
146	$(5,73)$	267	$(3,89)$	390	$(2,5)$	498	$(3,498)$
154	$(2,77)$	273	$(2,13)$	391	$(3,17)$		

Table 2.1. The values of D between 2 and 499 for which our methods show that $\mathcal{O}(\sqrt{D})$ is not a UFD.

Example 2.88. We wish to apply Theorem 2.87. For this we need primes congruent to 3 $(\bmod 8)$. The first such primes are $p = 3$ and $p = 11$.

(1) Consider $p = 3$. Then the hypotheses of Theorem 2.87 give the pair of simultaneous congruences

$$D \equiv 2 \pmod 8 \qquad \text{and} \qquad D \equiv 0 \pmod 3,$$

which, by the Chinese Remainder Theorem, is equivalent to the single congruence

$$D \equiv 18 \pmod{24}.$$

So for these values of D, by Theorem 2.87 we have that $\mathcal{O}(\sqrt{D})$ is not a UFD. Remembering that D must be square-free, the first few such positive values of D are $D = 42$, 66, 114, 138, and 186.

(2) Consider $p = 11$. Then the hypotheses of Theorem 2.87 give the pair of simultaneous congruences

$$D \equiv 2 \pmod 8 \qquad \text{and} \qquad D \equiv 0 \pmod{11},$$

which, by the Chinese Remainder Theorem, is equivalent to the single congruence

$$D \equiv 66 \pmod{88}.$$

So for these values of D, by Theorem 6.11 we have that $\mathcal{O}(\sqrt{D})$ is not a UFD. Remembering that D must be square-free, the first few such positive values of D are $D = 66$, 154, 330, 418, and 498.

Example 2.89. Table 2.1 is a table of values of D between 2 and 499 for which we can show that $\mathcal{O}(\sqrt{D})$ is not a UFD by using Theorem 2.78, Theorem 2.82, or Theorem 2.87. For each value of D we give a pair (q_1, q_2) that provides the argument. (In case $q_1 = 2$, we are using Theorem 2.78 with $q_2 = q$, as in Example 2.79. In case q_1 is an odd prime, we are using Theorem 2.82, as in Example 2.83 or Example 2.85, or Theorem 2.87, as in Example 2.88. In this last case we have simply set $q_2 = D$. For many values of D, there is more than one pair (q_1, q_2) that work. In those cases, we have simply chosen one.)

2.7 Summing Up

In the preceding sections of this chapter, we have shown that certain integral domains $\mathcal{O}(\sqrt{D})$ are or are not unique factorization domains. In this section we will sum up our work and also report on some interesting results that are beyond our ability to prove here.

We will state the results for the cases of imaginary quadratic fields ($D < 0$) and real quadratic fields ($D > 0$) separately.

First, imaginary quadratic fields.

Theorem 2.90. *Let $D < 0$, and let $R = \mathcal{O}(\sqrt{D})$.*

(1) If $D = -11, -7, -3, -2$, or -1, then R is a UFD.

(2) If $D \equiv 1, 2, 3, 6$, or $7 \pmod 8$ and $D \neq -1, -2$, or -7, then R is not a UFD.

(3) If $D \equiv 5 \pmod 8$ and $|D|$ is composite, then R is not a UFD.

Proof: (1) is Lemma 2.9; (2) is Theorem 2.68; and (3) is part of Proposition 2.70. □

Note that this theorem leaves an infinite number of cases open, those with $D \equiv 5 \pmod 8$, $|D|$ prime. Some of these cases we have dealt with in Proposition 2.72 (applied in Corollary 2.75).

The following is a very deep theorem.

Theorem 2.91. *There are exactly nine values of $D < 0$ for which $R = \mathcal{O}(\sqrt{D})$ is a UFD. They are $D = -1, -2, -3, -7, -11, -19, -43, -67$, and -163.*

Next, real quadratic fields.

Theorem 2.92. *Let $D > 0$, and let $R = \mathcal{O}(\sqrt{D})$.*

(1) If $D = 2, 3, 5, 6, 7, 11, 13, 17, 21$, or 29, then R is a UFD.

(2) If one of the following conditions holds,

(a) $D \equiv 1, 3, 6$, or $7 \pmod 8$ and (i) D is divisible by a prime congruent to $5 \pmod 8$ or (ii) D is divisible by a prime congruent to $3 \pmod 8$ and by a prime congruent to $7 \pmod 8$; or

(b) $D \equiv 2 \pmod 8$ and (i) D is divisible by a prime congruent to $5 \pmod 8$ or (ii) D is divisible by a prime congruent to $3 \pmod 8$,

then R is not a UFD.

Proof: (1) is Lemma 2.9 and (2) is Theorem 2.78 combined with Theorem 2.87. □

In Theorem 2.82 we were able to handle other values of D. See also Example 2.79, Example 2.83, Example 2.85, and Example 2.88.

Note that the results for real quadratic fields are much less complete than the results for imaginary quadratic fields. We have the following conjecture, which has been open for 200 years:

Conjecture 2.93 (Gauss). *There are an infinite number of values of $D > 0$ for which $\mathcal{O}(\sqrt{D})$ is a UFD.*

Remark 2.94. There is an effective procedure, due to Gauss, to decide whether $\mathcal{O}(\sqrt{D})$ is a UFD for any given value of D. To be precise, it computes the *class number* of $\mathcal{O}(\sqrt{D})$, and $\mathcal{O}(\sqrt{D})$ is a UFD precisely when the class number is 1.

Remark 2.95. Calculation shows that $R = \mathcal{O}(\sqrt{D})$ is a UFD for the following positive values of $D < 100$: $D = 2$, 3, 5, 6, 7, 11, 13, 14, 17, 19, 21, 22, 23, 29, 31, 33, 37, 38, 41, 43, 46, 47 ,53, 57, 59, 61, 62, 67, 69, 71, 73, 77, 83, 86, 89, 93, 94, and 97.

2.8 Exercises

Exercise 2.1. Complete the proof of Lemma 2.7. (You must prove Lemma 2.7 in the cases not explicitly done in the text, i.e., in the cases where $a \leq 0$ and $b > 0$; where $a \geq 0$ and $b < 0$; and where $a \leq 0$ and $b < 0$. You may do so by proving each of these cases from scratch, adapting the proof of the case $a \geq 0$ and $b > 0$ given in the text to these other cases, but even better would be a proof that simply reduces these other cases to the case $a \geq 0$ and $b > 0$ and uses the fact that we know Lemma 2.7 is true in that case.)

Exercise 2.2. Lemma 2.7 shows that for any integer a and any nonzero integer b, there exist integers q and r with $a = bq + r$ and $-|b| + 1 \leq r \leq |b| - 1$. Show that for any integer a and any nonzero integer b, there exist *unique* integers q and r with $a = bq + r$ and $0 \leq r \leq |b| - 1$.

Note by Lemma 2.14 that elements of an integral domain R will not in general have a unique gcd. Thus in the following problems we cannot write $LHS = RHS$ (where LHS (respectively RHS) stands for left-hand side (respectively right-hand side) of the equation). We thus write $LHS \cong RHS$ where by $\cong$ we mean *can be chosen to be equal to*. Also by Lemma 2.14, we see that the choice involves multiplication by some unit of R.

Exercise 2.3. Prove the following properties of a gcd of elements in an integral domain R:

(a) for any c, $\gcd(a, ca+b) \cong \gcd(a,b)$,

(b) for any $c \neq 0$, $\gcd(ac, bc) \cong c\gcd(a,b)$,

(c) if c divides both a and b, then $\gcd(a/c, b/c) \cong \gcd(a,b)/c$,

(d) if c is relatively prime to b, then $\gcd(ac, b) \cong \gcd(a,b)$.

In Exercises 2.4–2.11 R is assumed to be a PID. Do these exercises by using the results of Sections 2.1–2.3.

Exercise 2.4. Let α, β, γ, and δ be elements of R. If α divides γ and β divides δ, show that $\gcd(\alpha,\beta)$ divides $\gcd(\gamma,\delta)$.

Exercise 2.5. Let α and β be elements of R. Show that α and β are relatively prime if and only if α^2 and β^2 are relatively prime.

Exercise 2.6. Let α, β, and γ be nonzero elements of R. Show that

$$\gcd(\alpha,\beta,\gamma) \cong \gcd(\gcd(\alpha,\beta),\gamma).$$

(Hence the gcd of a finite set of elements of R can be found by successively finding the gcd of a pair of elements. In case R is a Euclidean domain this can be done by using Euclid's algorithm.)

Analogously to the gcd of elements of R, we can define an lcm of elements of R as follows:

> Let R be an integral domain and let $\{\alpha_i\}$ be a finite set of elements of R, not all of which are zero. Then an element λ of R is a *least common multiple* (lcm) of $\{\alpha_i\}$, $\lambda = \operatorname{lcm}(\{\alpha_i\})$ if
>
> (a) each α_i divides λ,
>
> (b) if ζ is any element of R that is divisible by α_i, then λ divides ζ.

Exercise 2.7. Prove the analog of Lemma 2.14:

(a) if λ is an lcm of $\{\alpha_i\}$ and ε is any unit of R, then $\lambda' = \lambda\varepsilon$ is also an lcm of $\{\alpha_i\}$,

(b) if λ and λ' are any two lcm's of $\{\alpha_i\}$, then $\lambda' = \lambda\varepsilon$ for some unit ε of R.

Exercise 2.8. Let α and β be elements of R, not both of which are zero. Suppose that α and β are relatively prime. Show that $\alpha\beta$ is an lcm of α and β.

Exercise 2.9. More generally, let α and β be any two elements of R, not both of which are zero. Let γ be a gcd of α and β. Show that $\lambda = \alpha\beta/\gamma$ is an lcm of α and β. Thus we see that if γ is a gcd of α and β, and λ is an lcm of α and β, then $\gamma\lambda \cong \alpha\beta$.

Exercise 2.10. Let α, β, and γ be any nonzero elements of R. Show that

$$\operatorname{lcm}(\alpha, \beta, \gamma) \cong \operatorname{lcm}(\operatorname{lcm}(\alpha, \beta), \gamma).$$

(Hence the lcm of a finite set of elements of R can be found by successively finding the lcm of a pair of elements. In case R is a Euclidean domain this can be done by using Euclid's algorithm to find gcd's and then using the result of Exercise 2.5.)

Exercise 2.11. Use the result of the preceding exercise and induction to show that every finite set of elements of R, not all of which are zero, has an lcm.

Exercise 2.12. Suppose that R is a UFD. In the notation of Proposition 2.59, let $\mu = p_1^{j_1} \cdots p_k^{j_k} q_1^{g_1} \cdots q_\ell^{g_\ell} r_1^{h_1} \cdots r_m^{h_m}$ where $j_i = max(e_i, f_i)$ for each $i = 1, \ldots, k$. Show that μ is an lcm of α and β. (In particular, α and β *have* an lcm.)

Exercises 2.3′–2.11′. Do Exercises 2.3–2.11 for a UFD R by using the results of Section 2.4. Exercises 2.3′–2.11′ are easier than Exercises 2.3–2.11, but that is because we are using more background—the prime factorization of elements of R.

Exercise 2.13. Give an example of an infinite set of elements in $\mathbb{Z}$, not all of which are zero, that does not have an lcm.

In the case $R = \mathbb{Z}$, the gcd of a set of elements is only defined up to multiplication by ± 1. We make the convention that in this case, we choose the gcd of a set of elements to be positive. Similarly, we choose the lcm of a set of elements to be positive.

Exercise 2.14. Let $\{a_1, \ldots, a_n\}$ be a set of integers, and let r be a rational number such that $a_i r$ is an integer, $i = 1, \ldots, n$.

(a) Suppose that $\{a_1, \ldots, a_n\}$ is relatively prime. Show that in fact r is an integer.

(b) More generally, let $\{a_1, \ldots, a_n\}$ have a gcd of d. Show that dr is an integer.

Exercise 2.15. Let $\{a_1/b_1, \ldots, a_n/b_n\}$ be a set of fractions in lowest terms. (By this we mean that a_i and b_i are relatively prime, for each i.)

(a) Suppose that $\{b_1, \ldots, b_n\}$ is pairwise relatively prime. Let $\ell = b_1 \cdots b_n$, and let $m_i = \ell/b_i$, $i = 1, \ldots, n$. Let $k = a_1 m_1 + \ldots + a_n m_n$. Show that k and ℓ are relatively prime. (Note that

$$(a_1/b_1) + \ldots + (a_n/b_n) = k/\ell,$$

so this shows that k/ℓ is a fraction in lowest terms.)

(b) Give an example of the following: a set of fractions in lowest terms $\{a_1/b_1, \ldots, a_n/b_n\}$ with $\{b_1, \ldots, b_n\}$ relatively prime (but not pairwise relatively prime), with $\ell = \text{lcm}(b_1, \ldots, b_n)$, $m_i = \ell/b_i$, $i = 1, \ldots, n$, $k = a_1 m_1 + \ldots + a_n m_n$, and k and ℓ *not* relatively prime. (Thus, in this case k/ℓ is not a fraction in lowest terms.)

Exercise 2.16. In each case, find the gcd of the following set of integers, and express the gcd as a linear combination of those integers:

(a) $\{19, 61\}$,

(b) $\{195, 37\}$,

(c) $\{391, 833\}$,

(d) $\{12345, 54321\}$,

(e) $\{65, 175, 233\}$,

(f) $\{1591, 1887, 2193\}$.

Exercise 2.17. Find the lcm of each of the sets of integers in Exercise 2.16.

Exercise 2.18. Let $R = \mathcal{O}(\sqrt{-1})$. Set $i = \sqrt{-1}$. In each case, find a gcd of the following sets of elements of R, and express that gcd as a linear combination of those elements:

(a) $\{13, 75\}$,

(b) $\{5 + i, 7 + 2i\}$,

(c) $\{17, 29 + 3i\}$,

(d) $\{3 + 2i, 2 + i\}$,

(e) $\{1 + i, 1 - i\}$,

(f) $\{5 + 5i, 14 + 8i, 9 + 7i\}$.

Exercise 2.19.

(a) Show that each of the following factorizations is a factorization into irreducibles in $\mathcal{O}(\sqrt{-14})$:

$$\begin{aligned} 47 &= (11 + 3\sqrt{-14}) \cdot (11 - 3\sqrt{-14}) = 13 \cdot 19, \\ 1745 &= (39 + 4\sqrt{-14}) \cdot (39 - 4\sqrt{-14}) = 5 \cdot 349. \end{aligned}$$

(b) Show that each of the following factorizations is a factorization into irreducibles in $\mathcal{O}(\sqrt{-17})$:

$$\begin{aligned} 121 &= 11 \cdot 11 = (2 + 3\sqrt{-17}) \cdot (2 - 3\sqrt{-17}), \\ 474 &= (7 + 5\sqrt{-17}) \cdot (7 - 5\sqrt{-17}) \\ &= 2 \cdot (13 + 2\sqrt{-17}) \cdot (13 - 2\sqrt{-17}). \end{aligned}$$

(c) Show that each of the following factorizations is a factorization into irreducibles in $\mathcal{O}(\sqrt{-23})$:

$$\begin{aligned} 89 &= 3 \cdot 163 = (11 + 4\sqrt{-23}) \cdot (11 - 4\sqrt{-23}), \\ 2049 &= 3 \cdot 683 = (41 + 4\sqrt{-23}) \cdot (41 - 4\sqrt{-23}). \end{aligned}$$

(d) Show that each of the following factorizations is a factorization into irreducibles in $\mathcal{O}(\sqrt{-26})$:

$$\begin{aligned} 1563 &= (17 + 7\sqrt{-26}) \cdot (17 - 7\sqrt{-26}) = 3 \cdot 521, \\ 8449 &= (5 + 18\sqrt{-26}) \cdot (5 - 18\sqrt{-26}) = 7 \cdot 17 \cdot 71, \\ 763 &= 7 \cdot 109 = (23 + 3\sqrt{-26}) \cdot (23 - 3\sqrt{-26}), \\ 136577 &= 7 \cdot 109 \cdot 179 = (23 + 3\sqrt{-26}) \cdot (23 - 3\sqrt{-26}) \cdot 179 \\ &= (369 + 4\sqrt{26}) \cdot (369 - 4\sqrt{26}). \end{aligned}$$

Exercise 2.20. Show that the following factorizations are factorizations into irreducibles in $\mathcal{O}(\sqrt{D})$ for the appropriate value of D:

(a) $55 = (9+\sqrt{26}) \cdot (9-\sqrt{26}) = 5 \cdot 11,$

(b) $-95 = (3+2\sqrt{26}) \cdot (3-2\sqrt{26}) = -5 \cdot 19,$

(c) $-21 = (3+\sqrt{30}) \cdot (3-2\sqrt{30}) = -3 \cdot 7,$

(d) $49 = (13+2\sqrt{30}) \cdot (13-2\sqrt{30}) = 7 \cdot 7,$

(e) $35 = (7+\sqrt{34}) \cdot (7-\sqrt{34}) = 5 \cdot 7,$

(f) $-25 = (3+\sqrt{34}) \cdot (3-\sqrt{34}) = -5 \cdot 5,$

(g) $14 = (7+\sqrt{35}) \cdot (7-\sqrt{35}) = 2 \cdot 7,$

(h) $-26 = (17+3\sqrt{35}) \cdot (17-3\sqrt{35}) = -2 \cdot 13,$

(i) $-21 = (7+\sqrt{70}) \cdot (7-\sqrt{70}) = -3 \cdot 7,$

(j) $-65 = (3+\sqrt{74}) \cdot (3-\sqrt{74}) = -5 \cdot 13,$

(k) $-77 = (1+\sqrt{78}) \cdot (1-\sqrt{78}) = -7 \cdot 11,$

(l) $-51 = (6+\sqrt{87}) \cdot (6-\sqrt{87}) = -3 \cdot 17.$

Exercise 2.21.

(a) Consider the following factorizations in $\mathcal{O}(\sqrt{-2})$:

$$51 = 3 \cdot 17 = (1+5\sqrt{-2}) \cdot (1-5\sqrt{-2}) = (7+\sqrt{-2})(7-\sqrt{-2}).$$

These *look like* three distinct factorizations into irreducibles, which would contradict $\mathcal{O}(\sqrt{-2})$ being a UFD. Show that in fact they are not factorizations into *irreducibles.* Find a factorization of 51 into irreducibles, and show how it yields these three factorizations of 51.

(b) Consider the following factorizations in $\mathcal{O}(\sqrt{-19})$:

$$\begin{aligned} 35 = 5 \cdot 7 &= (4+\sqrt{-19}) \cdot (4-\sqrt{-19}) \\ &= ((11+\sqrt{-19})/2) \cdot ((11-\sqrt{-19})/2). \end{aligned}$$

These *look like* three distinct factorizations into irreducibles, which would contradict $\mathcal{O}(\sqrt{-19})$ being a UFD. Show that in fact they are not factorizations into *irreducibles.* Find a factorization of 35 into irreducibles, and show how it yields these three factorizations of 35.

(c) Consider the following factorizations in $\mathcal{O}(\sqrt{6})$:

$$6 = \sqrt{6} \cdot \sqrt{6} = 2 \cdot 3.$$

These *look like* two distinct factorizations into irreducibles, which would contradict $\mathcal{O}(\sqrt{6})$ being a UFD. Show that in fact they are not factorizations into *irreducibles.* Find a factorization of 6 into irreducibles, show how it yields these two factorizations of 6, and use it to construct another factorization of 6.

(d) Consider the following factorizations in $\mathcal{O}(\sqrt{19})$:

$$-75 = -(1 + 2\sqrt{19}) \cdot (1 - 2\sqrt{19}) = -3 \cdot 5 \cdot 5.$$

These *look like* two distinct factorizations into irreducibles, which would contradict $\mathcal{O}(\sqrt{19})$ being a UFD. Show that in fact they are not factorizations into *irreducibles.* Find a factorization of -75 into irreducibles, show how it yields these two factorizations of -75, and use it to construct another factorization of -75.

Exercise 2.22.

(a) Let p be an odd prime. Set $D = -2p$. Show that

$$2p = -\sqrt{D} \cdot \sqrt{D} = 2 \cdot p$$

are two factorizations of $2p$ into irreducibles in $\mathcal{O}(\sqrt{D})$. (This applies to $D = -6, -10, -14, -22, -26, -34, \ldots$.)

(b) Let p be an odd prime and suppose that $D = 1 - 2p$ is square-free. Show that

$$2p = (1 - \sqrt{D})(1 + \sqrt{D}) = 2 \cdot p$$

are two factorizations of $2p$ into irreducibles in $\mathcal{O}(\sqrt{D})$. (This applies to $D = -5, -13, -21, -21, -33, -37, \ldots$.)

(c) Let p and q be primes with $pq \equiv 3 \pmod 4$ and suppose that $D = 1 - pq$ is square-free. Show that

$$pq = (1 - \sqrt{D})(1 + \sqrt{D}) = p \cdot q$$

are two factorizations of pq into irreducibles in $\mathcal{O}(\sqrt{D})$. (This applies to $D = -14, -34, -38, -94, -118, -142, \ldots$.)

(d) Let $p \geq 17$ and $q \geq 17$ be primes with $pq \equiv 1 \pmod 4$ and suppose that $D = (1 - pq)/4$ is square-free. Show that

$$pq = (1 - \sqrt{D})(1 + \sqrt{D}) = p \cdot q$$

are two factorizations of pq into irreducibles in $\mathcal{O}(\sqrt{D})$. (This applies to $D = -109, -123, -157, \ldots$.)

Exercise 2.23. Verify that Corollary 2.71 is true.

Exercise 2.24. Verify that Corollary 2.74 is true.

Exercise 2.25. Verify that Corollary 2.75 is true. (This requires the use of a computer.)

Exercise 2.26. Do the analogue of Example 2.83 for $p_2 = 17,\ 29$.

Exercise 2.27. Do the analogue of Example 2.85 for $p_2 = 17,\ 29$.

Exercise 2.28.

(a) Check the correctness of Table 2.1.

(b) Extend this table to values of D between 500 and 1000.

Exercise 2.29. Let R be an integral domain. Show that

$$\|f(X)\| = \deg(f(X)) \text{ for } f(X) \neq 0$$

is a norm on $R[X]$. (Here deg denotes the degree of a polynomial.) Note that the norm is not defined for the 0 polynomial.

Exercise 2.30. With the above definition of $\|\cdot\|$ on $R[X]$, show that, for any two nonzero polynomials $f(X)$ and $g(X)$,

(a) $\|f(X)g(X)\| = \|f(X)\| + \|g(X)\|$;

(b) $\|f(X) + g(X)\| \leq \max(\|f(X)\|, \|g(X)\|)$.

(The reason for not defining the norm of the 0 polynomial in $R[X]$ is to make these equations hold. If we defined the norm of the 0 polynomial to be 0, as you might think, we would have to make exceptions to (a) and (b) in order to make them hold.)

Exercise 2.31. Let R be a field. Show that $R[X]$ is a Euclidean domain with norm $\|\cdot\|$ as defined above. As a consequence, we conclude that $R[X]$ is a PID and hence a UFD.

Exercise 2.32.

(a) Show that $\mathbb{Z}[X]$ is *not* a Euclidean domain with the above norm.

(b) More generally, let R be any integral domain that is not a field. Show that $R[X]$ is *not* a Euclidean domain with the above norm.

In the case $R = \mathbf{Q}[X]$, the gcd of a set of elements is only defined up to multiplication by a nonzero rational number. We make the convention that in this case, we choose the gcd of a set of elements to be monic. A monic polynomial is one in which the highest power of "X" is equal to 1. Similarly, we choose the lcm of a set of elements to be monic.

Exercise 2.33. Find the gcd of each of the following sets of polynomials in $\mathbf{Q}[X]$, and express the gcd as a linear combination of those polynomials:

(a) $\{X^2 + X + 1, X + 1\}$;

(b) $\{X^3 + X + 1, X + 2\}$;

(c) $\{X^3 + 2X^2 + X + 2, X^4 + 5X^2 + 4\}$.

Exercise 2.34. Consider the following integral domain:

$$R = \{f(X) = a_0 + a_2X^2 + a_3X^3 + \ldots + a_nX^n \text{ in } \mathbf{Q}[X]\},$$

i.e., R consists of those polynomials in $\mathbf{Q}[X]$ that do not have an "X" term.

(a) Show that X^2 and X^3 are irreducible but not prime elements of R.

(b) This implies that R is *not* a UFD. Find an explicit element of R that has two distinct factorizations into irreducibles. (You may have already done (b) in doing (a).)

(c) Find a pair of elements of R that does not have a gcd.

Exercise 2.35. Let $R = \mathbb{Z}[X]$. Let

$$I = \{f(X) = a_nX_n + \ldots + a_0 \mid a_0 \text{ is even}\},$$

i.e., I consists of those polynomials whose constant term is even.

(a) Show that I is an ideal of R.

(b) (b) Show that $I = I_{\{2,X\}}$.

(c) Show that I is *not* a principal ideal of R.

Thus, we see that $\mathbb{Z}[X]$ is not a PID. (It turns out that $\mathbb{Z}[X]$ is a UFD, so $\mathbb{Z}[X]$ gives an example of a UFD that is not a PID.)

Chapter 3

The Gaussian Integers

In this chapter we will investigate $\mathcal{O}(\sqrt{-1})$, known as the *Gaussian integers*. We recall that

$$\begin{aligned}\mathcal{O}(\sqrt{-1}) &= \{a + b\sqrt{-1} \mid a \text{ and } b \text{ are integers}\} \\ &= \{a + bi \mid a \text{ and } b \text{ are integers}\},\end{aligned}$$

and we have shown that $\mathcal{O}(\sqrt{-1})$ *is* a UFD. We will begin by proving a justly famous theorem of Fermat: *every prime congruent to* $1 \pmod 4$ *can be written in as a sum of squares of two positive integers,* $p = x^2 + y^2$, *uniquely up to the order of the summands.* (For example, $5 = 2^2 + 1^2$, $13 = 3^2 + 2^2$, $17 = 4^2 + 1^2$, $29 = 5^2 + 2^2$, $37 = 6^2 + 1^2$, $41 = 5^2 + 4^2$). We shall present three proofs of this theorem in the first section of this chapter.

The first proof, due to Euler, is believed to be Fermat's original proof. (Fermat did not write his proof down, but left a hint as to his approach.) This is the longest and most difficult proof. (Actually, the proof we write down is a variant of Euler's proof, looking ahead to Chapter 4.)

The second proof is a twentieth-century proof due to Thue. It is of medium length and difficulty.

The third proof uses the fact that the Gaussian integers are a UFD, and, given that fact, is short and easy!

The fact that $\mathcal{O}(\sqrt{-1})$ is a UFD gives us unique factorization into primes, but by itself does not tell us what the primes are (or how to find a factorization). Using Fermat's theorem, we can concretely and completely answer this question, and we do so in the second section of this chapter.

3.1 Fermat's Theorem

We wish to represent numbers as sums of two squares. One case is easy to rule *out.*

Lemma 3.1. *No integer* $N \equiv 3 \pmod 4$ *can be written as a sum of two squares.*

Proof: First note that if z is even, $z = 2k$, then $z^2 = 4k^2 = 4(k^2) \equiv 0 \pmod 4$, while if z is odd, $z = 2k+1$, then $z^2 = 4k^2 + 4k + 1 = 4(k^2 + k) + 1 \equiv 1 \pmod 4$.

Now consider x^2+y^2. If x and y are both even, then $x^2 \equiv 0 \pmod 4$ and $y^2 \equiv 0 \pmod 4$, so $x^2 + y^2 \equiv 0 + 0 = 0 \pmod 4$. If x is even and y is odd, then $x^2 \equiv 0 \pmod 4$ and $y^2 \equiv 1 \pmod 4$, so $x^2 + y^2 \equiv 0 + 1 = 1 \pmod 4$. Similarly, if x is odd and y is even, $x^2 + y^2 \equiv 1 \pmod 4$. Finally, if x and y are both odd, then $x^2 \equiv 1 \pmod 4$ and $y^2 \equiv 1 \pmod 4$, so $x^2 + y^2 \equiv 1 + 1 = 2 \pmod 4$. Thus, in no case can we have $x^2 + y^2 \equiv 3 \pmod 4$. □

Now for the numbers we wish to rule *in.* We begin with an observation: if each of two numbers is representable as a sum of two squares, so is their product. For example, since $5 = 2^2 + 1^2$ and $13 = 3^2 + 2^2$, we can conclude that 65 is also a sum of two squares, and in fact $65 = 8^2 + 1^2 = 7^2 + 4^2$. That this is true is a simple algebraic fact.

Lemma 3.2. *If* $m = a^2 + b^2$ *and* $n = c^2 + d^2$, *then*

$$mn = e^2 + f^2$$

where $e = ac - bd$ *and* $f = ad + bc$.

Proof: We simply compute

$$\begin{aligned} e^2 + f^2 &= (ac - bd)^2 + (ad + bc)^2 \\ &= (a^2c^2 - 2acbd + b^2d^2) + (a^2d^2 + 2adbc + b^2c^2) \\ &= a^2c^2 + b^2d^2 + a^2d^2 + b^2c^2 \\ &= (a^2 + b^2)(c^2 + d^2), \end{aligned}$$

proving the result. □

It is convenient to introduce the following language.

Definition 3.3. The representation of mn as a sum of squares,

$$mn = e^2 + f^2$$

with $e = ac - bd$ and $f = ad + bc$ is obtained from the representations $m = a^2 + b^2$ and $n = c^2 + d^2$ of m and n as sums of squares by *composition*.

Fermat's proof contained several brilliant ideas. The first was to turn this lemma around and to show that if m and mn are sums of two squares, then, under proper conditions, so is n. The second was not to try to show directly that n is a sum of two squares, but instead to find some multiple mn of n that is, and to apply the first idea. Of course, to make that work Fermat had to know that m is a sum of two squares and for that he applied his "method of descent," a variant of mathematical induction.

Now let us look at Fermat's proof precisely. We will break it up into a number of steps.

Lemma 3.4.

(1) If p is a prime congruent to $1 \pmod 4$ then the congruence $x^2 + y^2 \equiv 0 \pmod p$ has a solution other than $x \equiv y \equiv 0 \pmod p$.

(2) If p is a prime congruent to $3 \pmod 4$ then the congruence $x^2 + y^2 \equiv 0 \pmod p$ only has the solution $x \equiv y \equiv 0 \pmod p$.

Proof:

(1) In this case we know, by Corollary B.33, that -1 is a quadratic residue $\pmod p$, i.e., there is an integer a with $a^2 \equiv -1 \pmod p$. Then, setting $x = a$ and $y = 1$,

$$x^2 + y^2 = a^2 + 1^2 \equiv -1 + 1 \equiv 0 \pmod p.$$

(2) We prove this by contradiction. Suppose $x^2 + y^2 \equiv 0 \pmod p$ with $x \not\equiv 0 \pmod p$. Then, by Theorem B.11, there is an integer a with $ax \equiv 1 \pmod p$. Then $(ax)^2 \equiv 1^2 \equiv 1 \pmod p$, so we have

$$\begin{aligned} x^2 + y^2 &\equiv 0 \pmod p \\ a^2(x^2 + y^2) &\equiv 0 \pmod p \\ (ax)^2 + (ay)^2 &\equiv 0 \pmod p \\ 1 + (ay)^2 &\equiv 0 \pmod p \\ (ay)^2 &\equiv -1 \pmod p, \end{aligned}$$

which would show -1 is a quadratic residue $\pmod p$. But, by Corollary B.30, we know that is not the case. □

Lemma 3.5. (Fermat). *Suppose $N = s^2 + t^2$ is a sum of two squares, and suppose p is a prime divisor of N that has a representation $p = a^2 + b^2$ as a sum of two squares. Then $M = N/p$ has a representation $M = c^2 + d^2$ such that the given representation $N = s^2 + t^2$ is obtained from the representations $p = a^2 + b^2$ and $M = c^2 + d^2$ by composition or from the representations $p = a^2 + (-b)^2$ and $M = c^2 + d^2$ by composition.*

Proof: Let us begin by *composing* the representations $N = s^2 + t^2$ and $p = a^2 + (-b)^2$, to obtain the representation

$$Np = (sa + tb)^2 + (-sb + ta)^2$$

and by *composing* the representations $N = s^2 + t^2$ and $p = a^2 + b^2$, to obtain the representation

$$Np = (sa - tb)^2 + (sb + ta)^2.$$

Now p divides N, by assumption, so p divides Nb^2. Certainly, p divides t^2p. Hence p divides the difference $t^2p - Nb^2$. But

$$t^2p - Nb^2 = t^2(a^2 + b^2) - (s^2 + t^2)b^2 = -s^2b^2 + t^2a^2 = (-sb + ta)(sb + ta).$$

Since p is a prime, it must divide one of the factors.

Case (I): p divides the first factor $-sb + ta$. Write $-sb + ta = pd$ and consider the first representation of Np,

$$Np = (sa + tb)^2 + (-sb + ta)^2.$$

Now p divides the left-hand side and the last term on the right-hand side, so it must divide the first term as well, so we may write $sa + tb = pc$. Then

$$Np = Mp^2 = (pc)^2 + (pd)^2 = p^2c^2 + p^2d^2,$$

so

$$M = c^2 + d^2$$

is a sum of two squares. Furthermore, we have just seen that

$$\begin{aligned} sa + tb &= pc \\ -sb + ta &= pd. \end{aligned}$$

Regard this as a system of two linear equations in the two unknowns s and t and solve. Doing so we obtain

$$s = ac - bd$$
$$t = ad + bc$$

and, referring to Definition 3.3, we see that $N = s^2 + t^2$ is obtained from $M = c^2 + d^2$ and $p = a^2 + b^2$ by composition.

Case (II): p divides the second factor $sb + ta$. Write $sb + ta = pd$ and consider the second representation of Np,

$$Np = (sa - tb)^2 + (sb + ta)^2.$$

Now p divides the left-hand side and the last term on the right-hand side, so it must divide the first term as well, so we may write $sa - tb = pc$. Then once again

$$Np = Mp^2 = (pc)^2 + (pd)^2 = p^2c^2 + p^2d^2,$$

so

$$M = c^2 + d^2$$

is a sum of two squares. Similarly, we have just seen that

$$sa - tb = pc$$
$$sb + ta = pd.$$

Regard this as a system of two linear equations in the two unknowns s and t and solve. Doing so we obtain

$$s = ac + bd$$
$$t = ad - bc$$

and, referring to Definition 3.3, we see that $N = s^2 + t^2$ is obtained from $M = c^2 + d^2$ and $p = a^2 + (-b)^2$ by composition.□

Theorem 3.6. (Fermat). *Every prime $p \equiv 1 \pmod 4$ can be as a sum of squares of positive integers, $p = x^2 + y^2$, and this representation is unique up to the order of x and y.*

Proof (Fermat/Euler): By Lemma 3.4, we can find integers x and y with $x^2 + y^2 \equiv 0 \pmod p$ and with x and y not divisible by p. Since this is a congruence, we may assume $1 \le x \le p-1$ and $1 \le y \le p-1$. Since $((p-1)-x)^2 \equiv x^2 \pmod p$ we may, replacing x by $p-x$ if necessary, assume that $1 \le x \le (p-1)/2$, and similarly we may assume that $1 \le y \le (p-1)/2$. Finally, if x and y have a common factor (which is certainly not p), we may divide x and y by that common factor, to obtain a pair of relatively prime integers x and y with $1 \le x \le (p-1)/2$, $1 \le y \le (p-1)/2$, and $x^2 + y^2$ a multiple of p. So we may assume x and y are relatively prime as well. Let $N = x^2 + y^2$ and $M = N/p$. Then M is an integer and $N = Mp$. Observe that, since $0 < x < p/2$ and $0 < y < p/2$,

$$N = x^2 + y^2 < (p/2)^2 + (p/2)^2 = p^2/2,$$

so $M < p/2$.

In particular, every prime factor of M is less than $p/2$. Let q be any prime factor of M. We claim $q \not\equiv 3 \pmod 4$. For suppose $q \equiv 3 \pmod 4$. Then $N = x^2 + y^2$ and N is a multiple of q, so $x^2 + y^2 \equiv 0 \pmod q$ and hence, by Lemma 3.4, $x \equiv y \equiv 0 \pmod q$. In other words, x and y are each divisible by q, which contradicts our assumption that x and y are relatively prime.

Thus we see that, for any prime factor q of N, $q = 2$ or $q \equiv 1 \pmod 4$. Also, let us observe that we may write $q = 2$ as a sum of two squares, $2 = 1^2 + 1^2$.

With this in hand, we prove the theorem by induction. Assume that every prime $q \equiv 1 \pmod 4$, $q < p$, can be written as a sum of two squares, and consider p. We have written $N = Mp$ as a sum of two squares, $N = x^2 + y^2$. But

$$N = (2 \cdot 2 \ldots \cdot 2 \cdot q_1 \cdot q_2 \cdots \cdot q_k)p$$

where we have a certain number of factors of 2 (perhaps none) and a certain number of odd prime factors q_1, q_2, …, q_k (perhaps none), with $q_1 < p$, $q_2 < p$, …, $q_k < p$. But now, since each of these factors is a sum of two squares, we may apply Lemma 3.5 repeatedly, eliminating one factor of M each time, until finally we obtain a representation of p as a sum of two squares, as claimed.

Now we must show the uniqueness part of the theorem.

Suppose $p = a^2+b^2 = s^2+t^2$. Apply Lemma 3.5 with $N = p$ to conclude that $p = s^2 + t^2$ is obtained from $p = a^2 + b^2$ and a representation of $M = p/p = 1$ by composition, or from $p = a^2 + (-b)^2$ and a representation of $M = p/p = 1$ by composition. But obviously the only representations of 1 are $1 = 1^2 + 0^2 = (-1)^2 + 0^2 = 0^2 + 1^2 = 0^2 + (-1)^2$. Composing a^2+b^2 and $a^2+(-b)^2$ with these yields the eight possibilities $p = a^2+b^2 = a^2 + (-b)^2 = (-a)^2 + b^2 = (-a)^2 + (-b)^2 = b^2 + a^2 = b^2 + (-a)^2 = (-b)^2 + a^2 = (-b)^2 + (-a)^2$, and we see that in exactly two of these cases we have $p = x^2 + y^2$ with x and y positive, and that these two cases differ from each other only in that x and y are interchanged. □

Remark 3.7. We have finished the proof by induction. Fermat phrased it "by descent," which is equivalent to induction. Fermat's phraseology is as follows: Suppose there is a prime $p \equiv 1 \pmod 4$ that cannot be written as a sum of two squares. In the notation of the proof, writing N as a sum of two squares, we see that there must be some smaller prime q that cannot be written as a sum of two squares, with $q = 2$ or $q \equiv 1 \pmod 4$, for all of the factors of M are of this form, and if each could be written as a sum of two squares, so could p, by Lemma 3.5. Now apply the same analysis to q instead of M to get a smaller prime q' that cannot be written as a sum of two squares. Continue in this fashion. Doing so, we obtain a descending sequence $p > q > q' > q'' > \dots$ of primes, each of which is either 2 or congruent to $1 \pmod 4$, and none of which can be written as a sum of two squares. This sequence must terminate at the smallest such prime, which is 2, and that is a contradiction, as $2 = 1^2 + 1^2$ is a sum of two squares.

We now present our second proof, and begin with a key lemma.

Lemma 3.8. (Thue). *Let p be a prime and let a be any integer relatively prime to p. Then there are integers x_0 and y_0 with $ax_0 \equiv y_0 \pmod p$ and $0 < |x_0| < \sqrt{p}$, $0 < |y_0| < \sqrt{p}$.*

Proof: Let $k = [\sqrt{p}]$, for convenience, and consider

$$\mathcal{S} = \{ax - y \mid 0 \leq x \leq k, 0 \leq y \leq k\}.$$

There are $k+1$ choices for x and $k+1$ choices for y, so the set $\mathcal{S}$ has $(k+1)^2 > p$ elements. Since there are only p congruence classes $\pmod p$, by the Pigeonhole Principle there must be two different elements of $\mathcal{S}$ that are congruent $\pmod p$,

$$ax_1 - y_1 \equiv ax_2 - y_2 \pmod p,$$

and so

$$a(x_1 - x_2) \equiv y_1 - y_2 \pmod p.$$

Set $x_0 = x_1 - x_2$ and $y_0 = y_1 - y_2$. Then certainly $ax_0 \equiv y_0 \pmod{p}$. We need to check that the other conditions on x_0 and y_0 are satisfied. First, since $0 \leq x_1 \leq k$ and $0 \leq x_2 \leq k$, the largest x_0 can be is $k - 0 = k$ and the smallest x_0 can be is $0 - k = -k$, so $|x_0| < \sqrt{p}$, and similarly $|y_0| < \sqrt{p}$. Also, since a is relatively prime to p, either x_0 and y_0 are both zero or both nonzero. But they cannot both be zero, as then $x_1 = x_2$ and $y_1 = y_2$, contradicting our choice of *different* elements of $\mathcal{S}$. □

Second Proof of Fermat's Theorem (Thue): Let $p \equiv 1 \pmod{4}$ and choose an integer a with $a^2 \equiv -1 \pmod{p}$. (Such an integer a exists by Corollary B.33).

Let $ax_0 \equiv y_0 \pmod{p}$ with x_0 and y_0 as in Lemma 3.8. Then

$$\begin{aligned} a^2x_0^2 &\equiv y_0^2 && \pmod{p} \\ -x_0^2 &\equiv y_0^2 && \pmod{p} \\ x_0^2 + y_0^2 &\equiv 0 && \pmod{p}, \end{aligned}$$

i.e., $x_0^2 + y_0^2$ is a multiple of p. But $0 < |x_0| < \sqrt{p}$ and $0 < |y_0| < \sqrt{p}$, so $0 < x_0^2 + y_0^2 < 2p$. Hence we must have $x_0^2 + y_0^2 = p$.

Now we must show uniqueness. Suppose $p = a^2 + b^2 = c^2 + d^2$ with $a, b, c, d \geq 0$. Note that we must have a and b relatively prime, as if they had a common factor r the prime p would be divisible by r^2. Similarly, c and d are relatively prime. Then, similarly to the proof of Lemma 3.5, we may apply composition and take absolute values to get two representations of p^2:

$$\begin{aligned} p^2 &= |ac + bd|^2 + |ad - bc|^2 \\ &= |ac - bd|^2 + |ad + bc|^2. \end{aligned}$$

Then (continuing as in the proof of Lemma 3.5) p divides

$$pd^2 - b^2p = (a^2 + b^2)d^2 - b^2(c^2 + d^2) = a^2d^2 - b^2c^2 = (ad - bc)(ad + bc),$$

so p must divide one of the two factors. Suppose p divides $ad - bc$ and consider the representation $p^2 = |ac + bd|^2 + |ad - bc|^2$. (If p divides $ad + bc$, consider the other representation and argue analogously.) Then p divides the left-hand side and the last term on the right-hand side, so must divide the first term on the right-hand side as well. Write $ac + bd = pu$ and $ad - bc = pv$. Then $p^2 = (pu)^2 + (pv)^2 = p^2(u^2 + v^2)$, i.e., $u^2 + v^2 = 1$. Since $ac + bd > 0, u > 0$, so we must have $u = 1$ and $v = 0$, i.e., $ac + bd = p$ and $ad - bc = 0$.

In particular, $ad = bc$. Then a divides the product bc. But and a and b are relatively prime, so by Euclid's Lemma a divides c. Similarly, c divides the product ad, and c and d are relatively prime, so by Euclid's Lemma c divides a. Hence a and c divide each other so $a = c$, and then $b = d$ as well. (In the other case we get $u = 0$ and $v = 1$, which yields the other order $a = d$ and $b = c$.) □

Now we present a lightning-fast proof of both existence and uniqueness using the fact that the Gaussian integers $\mathcal{O}(\sqrt{-1})$ is a UFD.

Third Proof of Fermat's Theorem:

Existence: Since $p \equiv 1 \pmod 4$, -1 is a quadratic residue $(\mathrm{mod}\, p)$, by Corollary B.33. Thus, there is an integer x with $x^2 \equiv -1 \pmod p$ and then $N = x^2 + 1$ is divisible by p, $N = Mp$ for some integer M. Note $x^2 + 1 = (x+i)(x-i)$ in $\mathcal{O}(\sqrt{-1})$. Consider the two factorizations of N,

$$N = (x+i)(x-i) = Mp.$$

Now p divides N but it certainly does not divide either $x+i$ or $x-i$ (as $(x+i)/p$ and $(x-i)/p$ are not in $\mathcal{O}(\sqrt{-1})$), so p is *not* a prime in $\mathcal{O}(\sqrt{-1})$. Now $\mathcal{O}(\sqrt{-1})$ is a UFD (Corollary 2.53), and in a UFD primes are irreducibles and vice versa (Lemma 2.55), so p is not irreducible in $\mathcal{O}(\sqrt{-1})$. Hence p has a factorization $p = \alpha\beta$ in $\mathcal{O}(\sqrt{-1})$ with neither α nor β a unit. Then $p^2 = \|p\| = \|\alpha\| \cdot \|\beta\|$ so $\|\alpha\| = \|\beta\| = p$ (as otherwise $\|\alpha\| = 1$ and α is a unit, or $\|\beta\| = 1$, and β is a unit). Write

$$\alpha = a + bi.$$

Then

$$p = \|\alpha\| = \alpha\overline{\alpha} = (a+bi)(a-bi) = a^2 + b^2,$$

and so p is written as a sum of squares, as claimed. (Also, we find that $\beta = \overline{\alpha} = a - bi$.)

Uniqueness: Suppose $p = a^2 + b^2 = c^2 + d^2$. Then

$$p = (a+bi)(a-bi) = (c+di)(c-di).$$

Now each of these four factors has norm p, and p is an ordinary prime, so each of these factors is irreducible (Lemma 2.64). In a UFD, primes are irreducible and vice versa (Lemma 2.55), so each of these factors is prime. Thus, since $a + b\sqrt{-1}$ is prime and divides the product, it must divide one of the factors on the right-hand side.

In general, if α divides β, $\beta = \alpha\gamma$, then $\|\beta\| = \|\alpha\| \cdot \|\gamma\|$, and here α and β each have norm p, so $\|\gamma\| = 1$, which implies γ is a unit (Lemma 1.14). But by Corollary 1.15 we already know all the units in $\mathcal{O}(\sqrt{-1})$. They are $\{\pm 1, \pm i\}$. Thus, we have the possibilities

$$c + di = (a + bi)\gamma \text{ with } \gamma = 1,\ -1,\ i, \text{ or } -i$$

or

$$c - di = (a + bi)\gamma \text{ with } \gamma = 1,\ -1,\ i, \text{ or } -i.$$

Solving for c and d, we see that these give the possibilities $(c, d) = (\pm a, \pm b)$ or $(\pm b, \pm a)$, where the signs can be chosen independently, so up to order and requiring both entries positive there is a unique solution. □

Remark 3.9. Combining the easy Lemma 3.1 and the trivial observation $2 = 1^2 + 1^2$ with Theorem 3.6, we see that Fermat's Theorem can be rephrased as follows:

Theorem (Fermat). *Let p be a prime. Then p can be written as a sum of squares of integers, $p = x^2 + y^2$ for some integers x and y, if and only if $p = 2$ or $p \equiv 1 \pmod 4$, in which case x and y are essentially unique.*

Here by "essentially unique" we mean unique up to sign and up to interchanging the order of x and y.

(We mention this not because it adds anything to what we have already proved but rather for comparison with the exercises for this chapter.)

For our purposes, Theorem 3.6 is all we need. But not only did Fermat show which primes could be represented as sums of two squares, he also showed which integers could be represented as sums of squares. Since this is also an interesting result, and since the rest of the work is relatively easy, we shall finish this section by showing that.

Theorem 3.10. (Fermat). *A positive integer n can be written as the sum of two squares if and only if for every prime $q \equiv 3 \pmod 4$ that divides n, the highest power of q dividing n is even.*

Proof: If $n = 1$ then $n = 1^2 + 0^2$ and we are done. For $n > 1$ let us factor n into primes,

$$n = 2^a p_1^{b_1} \dots p_k^{b_k} q_1^{c_1} \cdots q_\ell^{c_\ell}$$

where each prime p_i is congruent to 1 (mod 4) and each prime q_i is congruent to 3 (mod 4). Set $m = 2^a p_i^{b_1} \cdots p_k^{b_k}$, and $m' = q_1^{c_1} \cdots q_\ell^{c_\ell}$, so $n = mm'$.

Then $2 = 1^2 + 1^2$, and by Fermat's Theorem each p_i is a sum of two squares, so we may repeatedly compose representations (i.e., repeatedly apply Lemma 3.2) to obtain a representation $m = u^2 + v^2$. If each c_i is even, let

$$z = q_1^{c_1/2} \cdots q_\ell^{c_\ell/2}$$

and note that z is an integer with $z^2 = m'$. Then, setting $x = uz$ and $y = vz$,

$$x^2 + y^2 = (uz)^2 + (vz)^2 = (u^2 + v^2)z^2 = mm' = n,$$

so n is a sum of two squares.

On the other hand, suppose $n = x^2 + y^2$ is a sum of two squares, and let q be any prime congruent to 3 (mod 4) that divides n. Then n is a multiple of q, so $x^2 + y^2 = n$ gives $x^2 + y^2 \equiv 0 \pmod{q}$ and, by Lemma 3.4, that implies that x and y are each multiples of q. Let s and t be the highest powers of q that divide x and y respectively, so $x = q^s u$ and $y = q^t v$ with neither u nor v divisible by q. Let us assume $s \leq t$ (otherwise, simply interchange x and y). Then

$$n = x^2 + y^2 = (q^s u)^2 + (q^t v)^2 = q^{2s}(u^2 + q^{2(t-s)}v^2).$$

There are now two possibilities: If $s < t$, then u^2 is not divisible by q but $q^{2(t-s)}v^2$ is divisible by q, so their sum is not divisible by q. If $s = t$, then u and v are not divisible by q, and $q \equiv 3 \pmod{4}$, so again, by Lemma 3.4, $u^2 + v^2$ is not divisible by q. Thus, in either case the highest power of q dividing n is $2s$, which is even, as claimed. □

3.2 Factorization into Primes

We have two goals in this section. Our first is to find the primes in the Gaussian integers, and our second is to show how to factor an arbitrary Gaussian integer into primes. Recall from Definition 1.11 that if $\alpha = a + bi$ is a Gaussian integer, its *conjugate* is $\overline{\alpha} = a - bi$. (Note that $\alpha = \overline{\alpha}$ if and only if $\alpha = a$ is an ordinary integer.) We also observe that if α and β are Gaussian integers such that α divides β, then $\overline{\alpha}$ divides $\overline{\beta}$. This is a direct computation: if $\beta = \alpha\gamma$ with $\alpha = a + bi$, $\beta = c + di$, and $\gamma = e + fi$, so that $(c + di) = (a + bi)(e + fi)$, then also $(c - di) = (a - bi)(e - fi)$, so $\overline{\beta} = \overline{\alpha\gamma}$. In particular, if $\beta = b$ is an ordinary integer, then α divides β if and only if $\overline{\alpha}$ divides β.

Theorem 3.11. *The following is a complete list of primes in* $\mathcal{O}(\sqrt{-1})$*:*

(1) $1+i$ *and its associates* $-1+i$, $-1-i$, *and* $1-i$;

(2) *for any ordinary prime* $p \equiv 1 \pmod 4$, *write* p *as* $p = a^2 + b^2$ *with* a *and* b *integers;*

(a) $a+bi$ *and its associates* $-b+ai$, $-a-bi$, *and* $b-ai$;

(b) $a-bi$ *and its associates* $b+ai$, $-a+bi$, *and* $-b-ai$.

(3) *for any ordinary prime* $p \equiv 3 \pmod 4$, p *and its associates* pi, $-p$, *and* $-pi$.

Remark 3.12.

(1) Recall that the associates of α are $\alpha\varepsilon$ for any unit ε of $\mathcal{O}(\sqrt{-1})$. Since we determined in Corollary 1.15 that the units of $\mathcal{O}(\sqrt{-1})$ are ± 1 and $\pm i$, we have just multiplied the first prime in each list by these units to obtain the others.

(2) Note that $1+i$ and its conjugate $1-i$ *are* associates, but if a and b are as in Theorem 3.11(2) then $a+bi$ and its conjugate $a-bi$ are *not* associates.

Proof: In this theorem we are making two claims: first, that every Gaussian integer on this list is a prime, and second, that every prime in the Gaussian integers is on this list. Both of these claims are consequences of our earlier work.

If $\alpha = 1+i$ then $\|\alpha\| = 2$ and if $\alpha = a+bi$ as in (2a) or $\alpha = a-bi$ as in (2b) then $\|\alpha\| = p$. Thus, in either case $\|\alpha\|$ is an ordinary prime, so by Lemma 2.64(1) α is irreducible. If $p \equiv 3 \pmod 4$, then $\mathcal{O}(\sqrt{-1})$ has no element β of norm p (as if $\beta = x+yi$, $p = \|\beta\| = x^2+y^2$, and we know from Lemma 3.1 that this is impossible). Then, setting $\gamma = p$, we have $\|\gamma\| = p^2$, so γ is irreducible by Lemma 2.64(2). Finally, since $\mathcal{O}(\sqrt{-1})$ is a UFD, the primes and the irreducibles are the same (Lemma 2.55), so we have proved our first claim.

Now suppose α is prime, or equivalently, irreducible. Let $\alpha = a+bi$. Then $\overline{\alpha} = a-bi$ is also irreducible, or equivalently prime. (If we had $\overline{\alpha} = \beta\gamma$ then we would have $\alpha = \overline{\beta}\overline{\gamma}$). Let

$$N = \|\alpha\| = \alpha\overline{\alpha} = a^2 + b^2.$$

Factor N into ordinary primes, $N = p_1^{e_1} p_2 \cdots p_k^{e_k}$. Since α is a prime, and it divides the product $p_1^{e_1} \cdots p_k^{e_k}$, in $\mathcal{O}(\sqrt{-1})$, it must divide one of the factors, so suppose it divides p_1. Write $p_1 = \alpha\beta$. Then $\|\alpha\|$ divides $\|p_1\| = p_1^2$, so we must have $\|\alpha\| = 1$, p_1, or p_1^2. We cannot have $\|\alpha\| = 1$, as that would mean α is a unit.

Suppose $\|\alpha\| = p_1$. Now $\|\alpha\| = a^2 + b^2$, so that shows $p_1 = a^2 + b^2$. Then p_1 must either be 2 or a prime congruent to 1 (mod 4), by Lemma 3.1, so α is either as in (1), (2a), or (2b). (Furthermore, in this case $\beta = \overline{\alpha}$ and we have $p_1 = \alpha\overline{\alpha}$.)

Suppose $\|\alpha\| = p_1^2$. Then $\|\beta\| = 1$, and so β is a unit, and then α and p_1 are associates, $\alpha = \varepsilon p_1$ with ε a unit. Since, by assumption, α is a prime in $\mathcal{O}(\sqrt{-1})$, p_1 must also be a prime in $\mathcal{O}(\sqrt{-1})$. We have just ruled out $p_1 = 2$ or $p_1 \equiv 1 \pmod 4$, as we have just factored them as $\alpha\overline{\alpha}$ in those cases (so they are not irreducible and hence not prime), so we must have $p_1 \equiv 3 \pmod 4$, and α as is in (3). □

Example 3.13. We now show how to factor Gaussian integers into primes. There is a certain amount of trial and error involved, just as for factoring ordinary integers into primes. But we will see that there are only finitely many possibilities to try, so our method will eventually succeed.

We begin with some special cases.

Case 1 (a). $\alpha = 2$. Then $\alpha = -i(1+i)^2$. (Note $-i$ is a unit.)

Case 1 (b). $\alpha = p$, a prime congruent to 1 (mod 4). By trial and error, we find ordinary positive integers a and b with $a^2 + b^2 = p$, and then

$$\alpha = (a+bi)(a-bi).$$

(Note that there are only finitely many possibilities to check as a is a positive integer with $a < \sqrt{p}$.) For example, $5 = 2^2 + 1^2$ so $5 = (2+i)(2-i)$ in $\mathcal{O}(\sqrt{-1})$, and $13 = 2^2 + 3^2$ so $13 = (2+3i)(2-3i)$ in $\mathcal{O}(\sqrt{-1})$.

Case 1 (c). $\alpha = p$, a prime congruent to 3 (mod 4). Then α is prime, so its prime factorization is simply $\alpha = p$. For example, $7 = 7$ and $11 = 11$ in $\mathcal{O}(\sqrt{-1})$.

Case 2. α is an ordinary integer. Then factor α into ordinary primes and use case 1. For example, let $\alpha = 50$. Then

$$\begin{aligned}\alpha = 2 \cdot 5^2 &= ((-i)(1+i)^2)((2+i)(2-i))^2 \\ &= -i(1+i)^2(2+i)^2(2-i)^2.\end{aligned}$$

As a second example, let $\alpha = 44$. Then

$$\alpha = 2^2 \cdot 11 = ((-i)(1+i)^2)^2(11) = -1(1+i)^4(11).$$

As a third example, let $\alpha = 405$. Then

$$\alpha = 5 \cdot 7 \cdot 13 = (2+i)(2-i)(7)(2+3i)(2-3i).$$

Case 3. α is an arbitrary Gaussian integer, $\alpha = a+bi$. Let g be the ordinary gcd of the ordinary integers a and b, so $a = gc$ and $b = gd$ with c and d relatively prime. Then $\alpha = gc + gdi = g(c+di) = g\beta$ where $\beta = c + di$. We know how to factor g from Case 2, so we need only see how to factor β. To do this, we let $N = \|\beta\| = c^2 + d^2$. Then if γ is a prime dividing β, then $\|\gamma\|$ divides $\|\beta\| = N$. We know that N is not divisible by any prime $p \equiv 3 \pmod 4$. Also, if p is a prime $\equiv 1 \pmod 4$ that divides N, then either $\gamma = a + bi$ or $\overline{\gamma} = a - bi$ divides N, but not both. (We have shown that γ and $\overline{\gamma}$ are both primes, and that they are not associates. Thus, γ and $\overline{\gamma}$ are relatively prime. If γ and $\overline{\gamma}$ both divided β, then their product $p = \gamma\overline{\gamma}$ would divide β, i.e., p would divide $c + di$, which contradicts the fact that c and d are relatively prime.) Which of γ and $\overline{\gamma}$ works can only be determined by trial and error. But if γ, say, works, and e is the highest power of p dividing $\|\beta\|$, then β is divisible by γ^e.

Let us look at some examples:

(1) $\alpha = 2 + 5i$. Then $\beta = \alpha$ and $\|\beta\| = 29$ is prime, so β is prime and the prime factorization of α is $\alpha = 2 + 5i$.

(2) $\alpha = 4 + 3i$. Then $\beta = \alpha$ and $\|\beta\| = 25 = 5^2$. Recall that $5 = 2^2 + 1^2 = (2+i)(2-i)$, so β is divisible by either $(2+i)^2$ or $(2-i)^2$. The second of these works and we see α has the prime factorization $\alpha = i(2-i)^2$.

(3) $\alpha = 21 - 12i$. Then $\beta = 7 - 4i$ and $\|\beta\| = 65 = 5 \cdot 13$. Now $5 = (2+i)(2-i)$ and $13 = (2+3i)(2-3i)$. Trial and error shows that α has the prime factorization $\alpha = 3(2+i)(2-3i)$.

Note that our factorizations are only unique up to units, so there are many variants. For example, we also have the factorizations $2 = (1+i)(1-i)$, $5 = (1+2i)(1-2i)$, $13 = -i(2+3i)(3+2i)$, etc.

Remark 3.14. Let us now shift our point of view from factoring in the Gaussian integers and ask "what happens" to ordinary primes when we go from the ordinary integers to the Gaussian integers. Looking at Theorem 2.1, we can see three sorts of behavior:

(1) The ordinary prime 2 is (up to a unit) the square of a prime in $\mathcal{O}(\sqrt{-1})$: $2 = -i(1+i)^2$. The ordinary prime 2 is said to *ramify* in $\mathcal{O}(\sqrt{-1})$.

(2) An ordinary prime $p \equiv 1 \pmod 4$ is (up to a unit) a product of two distinct (i.e., nonassociated) primes in $\mathcal{O}(\sqrt{-1})$: $5 = (2+i)(2-i)$, $13 = (2+3i)(2-3i)$, etc. These ordinary primes are said to *split* in $\mathcal{O}(\sqrt{-1})$.

(3) An ordinary prime $p \equiv 3 \pmod 4$ is (up to a unit) still a prime in $\mathcal{O}(\sqrt{-1})$: $7 = 7$, $11 = 11$, etc. These ordinary primes are said to be *inert* in $\mathcal{O}(\sqrt{-1})$.

These three kinds of behavior are typical of what happens in general.

3.3 Exercises

Compare Exercises 3.1–3.5 with Remark 3.9, and Exercises 3.6–3.10 with Theorem 3.10. In Exercises 3.11, 3.12, and 3.13 "essentially" means up to sign and order and in the remaining exercises "essentially" means up to sign.

Exercise 3.1. Use the fact that $\mathcal{O}(\sqrt{-2})$ is a UFD and the fact that, for a prime $p \neq 2$, -2 is a quadratic residue $\pmod p$ if and only if $p \equiv 1$ or $3 \pmod 8$, as shown in Corollary B.37, to prove the following:

> Let p be a prime. Then p can be written as $p = x^2 + 2y^2$ for some integers x and y if and only if $p = 2$ or $p \equiv 1$ or $3 \pmod 8$, in which case x and y are essentially unique.

Exercise 3.2.

(a) Use the fact that $\mathcal{O}(\sqrt{-3})$ is a UFD and the fact that for an odd prime $p \neq 3$, -3 is a quadratic residue $\pmod p$ if and only if $p \equiv 1 \pmod 3$, by Corollary B.41, a corollary of the Law of Quadratic Reciprocity (Theorem B.40), to prove the following:

> Let p be a prime. Then $4p$ can be written as $4p = x^2 + 3y^2$ for some integers x and y if and only if p is odd and $p = 3$ or $p \equiv 1 \pmod 3$, in which case x and y are essentially unique.

(b) Let $\omega = (-1 + \sqrt{-3})/2$. Observe that if $\alpha = (a + b\sqrt{-3})/2$ for odd integers a and b, then either $\omega\alpha$ or $\overline{\omega}\alpha$ is of the form $c + d\sqrt{-3}$ with c and d integers. Use this to improve the result of part (a) to the following:

> Let p be a prime. Then p can be written as $p = x^2 + 3y^2$ for some integers x and y if and only if $p = 3$ or $p \equiv 1 \pmod 3$, in which case x and y are essentially unique.

Exercise 3.3.

(a) Use the fact that $\mathcal{O}(\sqrt{-7})$ is a UFD and the fact that for an odd prime $p \neq 7$, -7 is a quadratic residue $\pmod p$ if and only if $p \equiv 1$, 2, or $4 \pmod 7$, by Corollary B.41, a corollary of the Law of Quadratic Reciprocity (Theorem B.40), to prove the following:

> Let p be a prime. Then $4p$ can be written as $4p = x^2 + 7y^2$ for some integers x and y if and only if $p = 2$ or $p \equiv 1$, 2, or $4 \pmod 7$, in which case x and y are essentially unique.

(b) Observe that if x and y are both odd integers, then $x^2 + 7y^2 \equiv 0 \pmod 8$, while if p is an odd prime then $4p \equiv 4 \pmod 8$. Use this to improve the result of part (a) to the following:

> Let p be a prime. Then p can be written as $p = x^2 + 7y^2$ for some integers x and y if and only if p is odd and $p = 7$ or $p \equiv 1$, 2, or $4 \pmod 7$, in which case x and y are essentially unique.

Exercise 3.4.

(a) Use the fact that $\mathcal{O}(\sqrt{2})$ is a UFD and the fact that, for a prime $p \neq 2$, 2 is a quadratic residue $\pmod p$ if and only if $p \equiv 1$ or $7 \pmod 8$, as shown in Corollary B.37, to prove the following:

> Let p be a prime. Then p can be written as $p = x^2 - 2y^2$ for some integers x and y or $-p$ can be written as $-p = x^2 - 2y^2$ for some integers x and y if and only if $p = 2$ or $p \equiv 1$ or $7 \pmod 8$.

(b) Let $\varepsilon = 1 + \sqrt{2}$. Observe that $\varepsilon\bar{\varepsilon} = -1$. Use this to improve the result of part (a) to the following:

> Let p be a prime. Then p can be written as $p = x^2 - 2y^2$ for some integers x and y and $-p$ can be written as $-p = x^2 - 2y^2$ for some integers x and y if and only if $p = 2$ or $p \equiv 1$ or $7 \pmod 8$.

Exercise 3.5.

(a) Use the fact that $\mathcal{O}(\sqrt{5})$ is a UFD, the fact that 2 is not a quadratic residue $(\bmod 5)$, and the fact that for an odd prime $p \neq 5$, 5 is a quadratic residue $(\bmod p)$ if and only if $p \equiv 1$ or $4 \pmod 5$, by the Law of Quadratic Reciprocity (Theorem B.40), to prove the following:

> Let p be a prime. Then $4p$ can be written as $4p = x^2 - 5y^2$ for some integers x and y or $-4p$ can be written as $-4p = x^2 - 5y^2$ for some integers x and y if and only if p is odd and $p = 5$ or $p \equiv 1$ or $4 \pmod 5$.

(b) Let $\varepsilon_1 = (3 + \sqrt{5})/2$ and let $\varepsilon_2 = 2 + \sqrt{5}$. Observe that $\varepsilon_1\overline{\varepsilon_1} = 1$ and that $\varepsilon_2\overline{\varepsilon_2} = -1$. Use this to improve the result of part (a) to the following:

> Let p be a prime. Then p can be written as $p = x^2 - 5y^2$ for some integers x and y and $-p$ can be written as $-p = x^2 - 5y^2$ for some integers x and y if and only if p is odd and $p = 5$ or $p \equiv 1$ or $4 \pmod 5$.

Exercise 3.6. Prove the following:

> A positive integer n can be written as $n = x^2 + 2y^2$ for some integers x and y if and only if for every prime $q \equiv 5$ or $7 \pmod 8$ that divides n, the highest power of q dividing n is even.

Exercise 3.7. Prove the following:

> A positive integer n can be written as $n = x^2 + 3y^2$ for some integers x and y if and only if for every prime $q \equiv 2 \pmod 3$ that divides n, the highest power of q dividing n is even.

Exercise 3.8. Prove the following:

> A positive integer n can be written as $n = x^2 + 7y^2$ for some integers x and y if and only if for every prime q with either $q = 2$ or q odd and $q \equiv 3$, 5, or 6 $(\mathrm{mod}\, 7)$ that divides n, the highest power of q dividing n is even.

Exercise 3.9. Prove the following:

> A nonzero integer n can be written as $n = x^2 - 2y^2$ for some integers x and y if and only if for every prime $q \equiv 3$ or 5 $(\mathrm{mod}\, 8)$ that divides n, the highest power of q dividing n is even.

Exercise 3.10. Prove the following:

> A nonzero integer n can be written as $n = x^2 - 5y^2$ for some integers x and y if and only if for every prime $q \equiv 2$ or 3 $(\mathrm{mod}\, 5)$ that divides n, the highest power of q dividing n is even.

Exercise 3.11. Let n be a positive integer. Show that the number of essentially different ways of writing n as a sum of squares of two integers is equal to the number of essentially different ways of writing $2n$ as a sum of squares of two integers.

Exercise 3.12. Let p be a prime with $p \equiv 1 \pmod 4$.

(a) For a nonnegative integer k, show that p^k can be written essentially uniquely as a sum of squares of two relatively prime integers.

(b) For a nonnegative integer k, show that p^{2k} and p^{2k+1} can each be written as a sum of squares of two integers in $k+1$ essentially different ways.

Exercise 3.13.

(a) Let $p_1, \ldots, p_k$ be distinct primes, all of which are congruent to 1 modulo 4, and let n be the product $n = p_1 \cdots p_k$. Show that n can be written as a sum of squares of two integers in 2^{k-1} essentially distinct ways.

(b) Let p_1 and p_2 be distinct primes, both of which are congruent to 1 modulo 4, let k be a nonnegative integer, and let $n = p_1 p_2^k$. Show that n can be written as a sum of squares of two integers in $k+1$ essentially distinct ways.

Exercise 3.14. Let n be a positive integer. Show that the number of essentially different ways of writing n in the form $x^2 + 2y^2$ for integers x and y is equal to the number of essentially different ways of writing $2n$ in the form $x^2 + 2y^2$ for integers x and y.

Exercise 3.15. Let p be a prime with $p \equiv 1$ or $3 \pmod 8$.

(a) For a nonnegative integer k, show that p^k can be written essentially uniquely in the form $x^2 + 2y^2$ for relatively prime integers x and y.

(b) For a nonnegative integer k, show that p^{2k} and p^{2k+1} can each be written in the form $x^2 + 2y^2$ for integers x and y in $k + 1$ essentially different ways.

Exercise 3.16.

(a) Let $p_1, \ldots, p_k$ be distinct primes, all of which are congruent to 1 or 3 modulo 4, and let n be the product $n = p_1 \cdots p_k$. Show that n can be written in the form $x^2 + 2y^2$ for integers x and y in 2^{k-1} essentially distinct ways.

(b) Let p_1 and p_2 be distinct primes, both of which are congruent to 1 or 3 modulo 8, let k be a nonnegative integer, and let $n = p_1 p_2^k$. Show that n can be written in the form $x^2 + 2y^2$ for integers x and y in $k+1$ essentially distinct ways.

Clearly, Exercises 3.14, 3.15, and 3.16 are the analogs for $D = -2$ of Exercises 3.11, 3.12, and 3.13 for $D = -1$, and clearly, there are analogs of these exercises for $D = -3$ and $D = -7$. There are no such analogs for $D = 5$, as in this case there are infinitely many representations (a consequence of our study of Pell's equation in Chapter 4).

Chapter 4

Pell's Equation

In Remark 1.17 we saw that we had to consider *Pell's equation*

$$a^2 - b^2 D = 1,$$

when D is a square-free positive integer, $D \neq 1$.

In fact, we will consider this equation for any positive integer D that is not a square (since it requires no additional work) and show that it always has infinitely many solutions in integers a and b.

This was known to Fermat, who also developed a method for finding solutions. For example, Fermat knew that the smallest solution to $a^2 - 61b^2 = 1$ in positive integers is

$$a = 1766319049, \qquad b = 226153980.$$

Pell's equation has a long history. A method for solving it, called the *cakravala* method, was developed by Indian mathematicians in the seventh to twelfth centuries. Fermat (who was certainly unaware of this work) was the first western mathematician both to find a method and to prove that it always works. The full theory of Pell's equation is due to Lagrange, using the method of continued fractions. There is also a twentieth-century approach using Diophantine approximation. (Pell, however, had nothing to do with it. His name is attached to this equation because of Euler's mistaken belief that he did.)

The approach using Diophantine approximation gives a very quick proof that Pell's equation always has infinitely many solutions, but it has the serious disadvantage that it does not provide any method, other than trial and error, for finding them.

While the method of continued fractions gives the full theory, including an effective method for finding all solutions, it requires developing the theory of continued fractions first, and that would be a considerable digression for us.

Thus, we will present a variant of the cakravala method here, and furthermore we will prove that it always works. This method is undoubtedly very close to what Fermat had in mind, as you can tell from the similarity between the last chapter and this one. It is also quite beautiful and effective.

4.1 Representations and Their Composition

We let D be an arbitrary positive integer that is not a perfect square.

We shall prove all the results in this section by direct computation, but there is a deeper reason why they are true. (When a computation produces a particularly nice result, one should always ask why.) This reason can be found in the exercises to this chapter.

If a and b are integers with $m = a^2 - b^2 D$, we will say that $m = a^2 - b^2 D$ is a *representation* of m, or that (a, b) *represents* m.

If (a, b) represents m, then so does $\overline{(a, b)} = (a, -b)$. We shall call $\overline{(a, b)}$ the *conjugate* of (a, b).

If (a, b) represents m and c is any integer, then $c(a, b) = (ca, cb)$ represents $c^2 m$. On the other hand, if (a, b) represents m and c is any integer dividing both a and b, then $(a/c, b/c)$ represents m/c^2.

Finally, if (a, b) represents m and $d = \gcd(a, b)$, then $(a/d, b/d)$ represents m/d^2. In this case we write $(a, b)_{\text{red}} = (a/d, b/d)$ and call $(a, b)_{\text{red}}$ the *reduction* of (a, b). Of course, $(a, b)_{\text{red}} = (a, b)$ if and only if a and b are relatively prime. In this case, we say (a, b) is *reduced*.

We call the reader's attention to the similarity between Lemma 4.1 and Lemma 3.2, and to the similarity between Definition 4.2 and Definition 3.3.

Lemma 4.1. *Let (a, b) represent m and (c, d) represent n. Set*

$$e = ac + bcD \quad \textit{and} \quad f = ab + bc.$$

Then (e, f) represents mn.

Proof: We simply compute

$$\begin{aligned} e^2 - f^2 D &= (ac + bcD)^2 - (ad + bc)^2 D \\ &= (a^2c^2 + 2abcdD + b^2d^2D^2) - (a^2d^2 + 2adbc + b^2c^2)D \\ &= a^2c^2 + b^2d^2D^2 - a^2d^2D - b^2c^2D \\ &= (a^2 - b^2D)(c^2 - d^2D) = mn. \end{aligned}$$

□

Definition 4.2. The representation $mn = e^2 - f^2D$ of mn in Lemma 4.1 is obtained from the representations $m = a^2 - b^2D$ and $n = c^2 - d^2D$ by *composition*, denoted

$$(e, f) = (a, b) * (c, d).$$

We now give several properties of composition, which we verify by direct computation.

Lemma 4.3.

(1) *Composition is commutative:*

$$(a, b) * (c, d) = (c, d) * (a, b).$$

(2) *Composition is associative:*

$$((a, b) * (c, d)) * (e, f) = (a, b) * ((c, d) * (e, f)).$$

(3) *Composition commutes with conjugation:*

$$\overline{(a, b) * (c, d)} = \overline{(a, b)} * \overline{(c, d)}.$$

Proof:

(1) By definition,

$$(a, b) * (c, d) = (ac + bdD, ad + bc)$$

while

$$(c, d) * (a, b) = (ca + dbD, cb + ad)$$

and these are equal.

(2) By definition,

$$\begin{aligned}
&((a, b) * (c, d)) * (e, f) \\
&\quad = (ac + bdD, ad + bc) * (e, f) \\
&\quad = (ace + bdeD + (adf + bcf)D, acf + bdfD + ade + bce)
\end{aligned}$$

while

$$\begin{aligned}
&(a, b) * ((c, d) * (e, f)) \\
&\quad = (a, b) * (ce + dfD, cf + de) \\
&\quad = (ace + adfD + (bcf + bde)D, acf + ade + bce + bdfD)
\end{aligned}$$

and these are equal.

(3) By definition,

$$\overline{(a,b)*(c,d)} = \overline{(ac+bdD, ad+bc)} = (ac+bdD, -ad-bc)$$

while

$$\begin{aligned}\overline{(a,b)}*\overline{(c,d)} &= (a,-b)*(c,-d)\\ &= (ac+(-b)(-d)D, a(-d)+(-b)c)\end{aligned}$$

and these are equal. □

We now combine composition and reduction into a single operation.

Definition 4.4. The *reduced composition* $(a,b)*_r(c,d)$ of (a,b) and (c,d) is

$$(a,b)*_r(c,d) = ((a,b)*(c,d))_{\text{red}}.$$

(In other words, to obtain the reduced composition of (a,b) and (c,d), we first compose (a,b) and (c,d) and then reduce the result.)

We now give several properties of reduced composition.

Lemma 4.5.

(1) For any (a,b),

$$(a,b)*_r\overline{(a,b)} = \pm(1,0).$$

(2) For any (a,b) and (c,d),

$$(a,b)*_r(c,d) = (c,d)*_r(a,b).$$

(3) If (a,b) and (c,d) are reduced and $(a,b)*_r(c,d) = \pm(1,0)$, then $(c,d) = \pm\overline{(a,b)}$.

(4) For any (a,b), (c,d), and $t>0$

$$(a,b)*_r t(c,d) = (a,b)*_r(c,d).$$

Proof:

(1) By definition,

$$\begin{aligned}(a,b)*_r\overline{(a,b)} &= ((a,b)*\overline{(a,b)})_{\text{red}} = (a,b)*(a,-b)_{\text{red}}\\ &= (a^2-b^2D, 0)_{\text{red}} = \pm(1,0).\end{aligned}$$

(2) $(a,b) *_r (c,d) = ((a,b) * (c,d))_{\text{red}} = (c,d) * (a,b)_{\text{red}} = (c,d) *_r (a,b)$.

(3) If $(a,b) *_r (c,d) = \pm(1,0)$, then $(a,b) * (c,d) = (e,0)$ for some e, so in particular $ad + bc = 0$.

Then $ad = -bc$, so in particular a divides $-bc$. Now a and b are relatively prime, as (a,b) is assumed to be reduced. So, by Euclid's Lemma, a divides c, i.e., $c = ka$ for some a. Then $ad = -bc = -b(ka)$, so $d = -bk$. Thus $(c,d) = (ka, -kb) = k(a,-b)$. But (c,d) is assumed to be reduced, so $k = \pm 1$.

(4) Let $(e,f) = (a,b)*(c,d)$. Then $(te,tf) = (a,b)*t(c,d)$. Now $(e,f)_{\text{red}} = (a,b) *_r (c,d) = (e/s, f/s)$ where $s = \gcd(e,f)$, while $(te,tf)_{\text{red}} = (a,b) *_r t(c,d) = (te/s', tf/s')$ where $s' = \gcd(te,tf)$. But then we have that $s' = \gcd(te,tf) = t\gcd(e,f) = ts$, so

$$\begin{aligned} (a,b) *_r t(c,d) &= (te/s', tf/s') = (te/(ts), tf/(ts)) \\ &= (e/s, f/s) = (a,b) *_r (c,d). \end{aligned}$$ □

Corollary 4.6.

(1) For any (a,b), (c,d), and (e,f),

$$((a,b) * (c,d)) *_r (e,f) = ((a,b) *_r (c,d)) *_r (e,f),$$

and

$$((a,b) *_r (c,d)) *_r (e,f) = (a,b) *_r ((c,d) *_r (e,f)).$$

(2) If (a,b) is reduced and $(a,b) *_r (x,y) = (a,b)$, then $(x,y)_{\text{red}} = \pm(1,0)$.

Proof:

(1) The first claim follows immediately from Lemma 4.5(3). Let

$$(a,b) * (c,d) = (g,h) \text{ and } (a,b) *_r (c,d) = (i,j).$$

Then by definition $i = g/t$ and $j = h/t$, where $t = \gcd(g,h)$. Then $(g,h) = t(i,j)$. Then

$$\begin{aligned} ((a,b) * (c,d)) *_r (e,f) &= (g,h) *_r (e,f) \\ &= t(i,j) *_r (e,f) \\ &= (i,j) *_r (e,f) \\ &= ((a,b) *_r (c,d)) *_r (e,f). \end{aligned}$$

Now using this and the associativity of composition, we have

$$\begin{aligned}((a,b) *_r (c,d)) *_r (e,f) &= ((a,b) * (c,d)) *_r (e,f)\\ &= (((a,b) * (c,d)) * (e,f))_{\text{red}}\\ &= ((a,b) * ((c,d) * (e,f)))_{\text{red}}\\ &= (a,b) *_r ((c,d) * (e,f))\\ &= (a,b) *_r ((c,d) *_r (e,f)),\end{aligned}$$

proving the second claim.

(2) Suppose $(a,b) *_r (x,y) = (a,b)$. Then

$$\begin{aligned}\overline{(a,b)} *_r ((a,b) *_r (x,y)) &= \overline{(a,b)} *_r (a,b),\\ \overline{(a,b)} *_r ((a,b) *_r (x,y)) &= \pm(1,0),\\ (\overline{(a,b)} *_r (a,b)) *_r (x,y) &= \pm(1,0),\\ \pm(1,0) *_r (x,y) &= \pm(1,0),\\ (\pm(1,0) * (x,y))_{\text{red}} &= \pm(1,0),\\ (x,y)_{\text{red}} &= \pm(1,0).\end{aligned}$$

□

Let us call attention to what Corollary 4.6 says. The first claim is that if we wish to compute the reduced composition of (a,b), (c,d), and (e,f), we can either reduce at each stage or simply compute the composition at each stage and reduce at the end—it makes no difference. The same holds for any number of representations:

$$\begin{aligned}&(\ldots(((a_1,b_1) *_r (a_2,b_2)) *_r (a_3,b_3)) *_r \ldots (a_{n-1},b_{n-1})) *_r (a_n,b_n)\\ &\quad= (\ldots(((a_1,b_1) * (a_2,b_2)) * (a_3,b_3)) * \ldots (a_{n-1},b_{n-1})) *_r (a_n,b_n)\\ &\quad= ((\ldots(((a_1,b_1) * (a_2,b_2)) * (a_3,b_3)) * \ldots (a_{n-1},b_{n-1})) * (a_n,b_n))_{\text{red}}.\end{aligned}$$

The second claim is that reduced composition is associative, and the third claim is a cancellation result for reduced composition, up to sign.

In developing this theory of composition and reduction, we should not lose sight of our goal—to solve Pell's equation $a^2 - b^2D = 1$. In our language, this is finding a representation $1 = a^2 - b^2D$. Of course, 1 has the representations $1 = 1^2 - 0^2D = (-1)^2 - 0^2D$. We call these two representations *trivial representations* and all other representations of 1 *nontrivial representations*. So in fact, we are looking for nontrivial representations of 1.

Let us observe that if (a, b) represents 1, or if (a, b) represents -1, then (a, b) is automatically reduced. For if d is any common divisor of a and b, then d divides $a^2 - b^2D = \pm 1$, so $d = \pm 1$, i.e., a and b are relatively prime.

Our next lemma will give a key technique for finding nontrivial representations of 1, i.e., nontrivial solutions of Pell's equation.

Lemma 4.7. *Let (a, b) and (c, d) both be reduced and suppose that (a, b) and (c, d) represent the same integer m. Suppose also that $a \equiv c \pmod{m}$ and $b \equiv d \pmod{m}$. Let*

$$(e, f) = \overline{(a, b)} *_r (c, d).$$

Then (e, f) represents 1. Furthermore, if $(c, d) \neq \pm(a, b)$, then $(e, f) \neq \pm(1, 0)$.

Proof: Let $(E, F) = \overline{(a, b)} * (c, d) = (a, -b) * (c, d) = (ac - bdD, ad - bc)$. Then (E, F) represents m^2.

We claim that $\gcd(E, F) = m$.

Let us begin by seeing that each of E and F are divisible by m.

First F: Since $a \equiv c \pmod{m}$ and $d \equiv b \pmod{m}$, $ad \equiv cb \pmod{m}$, $ad - bc \equiv 0 \pmod{m}$, i.e., m divides $F = ad - bc$.

Next E: Since (a, b) represents m, $m = a^2 - b^2D$, so $a^2 - b^2D \equiv 0 \pmod{m}$, $a(a) - b(b)D \equiv 0 \pmod{m}$. But $c \equiv a \pmod{m}$ and $d \equiv b \pmod{m}$, so $ac - bdD \equiv 0 \pmod{m}$, i.e., m divides $E = ac - bdD$.

Thus we see that $(E/m, F/m)$ represents 1. But, as we observed, this automatically shows that $(E/m, F/m)$ is reduced, i.e., $\gcd(E/m, F/m) = 1$, and so $m = \gcd(E, F)$. But then

$$(e, f) = \overline{(a, b)} *_r (c, d) = (\overline{(a, b)} * (c, d))_{\mathrm{red}} = (E/m, F/m)$$

and (e, f) represents 1. Furthermore, the contrapositive of Lemma 4.5(3) gives that if $(c, d) \neq \pm(a, b)$, then $\overline{(a, b)} *_r (c, d) \neq \pm(1, 0)$, i.e., in our case $(e, f) \neq \pm(1, 0)$. □

In the next section, we will show that we can always find (a, b) and (c, d) satisfying the hypotheses of Lemma 4.7, and that will give us a single nontrivial solution of Pell's equation. But in fact, once we have a single nontrivial solution, we have infinitely many, as we see from the next lemma.

Lemma 4.8. *Let (e, f) represent 1 nontrivially (i.e., $(e, f) \neq \pm(1, 0)$). Let*

$$(e, f)^2 = (e, f) * (e, f),\ (e, f)^3 = (e, f)^2 * (e, f),\ (e, f)^4 = (e, f)^3 * (e, f), \ldots.$$

Then

$$\begin{array}{cccc} (e,f), & (e,f)^2, & (e,f)^3, & (e,f)^4,\dots \\ \overline{(e,f)}, & \overline{(e,f)}^2, & \overline{(e,f)}^3, & \overline{(e,f)}^4,\dots \end{array}$$

all represent 1 *nontrivially and are all distinct.*

Proof: If (e,f) represents 1, then $(e,f)*(e,f)$ represents $1\cdot 1$, i.e., $(e,f)^2$ represents 1, and then $(e,f)^2*(e,f)$ represents $1\cdot 1$, i.e., $(e,f)^3$ represents 1, etc., so $(e,f)^n$ represents 1 for every $n \geq 1$. But then $\overline{(e,f)}^n$ also represents 1 for every $n \geq 1$.

We must show that they are all nontrivial and all distinct.

Replacing (e,f) by $(e,-f)$, $(-e,f)$, or $(-e,-f)$, if necessary, we may assume that $e > 0$ and $f > 0$. Write $(e_k, f_k) = (e,f)^k$. We claim that $f_1, f_2, f_3, \dots$ are a strictly increasing sequence of positive integers (and hence that $e_1, e_2, e_3, \dots$ are a strictly increasing sequence of positive integers), which shows that $(e,f), (e,f)^2, (e,f)^3, \dots$ are all nontrivial and distinct. But also $\overline{(e,f)}^k = (e_k, -f_k)$, so all of these are nontrivial and distinct.

We show this claim by direct computation, using induction. By assumption, in case $k=1, (e_1, f_1) = (e,f)$ and e and f are positive integers. Now suppose e_k and f_k are positive integers. Then

$$(e_{k+1}, f_{k+1}) = (e_k, f_k) * (e,f) = (e_k e + f_k f D, e_k f + f_k e).$$

Since $e_k \geq 1$, $f \geq 1$, and $e \geq 1$, we see $f_{k+1} = e_k f + f_k e \geq 1 + f_k > f_k$ (and since $e \geq 1$, $f \geq 1$, and $f_k \geq 1$, we see $e_{k+1} = e_k e + f_k f D \geq e_k + D > e_k$), as claimed. □

4.2 Solving Pell's Equation

Rather than simply pull a rabbit out of a hat, we will proceed heuristically, seeing how one might find a method for solving Pell's equation. Then we will prove that this method actually works. At the outset, our approach will seem to have nothing to do with the last section, but we will soon see that our method involves a sequence of (carefully chosen) reduced compositions.

We wish to solve

$$a^2 - b^2 D = 1$$

and we see that this is equivalent to (assuming a and b are positive)

$$a^2 = b^2 D + 1 = b^2(D + 1/b^2),$$

so

$$a = b\sqrt{D + 1/b^2}$$
$$a/b = \sqrt{D + 1/b^2}.$$

Now for an arbitrary choice of b, $D + 1/b^2$ will not have a rational square root, so a/b will not be a rational number. We want to find a value of b for which it does, but there is no a priori guarantee that such a value of b exists. However, we do observe that for a solution (a, b), the ratio a/b will be close to $\sqrt{D}$, which we write as $a/b \sim \sqrt{D}$, and furthermore, as b gets larger the ratio a/b gets closer to $\sqrt{D}$.

So let's start with a guess for a and b that has a/b reasonably close to $\sqrt{D}$. Indeed, let us start with a pair (a, b) and set $e = a^2 - b^2D$, so (a, b) represents e. We would like to get $e = 1$, but let's settle for the moment for keeping $|e|$ small. (What "small" means turns out to be a delicate question, but we will save that for later.)

We would like to get a better guess (A, B) and we will try to do so (with hindsight gained from working many examples) by setting

$$B = a + bx$$

with x yet to be determined, and by choosing a suitable value of A. We want $A/B \sim \sqrt{D}$, and we have $a/b \sim \sqrt{D}$, so we look for $A/B \sim a/b$, i.e., $A \sim (a/b)B = (a/b)(a + bx) = a^2/b + ax$. We could try setting $A = a^2/b + ax$, but there are two problems with this: first, a^2/b is not an integer, and second, choosing $A = (a/b)B$ would not be making the situation any better, just keeping it the same (as then $A/B = a/b$).

We solve the first problem first. Note that from $a^2 - b^2D = 1$ we obtain $a^2/b = bD + 1/b$. So in our "try" for A above, we replace a^2/b with bD, which is pretty close to it, to obtain

$$A = ax + bD.$$

Then we set $E = A^2 - B^2D$ and hope that we have made things better. Unfortunately,

$$\begin{aligned}
E = A^2 - B^2D &= (ax + bD)^2 - (a + bx)^2D \\
&= (a^2x^2 + 2abxD + b^2D^2) - (a^2D + 2abxD + b^2x^2D) \\
&= a^2x^2 + b^2D^2 - a^2D - b^2x^2D \\
&= (a^2 - b^2D)(x^2 - D) \\
&= e(x^2 - D).
\end{aligned}$$

While we certainly should choose x near $\sqrt{D}$, so $|x^2 - D|$ is small, this is not an improvement, as $|E|$ is a multiple of $|e|$.

However, we are not done yet! We have flexibility in the choice of x, which we have not yet exploited.

Suppose A and B are divisible by $|e|$. Then $A/|e|$ and $B/|e|$ are integers, and $(A/|e|)^2 - (B/|e|)^2 D = E/e^2$ is an integer with $|E/e^2| < |E|$ (unless $e = 1$, in which case we are done, or $e = -1$, in which case, as we shall see, we are almost done). Guided by this observation, we shall choose x so that A and B are divisible by e, and set $a' = A/|e|$, $b' = B/|e|$, to find that $e' = (a')^2 - (b')^2 D = E/e^2 = |(x^2 - D)/e|$. Now e' is not necessarily equal to 1, but it cannot be too large—in particular, $|e'| \leq |x^2 - D|$. So we should try to choose x with $|x^2 - D|$ small, i.e., with x near $\sqrt{D}$.

Let us summarize this discussion in a lemma.

Lemma 4.9. *Let a and b be relatively prime, nonnegative integers and set $e = a^2 - b^2 D$. Then there is exactly one integer x with $\sqrt{D} - |e|/2 < x < \sqrt{D} + |e|/2$ such that $a + bx \equiv 0 \pmod{e}$. With this value of x, set*

$$a' = \left| \frac{ax + bD}{e} \right|,$$

$$b' = \left| \frac{a + bx}{e} \right|.$$

Then a' and b' are relatively prime, nonnegative integers.

Proof: There are two claims to prove:

(1) We can choose exactly one value of x as above.

(2) With this value of x, a' and b' are relatively prime, nonnegative integers.

We prove these in turn.

First, let us note that for any positive integer i, and any real number r, the inequality $r \leq x \leq r + i$ has exactly $i + 1$ integer solutions if r is an integer and exactly i integer solutions otherwise. If r is an integer, both $x = r$ and $x = r + i$ are integer solutions while if i is not an integer, neither is an integer solution. Thus, we see that the inequality $r < x < r + i$ has exactly $i - 1$ integer solutions if r is an integer and exactly i solutions otherwise.

In our situation, $\sqrt{D}$ is not a rational number, so $r = \sqrt{D} - |e|/2$ is certainly not an integer, and, setting $i = |e|$, we have $r + i = \sqrt{D} + |e|/2$, so the inequality $\sqrt{D} - |e|/2 < x < \sqrt{D} + |e|/2$ has exactly $i = |e|$ integer solutions. They must be consecutive integers $x_0, x_0 + 1, \ldots, x_0 + |e| - 1$ for some x_0.

With this in mind, we get to work. We claim that e and b are relatively prime. For suppose there was a prime p that divided both e and b. Then p would divide $e + b^2D = a^2$, and so p would divide a, contradicting our hypothesis that a and b are relatively prime.

Now, since e and b are relatively prime, the congruence

$$a + by \equiv 0 \pmod{e}$$

has exactly one solution $y_0 \pmod{e}$. But since $x_0, x_0 + 1, \ldots, x_0 + |e| - 1$ are $|e|$ consecutive integers, exactly one of them is congruent to $y_0 \pmod{e}$. Choose that one and call it x.

Now for the second claim.

Let $A = ax + bD$ and $B = a + bx$. By our choice of x, $B \equiv 0 \pmod{e}$ so $b' = |B/e|$ is a nonnegative integer. But also

$$aB - bA = a(a + bx) - b(ax + bD) = a^2 - b^2D = e \equiv 0 \pmod{e}.$$

Thus, $bA \equiv aB \pmod{e}$. But $B \equiv 0 \pmod{e}$, so $bA \equiv 0 \pmod{e}$. However, b and e are relatively prime, so $A \equiv 0 \pmod{e}$. Hence $a' = |A/e|$ is also a nonnegative integer.

It remains for us to show that a' and b' are relatively prime. Let $d' = \gcd(a', b')$. We want to prove $d' = 1$.

Now d' divides a' so $d'e$ divides A, and d' divides b' so $d'e$ divides B. Since $d'e$ divides both A and B, $d'e$ divides

$$xB - A = x(a + bx) - (ax + bD) = b(x^2 - D) = be,$$

so d' divides b. Similarly, $d'e$ divides

$$xA - BD = x(ax + bD) - (a + bx)D = a(x^2 - D) = ae,$$

so d' divides a. But a and b are assumed to be relatively prime, so $d' = 1$, and then a' and b' are also relatively prime. □

Now suppose a and b are in Lemma 4.9 and set $e = a^2 - b^2D$. Let a' and b' be as in Lemma 4.9 and set $e' = (a')^2 - (b')^2D$. We said in the above discussion that $|e|$ should be "small." To be precise, we want a bound on $|e|$ that ensures that if $|e|$ is "small," then $|e'|$ is also "small." For if we cannot ensure that, we lose control of the situation. The appropriate criterion is a delicate one, one that is not apparent a priori, but one that makes the proof work. We give it here.

Lemma 4.10. *In the situation of Lemma 4.9, let*

$$e = a^2 - b^2 D \quad \text{and} \quad e' = (a')^2 - (b')^2 D.$$

If $|e| < 2\sqrt{D}$, *then* $|e'| < 2\sqrt{D}$.

Proof: Following the notation of the proof of Lemma 4.9, let

$$A = ax + bD \quad \text{and} \quad B = a + bx,$$

so $a' = |A/e|$ and $b' = |B/e|$. Let $E = A^2 - B^2 D$. We compute, as before, that

$$E = (ax + bD)^2 - (a + bx)^2 D = e(x^2 - D),$$

and so

$$\begin{aligned} e' &= (A/e)^2 - (B/e)^2 D \\ &= (A^2 - B^2 D)/e \\ &= (1/|e|)(x^2 - D). \end{aligned}$$

We want to bound $|e'|$, so we must compute $x^2 - D$. We have

$$\sqrt{D} - |e|/2 < x < \sqrt{D} + |e|/2.$$

Since $|e| < 2\sqrt{D}$, we see that the lower bound $\sqrt{D} - |e|/2$ is positive (and so x is positive). Then

$$\begin{aligned} (\sqrt{D} - |e|/2)^2 < x^2 \qquad &< (\sqrt{D} + |e|/2)^2, \\ D - |e|\sqrt{D} + |e|^2/4 < x^2 \qquad &< D + |e|\sqrt{D} + |e|^2/4, \\ -|e|\sqrt{D} + |e|^2/4 < x^2 - D &< |e|\sqrt{D} + |e|^2/4. \end{aligned}$$

Now $x^2 - D$ may be positive or negative, so we must be careful about signs. There are two cases to consider:

(1) $x^2 - D$ is positive, so $0 < x^2 - D < |e|\sqrt{D} + |e|^2/4$.

(2) $x^2 - D$ is negative, so $-|e|\sqrt{D} + |e|^2/4 < x^2 - D < 0$.

In case (1),

$$|e'| = \left| \frac{x^2 - D}{e} \right| = \frac{x^2 - D}{|e|} < \sqrt{D} + |e|/4 < \sqrt{D} + 2\sqrt{D}/4 < 2\sqrt{D},$$

where we have used our hypothesis that $|e| < 2\sqrt{D}$, and in case (2),

$$|e'| = \left|\frac{x^2 - D}{|e|}\right| = \frac{-(x^2 - D)}{|e|} < \sqrt{D} - |e|/4 < 2\sqrt{D}.$$

Thus, in any case $|e'| < 2\sqrt{D}$, as claimed. □

Let us assemble these two results and relate them to composition.

Corollary 4.11. *Let a and b be nonnegative integers and suppose that (a, b) is reduced and represents e with $|e| < 2\sqrt{D}$. Then there is a unique integer x with $\sqrt{D} - |e|/2 < x < \sqrt{D} + |e|/2$ such that $a + bx \equiv 0 \pmod{e}$. (Note that x is positive.) For this value of x, let*

$$(a', b') = (a, b) *_r (x, 1).$$

Then (a', b') represents e' with $|e'| < 2\sqrt{D}$.

Proof: We follow the notation of the proof of Lemma 4.9 and Lemma 4.10. We have by Lemma 4.9 that

$$(A, B) = (ax + bD, a + bx)$$

and we compute that

$$(a, b) * (x, 1) = (ax + bD, a + bx),$$

so we immediately see that

$$(A, B) = (a, b) * (x, 1).$$

Then, by definition,

$$(a, b) *_r (x, 1) = ((a, b) * (x, 1))_{\text{red}} = (A, B)_{\text{red}} = (A/d, B/d)$$

where $d = \gcd(A, B)$. But we have seen that $|e|$ divides both A and B, and by the properties of the gcd,

$$d = \gcd(A, B) = |e| \gcd(A/|e|, B/|e|) = |e| \gcd(a', b') = |e|,$$

as we have seen that a' and b' are relatively prime. Hence

$$(a, b) *_r (x, 1) = (A/|e|, B/|e|) = (a', b')$$

as claimed.

Finally, if $|e| = |a^2 - b^2 D| < 2\sqrt{D}$, then also $|e'| = |(a')^2 - (b')^2 D| < 2\sqrt{D}$ by Lemma 4.10. □

We think of using Lemma 4.9 recursively. That is, we start with a pair (a, b), apply Lemma 4.9 to (a, b) to get a pair (a', b'), apply Lemma 4.9 again to (a', b') to get a pair (a'', b''), ... This works for Lemma 4.10, or for Corollary 4.11, in a similar way, but here we must start out with a pair (a, b) with $e = a^2 - b^2 D$ having $|e| < 2\sqrt{D}$. It may seem difficult to find such a pair, but in fact it is trivial—we choose $(a, b) = (1, 0)$!

Let us establish some notation.

Definition 4.12. Let (a, b) and (a', b') be as in Lemma 4.9. We write

$$(a', b') = \mathcal{P}(a, b).$$

If furthermore $(a'', b'') = \mathcal{P}(a', b')$ we write $(a'', b'') = \mathcal{P}^2(a, b)$, etc.

We let $(a_0, b_0) = (1, 0)$ and set

$$(a_i, b_i) = \mathcal{P}^i(1, 0)$$

for every $i > 0$. (Here "$\mathcal{P}$" stands for "Pell.")

We also set

$$e_i = a_i^2 - b_i^2 D.$$

Corollary 4.11 gives a value of x for which $(a_{i+1}, b_{i+1}) = (a_i, b_i) *_r (x, 1)$, and we denote that value of x by x_{i+1}, so

$$(a_{i+1}, b_{i+1}) = \mathcal{P}(a_i, b_i) = (a_i, b_i) *_r (x_{i+1}, 1).$$

Remark 4.13. Note, in particular, that x_1 is the unique integer with $\sqrt{D} - 1/2 < x_1 < \sqrt{D} + 1/2$, i.e., x_1 is the integer closest to $\sqrt{D}$. Then $(a_0, b_0) = (1, 0)$ and $(a_1, b_1) = (x_1, 1)$. Also note that each x_i is positive.

As a practical matter, to employ Lemma 4.9 we must search among the $|e|$ values of x with $\sqrt{D} - |e|/2 < x < \sqrt{D} + |e|/2$ for the one value of x that makes $a + bx \equiv 0 \pmod{e}$. But if we employ this method recursively, we may use the following relation.

Lemma 4.14. *In the above notation,*

$$x_{i+1} \equiv -x_i \pmod{e_i}.$$

Proof: We are looking for a value of x_{i+1} that in particular satisfies the congruence $a_i + b_i x_{i+1} \equiv 0 \pmod{e_i}$, a congruence with a unique solution mod e_i. But we know that

$$a_i = (a_{i-1}x_i + b_{i-1}D)/|e_{i-1}| \quad \text{and} \quad b_i = (a_{i-1} + b_{i-1}x_i)/|e_{i-1}|$$

and we then compute that

$$a_i + b_i(-x_i) = -b_{i-1}(x_i^2 - D)/|e_{i-1}| = \pm b_{i-1}e_i \equiv 0 \pmod{e_i},$$

as in the proof of Lemma 4.9 we computed that $e_i = (x_i^2 - D)/e_{i-1}$. Hence $x_{i+1} \equiv -x_i \pmod{e_i}$ is a solution to this congruence. But since this congruence has a unique solution, this must be the solution. □

It may seem that we have simply found a sequence of pairs (a_1, b_1), (a_2, b_2), (a_3, b_3), ... with $|e_1| < 2\sqrt{D}$, $|e_2| < 2\sqrt{D}$, $|e_3| < 2\sqrt{D}$, ..., while what we are looking for is a nontrivial solution to Pell's equation, i.e., a pair $(a, b) \neq \pm(1, 0)$ with $e = a^2 - b^2 D = 1$, so this is not good enough.

But in fact it is! In fact, we have found a solution (actually, infinitely many solutions) to Pell's equation!

Theorem 4.15. *Let D be any positive integer that is not a perfect square. Then Pell's equation $a^2 - b^2 D = 1$ has infinitely many solutions.*

Proof: To begin with, we claim that the pairs of integers

$$(a_0, b_0), \quad (a_1, b_1), \quad (a_2, b_2), \quad \ldots$$

are all distinct. To see this, suppose we choose any two nonnegative integers $s < t$. Recall that, by Corollary 4.11, and by Corollary 4.6(1),

$$\begin{aligned}
(a_{s+1}, b_{s+1}) &= (a_s, b_s) *_r (x_{s+1}, 1)\\
(a_{s+2}, b_{s+1}) &= (a_{s+1}, b_{s+1}) *_r (x_{s+2}, 1)\\
&= ((a_s, b_s) *_r (x_{s+1}, 1)) *_r (x_{s+2}, 1)\\
&= (a_s, b_s) *_r ((x_{s+1}, 1) *_r (x_{s+2}, 1))
\end{aligned}$$

and continuing in this way we see that

$$(a_t, b_t) = (a_s, b_s) *_r (X, Y)$$

where

$$(X, Y) = (x_{s+1}, 1) *_r (x_{s+2}, 1) * \cdots *_r (x_t, 1).$$

Note that each $x_i > 0$, so $Y \neq 0$ (by the formula for composition) and hence $(X, Y) \neq \pm(1, 0)$. Then, by the contrapositive of Corollary 4.6(2), $(a_t, b_t) \neq (a_s, b_s)$.

Now we shall show that Pell's equation has a single nontrivial solution.

For each i, let $\overline{a}_i \equiv a_i \pmod{e_i}$ with $0 \le \overline{a}_i < |e_i|$ and let $\overline{b_i} \equiv b_i \pmod{e_i}$ with $0 \le \overline{b}_i < |e_i|$. Consider the triples of integers

$$(e_0, \overline{a}_0, \overline{b}_0), \quad (e_1, \overline{a}_1, \overline{b}_1), \quad (e_2, \overline{a}_2, \overline{b}_2), \quad \ldots.$$

Let $k = [\sqrt{D}]$. Then there are at most $2k$ possibilities for e_i (as e_i is a nonzero integer between $-k$ and k). Also, there are at most k possibilities for $\overline{a}_i$ (as $0 \le \overline{a}_i < |e_i| \le k-1$) and at most k possibilities for $\overline{b}_i$ (as $0 \le \overline{b}_i < |e_i| \le k-1$). So there are at most $2k^3$ possibilities for $(e_i, \overline{a}_i, \overline{b}_i)$. But this sequence has infinitely many terms. So, by the Pigeonhole Principle, they cannot all be distinct, i.e., there must be a pair of nonnegative integers s and t with $s < t$ and $(e_s, \overline{a}_s, \overline{b}_s) = (e_t, \overline{a}_t, \overline{b}_t)$. Then $e_s = e_t$; call that common value e.

Now on the one hand $a_s \equiv a_t \pmod{e}$ and $b_s \equiv b_t \pmod{e}$, but on the other hand, as we saw in the first part of the proof, $(a_s, b_s) \ne (a_t, b_t)$. Thus we may apply Lemma 4.7 to conclude that, setting

$$(\alpha, \beta) = \overline{(a_s, b_s)} *_r (a_t, b_t),$$

then (α, β) represents 1, i.e.,

$$\alpha^2 - \beta^2 D = 1$$

and $(\alpha, \beta) \ne \pm(1, 0)$, so (α, β) is a nontrivial solution to Pell's equation.

Now that we have a single nontrivial solution to Pell's equation, we can get infinitely many.

The easiest way to do so is to simply quote Lemma 4.8, which tells us that the "powers" of (α, β), i.e.,

$$(\alpha, \beta), \quad (\alpha, \beta)^2 = (\alpha, \beta) * (\alpha, \beta), \quad (\alpha, \beta)^3 = (\alpha, \beta)^2 * (\alpha, \beta), \quad \ldots$$

are all distinct solutions of Pell's equation.

Alternatively, instead of appealing to Lemma 4.8, we may simply modify our proof that there is one nontrivial solution. Namely, since there are only finitely many possibilities for $(\overline{a}_i, \overline{b}_i, e_i)$ and there are infinitely many terms in this sequence, the Pigeonhole Principle tells us that there is some term that is repeated infinitely often. That is, there is a sequence of nonnegative integers $s_0 < s_1 < s_2 < \cdots$ with

$$(e_{s_0}, \overline{a}_{s_0}, \overline{b}_{s_0}) = (e_{s_1}, \overline{a}_{s_1}, \overline{b}_{s_1}) = (e_{s_2}, \overline{a}_{s_2}, \overline{b}_{s_2}) = \cdots.$$

Then, setting

$$(\alpha_i, \beta_i) = \overline{(a_{s_0}, b_{s_0})} *_r (a_{s_i}, b_{s_i})$$

for each $i = 1, 2, 3, \ldots$, we see that, by the same argument as above,

$$(\alpha_1, \beta_1), \quad (\alpha_2, \beta_2), \quad (\alpha_3, \beta_3), \quad \ldots$$

are all distinct nontrivial solutions of Pell's equation. □

4.3 Numerical Examples and Further Results

In this section we give a number of examples of the method we described in the last section, and describe some of its properties. We also give a few "tricks" for speeding up the computation. First, we have a number of tables giving the results of the method for various values of D. The values of D were chosen to provide an illustrative sample of the phenomena encountered.

For each value of D, the table gives a_i, b_i, $e_i = a_i^2 - Db_i^2$, $\overline{a}_i \equiv a_i \pmod{e_i}$, $\overline{b}_i \equiv b_i \pmod{e_i}$, and x_i, where

$$(a_{i+1}, b_{i+1}) = (a_i, b_i) *_r (x_{i+1}, 1), \quad \text{with} \quad (a_0, b_0) = (1, 0).$$

We observe that, in each case, the sequence $e_0, e_1, e_2, \ldots$ is periodic. This leads us to make the following definition.

Definition 4.16. For a fixed value and D, the period of the sequence $\{e_i\}$, i.e., the smallest value of k for which $e_{i+k} = e_i$ for every $i \geq 0$, is called the *small period* of D, denoted by $k = p(D)$.

In particular, we see that $1 = e_0 = e_{p(D)} = e_{2p(D)} \cdots$ so that $(a_{ip(D)}, b_{ip(D)})$ represents 1 for every i, giving a sequence of solutions to Pell's equation.

But we observe that the sequence of triples $(e_0, \overline{a}_0, \overline{b}_0)$, $(e_1, \overline{a}_1, \overline{b}_1)$, $(e_2, \overline{a}_2, \overline{b}_2)$, ... is also periodic, and so we make another definition.

Definition 4.17. For a fixed value of D, the period of the sequence $\{(e_i, \overline{a}_i, \overline{b}_i)\}$, i.e., the smallest value of k for which $(e_{i+k}, \overline{a}_{i+k}, \overline{b}_{i+k}) = (e_i, \overline{a}_i, \overline{b}_i)$ for every $i \geq 0$, is called the *large period* of D, denoted by $k = P(D)$.

Note in Tables 4.1–4.8 we have given at least one large period.

We also give in Table 4.9 the values of $p(D)$ and $P(D)$ for all square-free positive integers $D < 100$.

i	a_i	b_i	e_i	$\overline{a}_i$	$\overline{b}_i$	x_i
0	1	0	1	0	0	
1	4	1	3	1	1	4
2	11	3	4	3	3	5
3	18	5	−1	0	0	3
4	137	38	−3	2	2	4
5	393	109	−4	1	1	5
6	649	180	1	0	0	3
7	4936	1369	3	1	1	4
8	14159	3927	4	3	3	5
9	23382	6485	−1	0	0	3
10	177833	49322	−3	2	2	4
11	510117	141481	−4	1	1	5
12	842401	233640	1	0	0	3

Table 4.1. $D = 13$.

Remark 4.18. Recall we defined

$$(a_{i+1}, b_{i+1}) = \left(\left| \frac{a_i x_{i+1} + b_i D}{e_i} \right|, \left| \frac{a_i b_i x_{i+1}}{e_i} \right| \right) = (a_i, b_i) *_r (x_{i+1}, 1).$$

Suppose we had not used absolute values and instead had defined

$$(a'_{i+1}, b'_{i+1}) = \left(\frac{a_i x_{i+1} + b_i D}{e_i}, \frac{a_i b_i x_{i+1}}{e_i} \right).$$

Then we would have had $(a'_{i+1}, b'_{i+1}) = (a_i, b_i) *_r (x_{i+1}, 1)$ or $(a'_{i+1}, b'_{i+1}) = (a_i, b_i) *_r -(x_{i+1}, 1)$. This would not have changed the small period $p(D)$, but might have changed the large period. Call the new value $P'(D)$. Then it could only have changed $P'(D)$ by at most a factor of two, i.e., $P'(D) = P(D)$, $P'(D) = 2P(D)$, or $P'(D) = (1/2)P(D)$, and all three possibilities occur.

Remark 4.19. Since $1 = e_0 = e_{P(D)} = e_{2P(D)} = \cdots$ we see that $p(D)$ divides $P(D)$ (and also $p(D)$ divides $P'(D)$).

Now we give some results that enable us to speed up our search.

i	a_i	b_i	e_i	$\overline{a}_i$	$\overline{b}_i$	x_i
0	1	1	1	0	0	
1	4	1	−3	1	1	4
2	13	3	−2	1	1	5
3	61	14	−3	1	2	5
4	170	39	1	0	0	4
5	1421	326	−3	2	2	4
6	4433	1017	−2	1	1	5
7	20744	4759	−3	2	1	5
8	57799	13260	1	0	0	4
9	483136	110839	−3	1	1	4
10	1507207	345777	−2	1	1	5
11	7052899	1618046	−3	1	2	5
12	19651490	4508361	1	0	0	4
13	164264819	37684934	−3	2	2	4
14	512445947	117563163	−2	1	1	5
15	2397964916	550130881	−3	2	1	5
16	6681448801	1532829480	1	0	0	4

Table 4.2. $D = 19$.

i	a_i	b_i	e_i	$\overline{a}_i$	$\overline{b}_i$	x_i
0	1	0	1	0	0	
1	5	1	4	1	1	5
2	9	2	−3	0	2	3
3	32	7	−5	2	2	6
4	55	12	1	0	0	4
5	527	115	4	3	3	5
6	999	218	−3	0	2	3
7	3524	769	−5	4	4	6
8	6049	1320	1	0	0	4
9	57965	12649	4	1	1	5
10	109881	23978	−3	0	2	3
11	387608	84583	−5	3	3	6
12	665335	145188	1	0	0	4
13	6375623	1391275	4	3	3	5
14	12085911	2637362	−3	0	2	3
15	42633356	9303361	−5	1	1	6
16	73180801	15969360	1	0	0	4

Table 4.3. $D = 21$.

i	a_i	b_i	e_i	$\overline{a}_i$	$\overline{b}_i$	x_i
0	1	0	1	0	0	
1	5	1	−4	1	1	5
2	16	3	−5	1	3	7
3	27	5	4	3	1	3
4	70	13	−1	0	0	5
5	727	135	4	3	3	5
6	2251	418	5	1	3	7
7	3775	701	−4	3	1	3
8	9801	1820	1	0	0	5
9	101785	18901	−4	1	1	5
10	315156	58523	−5	1	3	7
11	528527	98145	4	3	1	3
12	1372210	254813	−1	0	0	5
13	14250627	2646275	4	3	3	5
14	44124091	8193638	5	1	3	7
15	73997555	13741001	−4	3	1	3
16	192119201	35675640	1	0	0	5
17	1995189565	370497401	−4	1	1	5
18	6177687896	1147167843	−5	1	3	7
19	10360186227	1923838285	4	3	1	3
20	26898060350	4994844413	−1	0	0	5
21	279340789727	51872282415	4	3	3	5
22	864920429531	160611691658	5	1	3	7
23	1450500069335	269351100901	−4	3	1	3
24	3765920568201	699313893460	1	0	0	5
25	39109705751345	7262490035501	−4	1	1	5
26	121095037822236	22486783999963	−5	1	3	7
27	203080369893127	37711077964425	4	3	1	3
28	527255777608490	97908939928813	−1	0	0	5
29	5475638145978027	1016800477252555	4	3	3	5
30	16954170215542571	3148310371686478	5	1	3	7
31	28432702285107115	5279820266120401	−4	3	1	3
32	73819574785756801	13707950903927280	1	0	0	5

Table 4.4. $D = 29$.

i	a_i	b_i	e_i	$\bar{a}_i$	$\bar{b}_i$	x_i
0	1	1	1	0	0	
1	6	1	5	1	1	6
2	11	2	−3	2	2	4
3	39	7	2	1	1	5
4	206	37	−3	2	1	5
5	863	155	−6	5	5	7
6	1520	273	1	0	0	5
7	17583	3158	5	3	3	6
8	33646	6043	−3	1	1	4
9	118521	21287	2	1	1	5
10	626251	112478	−3	1	2	5
11	2623525	471199	−6	1	1	7
12	4620799	829920	1	0	0	5
13	53452314	9600319	5	4	4	6
14	102283829	18370718	−3	2	2	4
15	360303801	64712473	2	1	1	5
16	1903802834	341933083	−3	2	1	5
17	7975515137	1432444805	−6	5	5	7
18	14047227440	2522956527	1	0	0	5
19	162495016977	29184966602	5	2	2	6
20	310942806514	55846976677	−3	1	1	4
21	1095323436519	196725896633	2	1	1	5
22	5787559989109	1039476459842	−3	1	2	5
23	24245563392955	4354631736001	−6	1	1	7
24	42703566796801	7669787012160	1	0	0	5

Table 4.5. $D = 31$.

i	a_i	b_i	e_i	$\bar{a}_i$	$\bar{b}_i$	x_i
0	1	0	1	0	0	
1	8	1	6	2	1	8
2	23	3	7	2	3	10
3	61	8	9	7	8	11
4	99	13	−1	0	0	7
5	1546	203	−6	4	5	8
6	4539	596	−7	3	1	10
7	12071	1585	−9	2	1	11
8	19603	2574	1	0	0	7
9	306116	40195	6	2	1	8
10	898745	118011	7	1	5	10
11	2390119	313838	9	7	8	11
12	3881493	509665	−1	0	0	7
13	60612514	7958813	−6	4	5	8
14	177956049	23366774	−7	5	4	10
15	473255633	62141509	−9	2	1	11
16	768555217	100916244	1	0	0	7
17	12001583888	1575885169	6	2	1	8
18	35236196447	4626739263	7	4	6	10
19	93707005453	12304332620	9	7	8	11
20	152177814459	19981925977	−1	0	0	7
21	2376374222338	312033222275	−6	4	5	8
22	6976944852555	916117740848	−7	6	2	10
23	18554460335327	2436320000269	−9	1	1	11
24	30131975818099	3956522259690	1	0	0	7

Table 4.6. $D = 58$.

i	a_i	b_i	e_i	$\overline{a}_i$	$\overline{b}_i$	x_i
0	1	0	1	0	0	
1	8	1	3	2	1	8
2	39	5	−4	3	1	7
3	164	21	−5	4	1	9
4	453	58	5	3	3	6
5	1523	195	4	3	3	9
6	5639	722	−3	2	2	7
7	29718	3805	−1	0	0	8
8	469849	60158	−3	1	2	8
9	2319527	296985	4	3	1	7
10	9747957	1248098	5	2	3	9
11	26924344	3447309	−5	4	4	6
12	90520989	11590025	−4	1	1	9
13	335159612	42912791	3	2	2	7
14	1766319049	226153980	1	0	0	8
15	27925945172	3575550889	3	2	1	8
16	137863406811	17651600465	−4	3	1	7
17	579379572416	74181952749	−5	1	4	9
18	1600275310437	204894257782	5	2	2	6
19	5380205503727	688864726095	4	3	3	9
20	19920546704471	2550564646598	−3	2	2	7
21	104982939026082	13441687959085	−1	0	0	8
22	1659806477712841	212516442698762	−3	1	2	8
23	8194049449538123	1049140525534725	4	3	1	7
24	34436004275865333	4409078544837662	5	3	2	9
25	95113963378057876	12178095108978261	−5	1	1	6
26	319777894410038961	40943363871772445	−4	1	1	9
27	1183997614262097968	151595360378111519	3	2	2	7
28	6239765965720528801	798920165762330040	1	0	0	8

Table 4.7. $D = 61$.

i	a_i	b_i	e_i	$\overline{a}_i$	$\overline{b}_i$	x_i
0	1	0	1	0	0	
1	10	1	9	1	1	10
2	19	2	−3	1	2	8
3	124	13	−3	1	1	10
4	849	89	−10	9	9	11
5	1574	165	1	0	0	9
6	30755	3224	9	2	2	10
7	59936	6283	−3	2	1	8
8	390371	40922	−3	2	2	10
9	2672661	280171	−10	1	1	11
10	4954951	519420	1	0	0	9
11	96816730	10149151	9	4	4	10
12	188678509	19778882	−3	1	2	8
13	1228887784	128822443	−3	1	1	10
14	8413535979	881978219	−10	9	9	11
15	15598184174	1635133995	1	0	0	9
16	304779035285	31949524124	9	8	8	10
17	593959886396	62263914253	−3	2	1	8
18	3868538353661	405533009642	−3	2	2	10
19	26485808589231	2776467153241	−10	1	1	11
20	49103078824801	5147401296840	1	0	0	9
21	959444306260450	100577091793201	9	7	7	10
22	1869785533696099	196006782289562	−3	1	2	8
23	12178157508437044	1276617785530573	−3	1	1	10
24	83377317025363209	8740317716424449	−10	9	9	11
25	154576476542289374	16204017647318325	1	0	0	9
26	3020330371328861310	316616653015472624	9	5	5	10
27	5886084266115433250	617029288383626923	−3	2	1	8
28	38336835968021460851	4018792383317234162	−3	2	2	10
29	262471767510034792701	27514517394837012211	−10	1	1	11
30	486606699052048124551	51010242406356790260	1	0	0	9

Table 4.8. $D = 91$.

D	$p(D)$	$P(D)$	D	$p(D)$	$P(D)$	D	$p(D)$	$P(D)$
2	2	2	37	2	2	69	6	24
3	1	1	38	2	2	70	4	8
5	2	2	39	2	2	71	6	24
6	2	2	40	2	2	72	2	2
7	2	2	41	6	12	73	10	30
8	1	1	42	2	2	74	6	36
10	2	2	43	6	36	75	2	4
11	2	2	44	4	8	76	8	48
12	2	2	45	4	4	77	4	8
13	6	6	46	8	96	78	2	4
14	2	2	47	2	2	79	2	2
15	1	1	48	1	1	80	1	1
17	2	2	50	2	2	82	2	2
18	2	2	51	2	2	83	2	2
19	4	8	52	4	4	84	2	2
20	2	2	53	8	8	85	8	8
21	4	16	54	4	16	86	6	72
22	4	8	55	4	8	87	2	2
23	2	2	56	2	2	88	4	4
24	1	1	57	4	24	89	10	10
26	2	2	58	8	24	90	2	2
27	2	2	59	4	16	91	5	30
28	4	8	60	2	4	92	6	180
29	8	8	61	14	28	93	6	12
30	2	2	62	2	2	94	10	120
31	6	24	63	1	1	95	2	4
32	2	4	65	2	2	96	2	4
33	2	4	66	2	2	97	12	180
34	2	2	67	8	48	98	2	2
35	1	1	68	2	2	99	1	1

Table 4.9. Small and large periods for values of D between 2 and 99.

Lemma 4.20.

(1) Suppose (a, b) represents -1. Then

$$(a, b) * (a, b) = (a, b) *_r (a, b)$$

represents 1.

(2) Suppose that (a, b) represents ± 2. Then

$$(a, b) *_r (a, b) = \tfrac{1}{2}((a, b) * (a, b))$$

represents 1.

Proof:

(1) This is immediate.

(2) If (a, b) represents ± 2 then it must be reduced. Then $(a, b) * (a, b)$ represents 4. But

$$\begin{aligned}(a, b) * (a, b) &= (a^2 + b^2 D, 2ab) \\ &= (a^2 - b^2 D + 2b^2 D, 2ab) \\ &= (\pm 2 + 2b^2 D, 2ab) \\ &= 2(\pm 1 + b^2 D, ab),\end{aligned}$$

so $(a, b) *_r (a, b) = (\pm 1 + b^2 D, ab)$ represents 1. □

Lemma 4.21. *Let (a, b) and (c, d) both be reduced and suppose that, for some integer m, (a, b) represents m and (c, d) represents $-m$. Suppose also that $a \equiv c \pmod{m}$ and $b \equiv d \pmod{m}$. Then $\overline{(a, b)} *_r (c, d)$ represents -1.*

Proof: This is almost identical to the proof of Lemma 4.7. $(E, F) = \overline{(a, b)} * (c, d)$ represents $-m^2$ and then $\gcd(E, F) = |m|$, so $(e, f) = \frac{1}{|m|}(\overline{(a, b)} * (c, d)) = \overline{(a, b)} *_r (c, d)$ represents -1. □

We also make the trivial but useful observation that if (a, b) represents m, then $(a, -b)$, $(-a, b)$, and $(-a, -b)$ also represent m.

Now let us look at some examples (and we refer the reader to Tables 4.1–4.8).

Example 4.22.

(1) $D = 13$: $(a_0, b_0) = (1, 0)$ with $e_0 = 1$ and we compute $(a_1, b_1) = (4, 1)$ with $e_1 = 3$, $(a_2, b_2) = (11, 3)$ with $e_2 = 4$, and $(a_3, b_3) = (18, 5)$ with $e_3 = -1$. Then $(a_3, b_3) *_r (a_3, b_3) = (649, 180) = (a_6, b_6)$ represents 1.

(2) $D = 19$: $(a_0, b_0) = (1, 0)$ with $e_0 = 1$ and we compute $(a_1, b_1) = (4, 1)$ with $e_1 = -3$, and $(a_2, b_2) = (13, 3)$ with $e_2 = -2$. Then $(a_2, b_2) *_r (a_2, b_2) = (170, 39) = (a_4, b_4)$ represents 1.

(3) $D = 29$: $(a_0, b_0) = 1$ with $e_0 = 1$ and we compute $(a_1, b_1) = (5, 1)$ with $e_1 = -4$, $(a_2, b_2) = (16, 3)$ with $e_2 = -5$, and $(a_3, b_3) = (27, 5)$ with $e_3 = -4$. We cannot apply Lemma 4.21 directly because, while $e_3 = -e_1$ and $\bar{b}_3 = \bar{b}_1$, $\bar{a}_3 \neq \bar{a}_1$. But if we replace (a_1, b_1) by $\overline{(-5, 1)}$ then $-5 \equiv 3 \pmod 4$ and we can use Lemma 4.21. Then $\overline{(-5, 1)} *_r (27, 5) = (-70, -13)$, so $(70, 13) = (a_4, b_4)$ represents -1, and then $(70, 13) * (70, 13) = (9801, 1820) = (a_8, b_8)$ represents 1.

(4) $D = 61$: $(a_0, b_0) = (1, 0)$ with $e_0 = 1$ and we compute $(a, b) = (8, 1)$ with $e_1 = 3$, $(a_2, b_2) = (39, 5)$ with $e_2 = -4$, $(a_3, b_3) = (164, 21)$ with $e_3 = -5$, and $(a_4, b_4) = (453, 58)$ with $e_4 = 5$. Although $e_4 = -e_3$ we cannot apply Lemma 4.21 (even after a sign change). But $(a_5, b_5) = (1523, 195)$ with $e_5 = 4$, so using (a_2, b_2) and (a_5, b_5) and a sign change, we find that $\overline{(39, -5)} *_r (1523, 195) = (29718, 3805) = (a_7, b_7)$ represents -1 and then $(29718, 3805) * (29718, 3805) = (1766319049, 226153980) = (a_{14}, b_{14})$ represents 1.

(5) $D = 91$: $(a_0, b_0) = (1, 0)$ with $e_0 = 1$ and we compute $(a_1, b_1) = (10, 1)$ with $e_1 = 9, (a_2, b_2) = (19, 2)$ with $e_2 = -3$, and $(a_3, b_3) = (124, 13)$ with $e_3 = -3$. Then, using (a_2, b_2) and (a_3, b_3) and a sign change, we find that $\overline{(19, -2)} *_r (124, 13) = (1574, 165) = (a_5, b_5)$ represents 1.

4.4 Units in $\mathcal{O}(\sqrt{D})$

In this section we assume that D is a square-free positive integer, $D \neq 1$. Our objective is to find *all* units of $R = \mathcal{O}(\sqrt{D})$.

Definition 4.23. Let D be a square-free positive integer, $D \neq 1$, and let $R = \mathcal{O}(\sqrt{D})$. Among all units $\varepsilon = c + d\sqrt{D}$ of R, let ε_0 be defined as follows: $\varepsilon_0 = c_0 + d_0\sqrt{D}$ where d_0 is the minimum positive value of d and c_0 is positive. In almost all cases this determines c_0 uniquely, but if not, we choose c_0 to be the smallest positive value of c for our given choice of d_0. The unit ε_0 is called the *fundamental unit* of R.

Let us observe that this definition makes sense. First note that, since Pell's equation $x^2 - y^2D = 1$ has a nontrivial solution, there *is* a unit $c + d\sqrt{D}$ of R with c and d both positive (and in fact, there are infinitely many such units). Next note that $\varepsilon = c + d\sqrt{D} = (a + b\sqrt{D})/2$ where a and b are both even integers, if $D \equiv 2$ or $3 \pmod 4$, or where a and b are either both even integers or both odd integers, if $D \equiv 1 \pmod 4$, and in each case the possible values of b (and if necessary, the possible values of a) are well-ordered, as we require that b and a be positive.

Lemma 4.24. *Let D be a square-free positive integer, $D \neq 1$, and let $R = \mathcal{O}(\sqrt{D})$. The fundamental unit ε_0 of R is the smallest unit of R that is greater than* 1.

Proof: First let us observe that it is not a priori clear that there is a smallest unit greater than 1, as there could be an infinite sequence of units $\varepsilon, \varepsilon', \varepsilon'', \dots$ with $\varepsilon > \varepsilon' > \varepsilon'' > \dots > 1$. But in fact there is such a smallest unit.

Note that the statement of the lemma is equivalent to the following statement:

Let ε be any unit of R with $\varepsilon > 1$. Then $\varepsilon \geq \varepsilon_0$.

We leave the proof of this to the exercises. □

Theorem 4.25. *Let D be a square-free positive integer, $D \neq 1$, and let $R = \mathcal{O}(\sqrt{D})$. Then the units of R are*

$$\{\dots, \pm\varepsilon_0^{-3}, \pm\varepsilon_0^{-2}, \pm\varepsilon_0^{-1}, \pm 1, \pm\varepsilon_0, \pm\varepsilon_0^2, \pm\varepsilon_0^3, \dots\}.$$

Proof: Let $\varepsilon = c + d\sqrt{D}$ be any unit of R. Note that $\overline{\varepsilon} = c - d\sqrt{D}$, $-\varepsilon = -c - d\sqrt{D}$, and $-\overline{\varepsilon} = -c + d\sqrt{D}$ are all units of R. Also note that $\overline{\varepsilon} = \pm\varepsilon^{-1}$ (+ if $\varepsilon\overline{\varepsilon} = 1$ and − if $\varepsilon\overline{\varepsilon} = -1$). Thus in order to prove the theorem, it suffices to prove if $\varepsilon = c + d\sqrt{D}$ is any unit of R with c and d positive, then $\varepsilon = \varepsilon_0^k$ for some positive integer k.

Since $\varepsilon = c + d\sqrt{D}$ with c and d positive, we see that $\varepsilon > 1$. Also, $\varepsilon_0 > 1$, so the sequence 1, ε_0, ε_0^2, ε_0^3, ... of nonnegative powers of ε_0 is a (strictly) increasing sequence that diverges to $+\infty$. Hence there is some positive integer k with $\varepsilon_0^{k-1} < \varepsilon \leq \varepsilon_0^k$. But then $1 < \varepsilon' \leq \varepsilon_0$ where $\varepsilon' = \varepsilon/(\varepsilon_0^{k-1})$.

$D = 2$	$\varepsilon_0 = 1 + \sqrt{2}$	$\varepsilon_{\text{Pell}} = 3 + 2\sqrt{2}$	$\varepsilon_{\text{Pell}} = \varepsilon_0^2$
$D = 3$	$\varepsilon_0 = 2 + \sqrt{3}$	$\varepsilon_{\text{Pell}} = 2 + \sqrt{3}$	$\varepsilon_{\text{Pell}} = \varepsilon_0$
$D = 5$	$\varepsilon_0 = (1 + \sqrt{5})/2$	$\varepsilon_{\text{Pell}} = 9 + 4\sqrt{5}$	$\varepsilon_{\text{Pell}} = \varepsilon_0^6$
$D = 6$	$\varepsilon_0 = 5 + 2\sqrt{6}$	$\varepsilon_{\text{Pell}} = 5 + 2\sqrt{6}$	$\varepsilon_{\text{Pell}} = \varepsilon_0$
$D = 7$	$\varepsilon_0 = 8 + 3\sqrt{7}$	$\varepsilon_{\text{Pell}} = 8 + 3\sqrt{7}$	$\varepsilon_{\text{Pell}} = \varepsilon_0$
$D = 10$	$\varepsilon_0 = 3 + \sqrt{10}$	$\varepsilon_{\text{Pell}} = 19 + 6\sqrt{10}$	$\varepsilon_{\text{Pell}} = \varepsilon_0^2$
$D = 11$	$\varepsilon_0 = 10 + 3\sqrt{11}$	$\varepsilon_{\text{Pell}} = 10 + 3\sqrt{11}$	$\varepsilon_{\text{Pell}} = \varepsilon_0$
$D = 13$	$\varepsilon_0 = (3 + \sqrt{13})/2$	$\varepsilon_{\text{Pell}} = 649 + 180\sqrt{13}$	$\varepsilon_{\text{Pell}} = \varepsilon_0^6$
$D = 14$	$\varepsilon_0 = 15 + 4\sqrt{14}$	$\varepsilon_{\text{Pell}} = 15 + 4\sqrt{14}$	$\varepsilon_{\text{Pell}} = \varepsilon_0$
$D = 15$	$\varepsilon_0 = 4 + \sqrt{15}$	$\varepsilon_{\text{Pell}} = 4 + \sqrt{15}$	$\varepsilon_{\text{Pell}} = \varepsilon_0$
$D = 17$	$\varepsilon_0 = 4 + \sqrt{17}$	$\varepsilon_{\text{Pell}} = 33 + 8\sqrt{17}$	$\varepsilon_{\text{Pell}} = \varepsilon_0^2$
$D = 19$	$\varepsilon_0 = 170 + 39\sqrt{19}$	$\varepsilon_{\text{Pell}} = 170 + 39\sqrt{19}$	$\varepsilon_{\text{Pell}} = \varepsilon_0$
$D = 21$	$\varepsilon_0 = (5 + \sqrt{21})/2$	$\varepsilon_{\text{Pell}} = 55 + 12\sqrt{21}$	$\varepsilon_{\text{Pell}} = \varepsilon_0^3$
$D = 22$	$\varepsilon_0 = 197 + 42\sqrt{22}$	$\varepsilon_{\text{Pell}} = 197 + 42\sqrt{22}$	$\varepsilon_{\text{Pell}} = \varepsilon_0$
$D = 23$	$\varepsilon_0 = 24 + 5\sqrt{23}$	$\varepsilon_{\text{Pell}} = 24 + 5\sqrt{23}$	$\varepsilon_{\text{Pell}} = \varepsilon_0$
$D = 33$	$\varepsilon_0 = 23 + 4\sqrt{33}$	$\varepsilon_{\text{Pell}} = 23 + 4\sqrt{33}$	$\varepsilon_{\text{Pell}} = \varepsilon_0$
$D = 34$	$\varepsilon_0 = 35 + 6\sqrt{34}$	$\varepsilon_{\text{Pell}} = 35 + 6\sqrt{34}$	$\varepsilon_{\text{Pell}} = \varepsilon_0$
$D = 37$	$\varepsilon_0 = 6 + \sqrt{37}$	$\varepsilon_{\text{Pell}} = 735 + 12\sqrt{37}$	$\varepsilon_{\text{Pell}} = \varepsilon_0^2$
$D = 141$	$\varepsilon_0 = 95 + 8\sqrt{141}$	$\varepsilon_{\text{Pell}} = 95 + 8\sqrt{141}$	$\varepsilon_{\text{Pell}} = \varepsilon_0$

Table 4.10. Relation between ε_0 and $\varepsilon_{\text{Pell}}$ for selected values of D.

But ε' is a unit as $(\varepsilon')^{-1} = \varepsilon^{-1}\varepsilon_0^{k-1}$. In other words, ε' is a unit that is greater than 1 and less than or equal to ε_0, so by Lemma 4.10 we must have $\varepsilon' = \varepsilon_0$, which gives $\varepsilon = \varepsilon_0^k$. □

Remark 4.26. For a given value of D, let $\varepsilon_{\text{Pell}}$ be the unit $\varepsilon_{\text{Pell}} = a + b\sqrt{D}$ obtained from the smallest solution to Pell's equation $a^2 - b^2D = 1$ in positive integers a and b. Then sometimes $\varepsilon_{\text{Pell}}$ is the fundamental unit ε_0 of $\mathcal{O}(\sqrt{D})$ and sometimes not. Table 4.10 is a table of ε_0, $\varepsilon_{\text{Pell}}$, and the relation between them for selected values of D.

4.5 Exercises

In Exercises 4.1 and 4.2, D is a positive integer that is not a perfect square. In the remaining exercises, D is a square-free positive integer, $D \neq 1$.

Exercise 4.1.

(a) Use our method to find a solution of Pell's equation $a^2 - b^2D = 1$ for various small values of D.

(b) In particular, use our method to find, by hand computation, the smallest solution $a = 176631909$ and $b = 226153980$ of Pell's equation $a^2 - 61b^2 = 1$ in positive integers a and b. (Compare Table 4.7 and Example 4.22.) If Fermat could do this computation by hand, so can you.

Exercise 4.2. Write a computer program to implement our method of solving Pell's equation for relatively small values of a, b, and D, and verify parts of Tables 4.1–4.9.

Exercise 4.3. The *Archimedes cattle problem* was a famous problem posed by Archimedes. Search the Internet for this problem. (You will find lots of references.) The heart of this problem is solving Pell's equation $a^2 - 4729494b^2 = 1$. Write a computer program to implement our method of solving Pell's equation that can handle relatively large values of a, b, and D and apply it to the case $D = 4729494$. Show that at step 60 it produces the following solution:

$$a = 109931986732829734979866232821433543901088049,$$
$$b = 50549485234315033074477819735540408986340.$$

The next few problems explain why many of the results in Section 4.1 are true, by relating these results to arithmetic in the field $\mathbf{Q}(\sqrt{D})$. Given a pair of rational numbers (a, b), let (a, b) correspond to the element $\alpha = a + b\sqrt{D}$ of $\mathbf{Q}(\sqrt{D})$. We write this as $(a, b) \leftrightarrow \alpha = a + b\sqrt{D}$.

Exercise 4.4.

(a) Show that if $(a, b) \leftrightarrow \alpha$ then $\overline{(a, b)} \leftrightarrow \overline{\alpha}$, so conjugation of representations corresponds to conjugation in $\mathbf{Q}(\sqrt{D})$.

(b) Show that if $(a, b) \leftrightarrow \alpha$ and $(c, d) \leftrightarrow \beta$, then $(a, b) * (c, d) = \alpha\beta$. Thus, composition of representations (as defined in Definition 4.2) corresponds to multiplication in $\mathbf{Q}(\sqrt{D})$.

Exercise 4.5. Show that the fact that composition of representations is commutative, associative, and commutes with conjugation follows from the fact that multiplication in $\mathbf{Q}(\sqrt{D})$ is commutative, associative, and commutes with conjugation, thereby proving Lemma 4.3.

Exercise 4.6.

(a) If $(a, b) \leftrightarrow \alpha$, show that (a, b) represents $\mathrm{N}(\alpha)$, the norm of α.

(b) Use the fact that $\mathrm{N}(\alpha\beta) = \mathrm{N}(\alpha)\,\mathrm{N}(\beta)$ for any two elements α and β of $\mathbf{Q}(\sqrt{D})$ to prove Lemma 4.1.

Exercise 4.7. Let $(a, b) \leftrightarrow \alpha$. Observe that, if $\alpha \neq \pm 1$, then the powers of α are all distinct. Use this to prove Lemma 4.8.

Exercise 4.8. We have often had occasion to compute $\overline{(a, b)} * (c, d)$. If $(a, b) \leftrightarrow \alpha$ and $(c, d) \leftrightarrow \beta$, show that $\overline{(a, b)} * (c, d) \leftrightarrow \mathrm{N}(\alpha)\beta/\alpha$, so composition with $\overline{(a, b)}$ corresponds to multiplication by $\mathrm{N}(\alpha)/\alpha$.

Exercise 4.9. If $(a, b) \leftrightarrow \alpha$, $(x, 1) \leftrightarrow \chi$, and $(a', b') \leftrightarrow \chi'$, show that $\alpha' = \pm\chi/\overline{\alpha}$, so reduced composition of $(x, 1)$ with (a, b) corresponds (up to sign) to division of χ by $\overline{\alpha}$. (Compare Corollary 4.11.)

The next few problems give the proof of Lemma 4.24.

Exercise 4.10. Let $\varepsilon = a + b\sqrt{D}$ be a unit of $R = \mathcal{O}(\sqrt{D})$. Observe that $|\overline{\varepsilon}| = 1/|\varepsilon|$. Use this observation to prove that, for a unit $\varepsilon = a + b\sqrt{D}$ of R, $\varepsilon > 1$ if and only if a and b are both positive.

Exercise 4.11. Let $\varepsilon = a + b\sqrt{D}$ be a unit of $R = \mathcal{O}(\sqrt{D})$. Observe that $a = \pm\sqrt{e + b^2 D}$, where $e = \varepsilon\overline{\varepsilon} = \pm 1$. Use this observation to prove that if $\varepsilon_1 = a_1 + b_1\sqrt{D}$ and $\varepsilon_2 = a_2 + b_2\sqrt{D}$ are units of $\mathcal{O}(\sqrt{D})$, with a_1, b_1, a_2, and b_2 all positive, and with $b_1 < b_2$, or with $b_1 = b_2$ and $a_1 < a_2$, then $\varepsilon_1 < \varepsilon_2$.

Exercise 4.12. Show that the only case in which it is possible to have distinct units $\varepsilon_1 = a_1 + b\sqrt{D}$ and $\varepsilon_2 = a_1 + b\sqrt{D}$ with a_1, b, and a_2 all positive is for $D = 5$ when $\varepsilon_1 = (1 + \sqrt{5})/2$ and $\varepsilon_2 = (3 + \sqrt{5})/2$, or vice versa.

Exercise 4.13. Use the results of Exercises 4.10 and 4.11 to prove Lemma 4.24.

Exercise 4.14. Let $R = \mathcal{O}(\sqrt{D})$. Suppose that $D \equiv 1 \pmod 8$. Show that every unit ε of R is of the form $\varepsilon = a + b\sqrt{D}$ with a and b integers. (Of course, this is automatically true for $D \equiv 2$, 3, 6, or 7 $\pmod 8$ as in those cases, every element of R is of that form. In case $D \equiv 5 \pmod 8$, all units of R may or may not be of this form, as we see from Table 4.10.)

Exercise 4.15. Let $R = \mathcal{O}(\sqrt{D})$. Suppose that $D \equiv 3$, 6, or 7 $\pmod 8$. Show that every unit ε of $R = \mathcal{O}(\sqrt{D})$ has norm $\mathrm{N}(\varepsilon) = 1$.

Exercise 4.16.

(a) Verify the entries in Table 4.10.

(b) Extend this table to include all values of D between 26 and 47.

In the next two exercises we adopt the notation of Remark 4.26.

Exercise 4.17. Let $R = \mathcal{O}(\sqrt{D})$.

(a) Suppose that $D \equiv 3$, 6, or 7 $(\bmod 8)$. Show that $\varepsilon_{\text{Pell}} = \varepsilon_0$.

(b) Suppose that $D \equiv 1$ or 2 $(\bmod 8)$. Show that $\varepsilon_{\text{Pell}} = \varepsilon_0^k$ for $k = 1$ or 2.

(c) Suppose that $D \equiv 5 \pmod 8$. Show that $\varepsilon_{\text{Pell}} = \varepsilon_0^k$ for $k = 1$, 2, 3, or 6. (All of these possibilities occur, as illustrated in Table 4.10.)

Exercise 4.18. Let $R = \mathcal{O}(\sqrt{D})$.

(a) Suppose that $a^2 - b^2 D = -4$ has a solution in odd integers a and b. Show that $\varepsilon_{\text{Pell}} = \varepsilon_0^6$, and conversely.

(b) Suppose that $a^2 - b^2 D = -4$ does not have a solution in odd integers a and b, but that $a^2 - b^2 D = 4$ has a solution in odd integers a and b. Show that $\varepsilon_{\text{Pell}} = \varepsilon_0^3$, and conversely.

(c) Suppose that $a^2 - b^2 D = -4$ does not have a solution in odd integers a and b, and that $a^2 - b^2 D = 4$ does not have a solution in odd integers a and b, but that $a^2 - b^2 D = -1$ has a solution in integers a and b. Show that $\varepsilon_{\text{Pell}} = \varepsilon_0^2$, and conversely.

(d) Suppose that $a^2 - b^2 D = -4$ does not have a solution in odd integers a and b, that $a^2 - b^2 D = 4$ does not have a solution in odd integers a and b, and that $a^2 - b^2 D = -1$ does not have a solution in integers a and b. Show that $\varepsilon_{\text{Pell}} = \varepsilon_0$, and conversely.

Chapter 5

Towards Algebraic Number Theory

In Chapters 1 through 4 we considered the quadratic fields $\mathbf{K} = \mathbf{Q}(\sqrt{D})$ and the rings of integers $R = \mathcal{O}(\sqrt{D})$, and we kept our discussion as elementary as possible while still proving interesting results. But this consideration was just the tip of an iceberg. Our aim in this chapter is to indicate how these results generalize. The field of mathematics that these generalize to is known as *algebraic number theory*. As a matter of historical fact, our development here parallels the development of algebraic number theory. Historically, quadratic fields were considered first, and their investigation motivated the investigation of more general *algebraic number fields*.

Our goal here is to provide a guide to some of the high points of algebraic number theory, especially those most related to the topics we have studied. In order to do so, we will have to introduce some more advanced concepts than we have done so far, so this chapter will require more background of the reader. Also, as this is a guide and not a complete treatment, we will not prove the more advanced results we state. But we hope this chapter will motivate the reader to go on and study algebraic number theory further.

However, we will revisit quadratic fields and prove a number of results, and do a number of examples, in them. These will provide concrete examples of this general theory, as well as being very interesting in themselves.

There is one notational point that we should mention up front. We will be intensively studying ideals in this chapter. Standard mathematical notation, which we shall adopt below, gives parentheses special meaning in the context of ideals. Parentheses are of course also the standard way to group items for multiplication. In order to avoid confusion, we shall instead, throughout this chapter, except in the last section, use square brackets to group items for multiplication.

5.1 Algebraic Numbers and Algebraic Integers

We begin by defining algebraic numbers.

Definition 5.1. A complex number α is *algebraic* if α is a root of a polynomial $f(X)$ with rational coefficients. Otherwise α is *transcendental.*

It follows from properties of polynomials that if α is algebraic, there is a unique monic polynomial of lowest degree having α as a root. (A polynomial $f(X)$ is monic if the coefficient of the highest power of X is 1.) We call this polynomial the *minimum polynomial* of α and denote it by $m_\alpha(X)$.

Example 5.2.

(1) If r is a rational number then $m_r(X) = X - r$, a polynomial of degree 1. More interestingly, $m_{\sqrt{2}}(X) = X^2 - 2$ and $m_{\sqrt{3}}(X) = X^2 - 3$. Also, $m_{\sqrt{2}\sqrt{3}}(X) = X^2 - 6$ and $m_{\sqrt{2}+\sqrt{3}}(X) = X^4 - 10X^2 + 1$. Finally, $m_{\sqrt[3]{2}}(X) = X^3 - 2$.

(2) It is relatively easy to show that there are transcendental numbers, but not so easy to show that any particular number is transcendental. It is a famous theorem of Hermite that e is transcendental and a famous theorem of Lindemann that π is transcendental, and a case of the famous Gelfond-Schneider theorem that $2^{\sqrt{2}}$ is transcendental.

Let $\mathbf{K}$ be any subfield of the field of complex numbers $\mathbf{C}$. Then $\mathbf{K}$ is a vector space over $\mathbf{Q}$ and its dimension is called the *degree* of $\mathbf{K}$, denoted $\deg_{\mathbf{K}/\mathbf{Q}}$.

Definition 5.3. A subfield $\mathbf{K}$ of $\mathbf{C}$ is an *algebraic number field* if its degree $\deg_{\mathbf{K}/\mathbf{Q}}$ is finite.

At first glance it may not be obvious what this has to do with $\mathbf{K}$ being "algebraic," but here is the connection.

Lemma 5.4. *Let $\mathbf{K}$ be an algebraic number field. Then every element α of $\mathbf{K}$ is an algebraic number.*

Proof: Let $n = \deg_{\mathbf{K}/\mathbf{Q}}$ and consider $\{1, \alpha, \ldots, \alpha^n\}$. This is a set of $n+1$ elements in an n-dimensional vector space, so it must be linearly dependent. Thus, there are rational numbers $a_0, a_1, \ldots, a_n$ with $a_n\alpha^n + \ldots + a_1\alpha + a_0 = 0$, and then α is a root of the polynomial $f(X) = a_nX^n + \ldots + a_1X + a_0$. □

Now that we know what an algebraic number field is, we need to know what an algebraic integer is.

Definition 5.5. Let $\mathbf{K}$ be an algebraic number field. An element α of $\mathbf{K}$ is an *algebraic integer* if $m_\alpha(X)$ is a polynomial with integer coefficients.

Lemma 5.6. *Let $\mathbf{K}$ be an algebraic number field. If α and β are algebraic integers in $\mathbf{K}$, then $\alpha + \beta$ and $\alpha\beta$ are algebraic integers in $\mathbf{K}$.*

Note that this lemma is far from obvious. It is a nontrivial fact that if $m_\alpha(X)$ has integer coefficients and $m_\beta(X)$ has integer coefficients, then $m_{\alpha+\beta}(X)$ and $m_{\alpha\beta}(X)$ have integer coefficients, but it turns out to be true. But having this lemma we may make the following definition.

Definition 5.7. The *ring of integers* $\mathcal{O}(\mathbf{K})$ of $\mathbf{K}$ is

$$\mathcal{O}(\mathbf{K}) = \{\alpha \text{ in } \mathbf{K} \mid \alpha \text{ is an algebraic integer}\}.$$

Example 5.8. Referring back to Example 5.2, we see that $\sqrt{2}$, $\sqrt{3}$, $\sqrt{2}\sqrt{3}$, $\sqrt{2}+\sqrt{3}$, and $\sqrt[3]{2}$ are all algebraic integers.

Remark 5.9. For any algebraic number field $\mathbf{K}$, $\mathcal{O}(\mathbf{K}) \cap \mathbf{Q} = \mathbb{Z}$. For if r is in $\mathbf{Q}$, then $m_r(X) = X - r$, and this polynomial has integer coefficients if and only if r is in $\mathbb{Z}$.

Remark 5.10. It is important to distinguish between the integers in $\mathbf{Q}$ and the (algebraic) integers in $\mathbf{K}$. We shall always use the term "integer" to refer to an element of $\mathbb{Z}$ and the term "algebraic integer" to refer to an element of $\mathcal{O}(\mathbf{K})$. Also, we will always use the term "prime" to refer to a prime number in $\mathbb{Z}$. We should point out that it is often the case in the literature that, in order to emphasize the distinction, the elements of $\mathbb{Z}$ are called the "rational integers" and the primes in $\mathbb{Z}$ are called the "rational primes."

Lemma 5.11. *Let $\mathbf{K}$ be an algebraic number field and let α be any element of $\mathbf{K}$. Then there is an integer m such that $m\alpha$ is an algebraic integer.*

Proof: Let $m_\alpha(X) = X^d + a_{d-1}X^{d-1} + \ldots + a_1X + a_0$ and let each a_i be the rational number $a_i = r_i/s_i$. Let $m = \operatorname{lcm}(s_0, \ldots, s_{d-1})$ and let $\beta = m\alpha$. Then $\alpha = \beta/m$, so $0 = m_\alpha(\alpha) = m_\alpha(\beta/m) = [\beta/m]^d + a_{d-1}[\beta/m]^{d-1} + \ldots + a_1[\beta/m] + a_0$. Multiplying through by m^d we see that $\beta^d + a_{d-1}m\beta^{d-1} + \ldots + a_1m^{d-1}\beta + a_0m^d = 0$, so $m_\beta(X) = X^d + a_{d-1}mX^{d-1} + \ldots + a_1m^{d-1}X + a_0m^d$ and all of the coefficients of $m_\beta(X)$ are integers. $\square$

Using this lemma we may find the additive structure of $\mathcal{O}(\mathbf{K})$.

Proposition 5.12. *Let* $\mathbf{K}$ *be an algebraic number field. Then* $\mathcal{O}(\mathbf{K})$ *is a free* $\mathbb{Z}$*-module of rank* $\deg_{\mathbf{K}/\mathbf{Q}}$.

Proof: Let $\deg_{\mathbf{K}/\mathbf{Q}} = n$ and let $\{\alpha_1, \ldots, \alpha_n\}$ be a basis for $\mathbf{K}$ as a vector space over $\mathbf{Q}$. Then, replacing α_i by $\beta_i = m_i\alpha_i$ for a suitable integer m_i, we have a new basis of $\mathbf{K}$ given by $\{\beta_1, \ldots, \beta_n\}$ with each β_i in $\mathcal{O}(\mathbf{K})$. Thus $\mathcal{O}(\mathbf{K})$ contains a free $\mathbb{Z}$-module of rank n.

On the other hand, if $\{\gamma_1, \ldots, \gamma_{n+1}\}$ is any set of $n+1$ elements in $\mathcal{O}(\mathbf{K})$, there is a linear relation $c_1\gamma_1 + \ldots + c_{n+1}\gamma_{n+1} = 0$ with each c_i in $\mathbf{Q}$. Multiplying through by the lcm of the denominators of the c_i's gives a linear relation $d_1\gamma_1 + \ldots + d_{n+1}\gamma_{n+1} = 0$ with each d_i in $\mathbb{Z}$, so $\mathcal{O}(\mathbf{K})$ does not contain a free $\mathbb{Z}$-module of rank $n+1$.

Hence $\mathcal{O}(\mathbf{K})$ must be a free $\mathbb{Z}$-module of rank n. □

Example 5.13. An algebraic number field $\mathbf{K}$ of degree 2 is called a *quadratic field*. Any quadratic field $\mathbf{K}$ must be $\mathbf{Q}(\sqrt{D})$ for some square-free integer $D \neq 1$. We let $\mathcal{O}(\sqrt{D})$ denote $\mathcal{O}(\mathbf{Q}(\sqrt{D}))$.

If $D \equiv 2$ or $3 \pmod 4$, then

$$\mathcal{O}(\sqrt{D}) = \{a + b\sqrt{D} \mid a \text{ and } b \text{ are integers }\}.$$

In this case, we see that $\mathcal{O}(\sqrt{D})$ is a free $\mathbb{Z}$-module of rank 2 with basis $\{1, \sqrt{D}\}$.

If $D \equiv 1 \pmod 4$, then

$$\mathcal{O}(\sqrt{D}) = \{[a + b\sqrt{D}]/2 \mid a \text{ and } b \text{ are integers, and either they are both even or they are both odd}\}.$$

In this case, we see that $\mathcal{O}(\sqrt{D})$ is a free $\mathbb{Z}$-module of rank 2 with basis $\{1, [1 + \sqrt{D}]/2\}$.

Remark 5.14. We defined an algebraic integer α in Definition 5.5 as an algebraic number for which the polynomial $m_\alpha(X)$ is a monic polynomial with integer coefficients. Of course, $m_\alpha(X)$ is a particular polynomial with $m_\alpha(\alpha) = 0$. There is an alternate definition of an algebraic integer: α is an algebraic integer if there is *some* monic polynomial with integer coefficients $f(X)$ with $f(\alpha) = 0$. It is a theorem (a consequence of Gauss's Lemma for polynomials) that these two definitions are equivalent.

5.2 Ideal Theory

Let us return to the consideration of a general integral domain R. We have defined the notion of an ideal of R in Definition 2.28, and the notion of a principal ideal of R in Definition 2.29.

First we introduce a bit of (standard) notation. We have denoted the principal ideal generated by α as I_α. This notation has the disadvantage that the interesting information is relegated to the subscript. So, instead, we will denote it by (α). In Definition 2.34, we denoted the ideal generated by the set $\mathcal{A} = \{\alpha_1, ..., \alpha_k\}$ by $I_\mathcal{A}$. This notation has the same disadvantage, so we will adopt the same solution, denoting it instead by $(\alpha_1, \ldots, \alpha_k)$.

We certainly know how to multiply elements of R and we certainly know what it means for one element of R to divide another. Guided by this knowledge, we shall seek the analogs for ideals.

Consider the principal ideal (α) consisting of multiples of the element α and the principal ideal (β) consisting of multiples of the element β. Let $\gamma = \alpha\beta$ be the product of these two elements, and consider the principal ideal (γ). We would like to have a definition of multiplication of ideals that makes $(\gamma) = (\alpha)(\beta)$. Let us see what that might be.

Let α' be an arbitrary element of the principal ideal (α), so $\alpha' = \alpha\lambda$ for some element λ of R. Similarly, let β' be an arbitrary element of the principal ideal (β), so $\beta' = \beta\mu$ for some element μ of R. Then

$$\alpha'\beta' = (\alpha\lambda)(\beta\mu) = (\alpha\beta)(\lambda\mu) = \gamma\nu \text{ where } \nu = \lambda\mu.$$

Thus, we conclude that $\alpha'\beta'$ is an element of (γ). But we can also see that every element of (γ) is of this form. For if γ' is an arbitrary element of (γ), then $\gamma' = \gamma\nu$ for some element ν of R, and then

$$\gamma' = \gamma\nu = (\alpha\beta)\nu = \alpha(\beta\nu) \text{ (or } = (\alpha\nu)\beta\text{)}$$

is the product of an element of (α) and an element of (β). Guided by this, we are tempted to define the product of two ideals as follows: Let I and J be ideals of R. Then their product $K = IJ$ is the ideal consisting of all elements of R of the form $\alpha'\beta'$ where α' is any element of I and β' is any element of J. But examining the definition of an ideal (Definition 2.28) we see that this does not quite work (although, as you can verify, it does work for principal ideals).

Definition 2.28 tells us that if K is an ideal, it must satisfy two properties. The second property is that if γ' is any element of K and ν is any element of R, then $\gamma'\nu$ must also be an element of K. This is no problem: since

γ' is in IJ, it is of the form $\gamma' = \alpha'\beta'$ and so $\gamma'\nu = (\alpha'\beta')\nu = \alpha'\beta''$ where $\beta'' = \beta'\nu$. Then α' is certainly an element of I and β'' is an element of J (since J is an ideal), so $\gamma'\nu$ is of the required form. However, the first property does not hold, in general. Suppose γ_1' and γ_2' are two elements of the required form, $\gamma_1' = \alpha_1'\beta_1'$ and $\gamma_2' = \alpha_2'\beta_2'$. Then $\gamma_3' = \gamma_1' + \gamma_2' = \alpha_1'\beta_1' + \alpha_2'\beta_2'$, and there is no reason to expect that this sum can be written in the form $\alpha_3'\beta_3'$.

So instead we modify our definition in the simplest possible way in order to force the first property to work as well.

Definition 5.15. Let I and J be two ideals of R. Their *product* is the ideal IJ given by

$$IJ = \{\sum \alpha_i\beta_i \mid \text{ each } \alpha_i \text{ is in } I_1 \text{ and each } \beta_i \text{ is in } I_2, \text{ and the sum is finite}\}.$$

Of course, we must check that this does define an ideal, and this is indeed the case. The proof of this is similar to the proof of Lemma 2.35.

We leave the proof of the following useful lemma to the reader.

Lemma 5.16. *Let R be an integral domain and let $I = (\alpha_1, \ldots, \alpha_i)$ and $J = (\beta_1, \ldots, \beta_j)$ be ideals in R.*

(1) $I \subseteq J$ if and only if α_k is an element of J for each $k = 1, \ldots, i$. Consequently, $I = J$ if and only if α_k is an element of J for each $k = 1, \ldots, i$ and β_k is an element of I for each $k = 1, \ldots, j$.

(2) Let $K = IJ$ be the product of the ideals I and J. Then

$$K = (\alpha_1\beta_1, \ldots, \alpha_1\beta_j, \alpha_2\beta_1, \ldots, \alpha_2\beta_j, \ldots, \alpha_i\beta_1, \ldots, \alpha_i\beta_j).$$

Remark 5.17. Note that $R = (1)$ and that $IR = I$ for any ideal I of R. Also note that multiplication of ideals is certainly commutative, and that Lemma 5.16(2) implies that multiplication of ideals is also associative.

Now let us think about the analog of divisibility. Suppose that the element α of R divides the element β of R. Then β is a multiple of α, and then every multiple of β is a multiple of α. In other words, the principal ideal (α) contains the principal ideal (β). So, here we do not need to make another definition. We simply regard the analog of divisibility as being that the ideal I contains the ideal J, $I \supseteq J$. Now recall from Definition 2.45 that an element α of R is a prime if α dividing $\beta\gamma$ implies that α divides β or α divides γ. With this analogy and definition in mind, we can define a prime ideal in general.

Definition 5.18. A proper ideal P of R is a *prime ideal* if, whenever J and K are ideals of P with $P \supseteq JK$, then $P \supseteq J$ or $P \supseteq K$.

We just noted that the analog of α dividing β for two ideals I and J was that $I \supseteq J$. But we could have asked for something stronger. Namely, if α divides β, then $\beta = \alpha\gamma$ for some γ. Thus, passing to ideals, we might ask that the analog be that there is an ideal K with $J = IK$. This is indeed stronger in general, as the next proposition shows.

Proposition 5.19. *Let I and J be ideals in an integral domain R. If there is an ideal K with $J = IK$, then $I \supseteq J$.*

Proof: By Definition 5.15, $J = IK$ consists of linear combinations of elements of the form $\alpha\beta$, with α in I and β in K. But by the definition of an ideal, every element of this form, and also every linear combination of elements of this form, is in I. □

We need to introduce one further property of ideals.

Definition 5.20. Let I be a proper ideal in an integral domain R. If $J \supseteq I$ for some ideal J implies that $J = I$ or $J = R$, then I is a *maximal* ideal.

Being maximal is a stronger condition than being prime, as the next proposition shows.

Proposition 5.21. *Let I be an ideal in an integral domain R. If I is a maximal ideal, then I is a prime ideal.*

Proof: Let I be a maximal ideal, and suppose that $I \supseteq JK$. We need to show that $I \supseteq J$ or $I \supseteq K$. We prove this by contradiction.

Assume this is not the case. Then there is an element α of J that is not in I, and an element β of K that is not in I. But, since $I \supseteq JK$, the element $\gamma = \alpha\beta$ is in I. Let I' be the ideal generated by α and I, $I' = \{\alpha\zeta + \delta \mid \zeta \text{ in } R, \delta \text{ in } I\}$. Certainly $I' \supseteq I$, and indeed $I' \supset I$, as α is in I' but not in I. Hence, by the definition of a maximal ideal, $I' = R$. In particular, 1 is in I', so $1 = \alpha\zeta_0 + \delta_0$ for some element ζ_0 of R and some element δ_0 of I. But then

$$\beta = \beta(\alpha\zeta_0 + \delta_0) = (\alpha\beta)\zeta_0 + \delta_0\beta = \gamma\zeta_0 + \delta_0\beta$$

is an element of I by the definition of an ideal, which is a contradiction. □

5.3 Dedekind Domains

We now single out a key class of integral domains, the Dedekind domains. The definition of a Dedekind domain involves three conditions. We will begin by stating the definition and then we will explain what these conditions mean.

Definition 5.22. An integral domain R is a *Dedekind domain* if

(1) every prime ideal of R is maximal;

(2) R satisfies the ascending chain condition;

(3) R is integrally closed in its quotient field.

We know what the first condition means. In Proposition 5.21 we proved that for an integral domain R, every maximal ideal R is prime. Here we want the converse to be true as well, so that the prime ideals are exactly the maximal ideals.

Lemma 5.23. *Every prime ideal of* $\mathbb{Z}$ *is maximal.*

Proof: We know that every ideal I of $\mathbb{Z}$ consists of the multiples of some integer i, $I = (i)$.

Suppose that $I = (i)$ is not maximal. Then there is a proper ideal J of $\mathbb{Z}$ with $J \supset I$. Then we must have $J = (j)$ for some integer $j \neq \pm 1$ and $j \neq \pm i$. Since $I \subseteq J$, $i = jk$ for some integer k, with $k \neq \pm 1$ and $k \neq \pm i$. Let $K = (k)$. Then $I = JK \supseteq JK$ but $I \not\supseteq J$ and $I \not\supseteq K$, so I is not a prime ideal. □

Definition 5.24. An integral domain R satisfies the *ascending chain condition* (ACC), or is *noetherian*, if every sequence of ideals $I_1 \subset I_2 \subset I_3 \ldots$ of R is finite.

Lemma 5.25. $\mathbb{Z}$ *satisfies the ascending chain condition.*

Proof: Let $I_1 \subset I_2 \subset I_3 \ldots$ be a sequence of ideals in $\mathbb{Z}$. Then $I_1 = (i_1)$, $I_2 = (i_2)$, $I_3 = (i_3), \ldots$, with $|i_1| > |i_2| > |i_3| > \ldots$. Thus $\{|i_1|, |i_2|, |i_3|, \ldots\}$ is a strictly decreasing sequence of nonnegative integers, which must be finite. □

The last condition in the definition of a Dedekind domain is considerably more subtle. To begin with, we must define the quotient field of an integral domain R. This definition mimics the construction of the rational numbers $\mathbf{Q}$ from the integers $\mathbb{Z}$.

Definition 5.26. Let R be an integral domain. Its quotient field $\mathbf{F}$ is given by

$$\mathbf{F} = \{a/b \mid a \text{ and } b \text{ are in } R, \text{ and } b \neq 0\},$$

with $a/b = c/d$ in $\mathbf{F}$ if $ad = bc$ in R. We define addition and multiplication in $\mathbf{F}$ by

$$a/b + c/d = [ad + bc]/[bd], \qquad [a/b][c/d] = [ac]/[bd],$$

and we regard R as a subset of $\mathbf{F}$ by identifying a with $a/1$.

(It is easy to check that $\mathbf{F}$ is a field, and that the "usual" laws of fractions work in $\mathbf{F}$. The reader may worry that the definition of $\mathbf{F}$ is circular in that we are already assuming that we can divide elements of R in defining $\mathbf{F}$, but this is not the case. In the definition of $\mathbf{F}$ "/" is just a symbol, but once $\mathbf{F}$ is defined, we can give it its usual interpretation, so that $b[a/b] = a$.)

Example 5.27.

(1) If $R = \mathbb{Z}$, its quotient field is $\mathbf{Q}$.

(2) Let $\mathbf{K}$ be an algebraic number field and let $R = \mathcal{O}(\mathbf{K})$ be the ring of integers of $\mathbf{K}$. Then $\mathbf{K}$ is the quotient field of R. To see this, let α be any element of $\mathbf{K}$. By Lemma 5.11, there is an integer n with $\beta = n\alpha$ in R, and then $\alpha = \beta/n$, and n is in R by Remark 5.9.

To state the last condition properly, we need to slightly generalize the notion of an algebraic integer. To this end, let Z be any subring of $\mathbf{Q}$. (Note that $Z \supseteq \mathbb{Z}$ as Z must contain 1.) Then an element α of an algebraic number field $\mathbf{K}$ is *Z-integral* if its minimum polynomial $m_\alpha(X)$ has all of its coefficients in Z.

Definition 5.28. Let R be a subring of an algebraic number field $\mathbf{K}$ and let $Z = R \cap \mathbf{Q}$. The *integral closure* of R in $\mathbf{K}$ is $S = \{\alpha \text{ in } \mathbf{K} \mid \alpha \text{ is } Z\text{-integral}\}$. R is *integrally closed* in $\mathbf{K}$ if $R = S$.

Lemma 5.29. *For any algebraic number field $\mathbf{K}$, $R = \mathcal{O}(\mathbf{K})$ is integrally closed in $\mathbf{K}$.*

Proof: By Remark 5.9, $Z = R \cap \mathbf{Q} = \mathbb{Z}$, so an element α of $\mathbf{K}$ is Z-integral if and only if it is $\mathbb{Z}$-integral. But this is exactly the definition of an algebraic integer. □

This definition seems like virtually a tautology, and to understand it we should see how it can possibly go wrong.

Example 5.30.

(1) Let $\mathbf{K} = \mathbf{Q}(\sqrt{D})$. Fix an integer $m > 1$ and let

$$R' = \{a + b\sqrt{D} \mid a \text{ and } b \text{ are integers and } b \text{ is divisible by } m\}.$$

Note that for any element β of $R = \mathcal{O}(\mathbf{K})$, $\beta' = 2m\beta$ is in R'. (The factor of 2 is to take care of the case when $D \equiv 1 \pmod 4$.) As we have seen in Example 5.27, for any element α of $\mathbf{K}$, there is an element β of R and an integer n with $\alpha = \beta/n$. But then $\alpha = \beta'/[2mn]$, so $\mathbf{K}$ is also the quotient field of R'. But of course in this case $S = R \supset R'$. (This example may seem artificial but in fact is not. Let $D' = m^2 D$. Then $\mathbf{Q}(\sqrt{D'}) = \mathbf{Q}(\sqrt{D})$ and we might naturally have been led to R' if we had considered $\mathbf{Q}(\sqrt{D'})$.)

(2) Here is a very important example. Let $\mathbf{K} = \mathbf{Q}(\sqrt{D})$ where $D \equiv 1 \pmod 4$ and let

$$R' = \{a + b\sqrt{D} \mid a \text{ and } b \text{ are integers}\}.$$

Then for any element β of $R = \mathcal{O}(\mathbf{K})$, $\beta' = 2m\beta$ is in R', so, by the same argument as above, we see that $\mathbf{K}$ is the quotient field of R' but $S = R \supset R'$. Thus, while in this case one might have naively thought of R' as the right ring to consider, it is *not*.

Definition 5.31. Let R be an integral domain with quotient field $\mathbf{K}$. A subset I of $\mathbf{K}$ is a *fractional ideal* of $\mathbf{K}$ if there is an element γ_0 of R such that $J = \gamma_0 I$ is an ideal of R, where $J = \gamma_0 I = \{\gamma_0\beta \mid \beta \text{ in } I\}$.

Observe that every ideal I of R is a fractional ideal of $\mathbf{K}$, as in this case we may choose $\gamma_0 = 1$. Also observe that the definition of a fractional ideal is a very natural one, as in this case $I = [1/\gamma_0]J$. From this we can easily see that if I_1 and I_2 are fractional ideals of $\mathbf{K}$, then $I_1 I_2$ is also a fractional ideal of $\mathbf{K}$. Finally, we can easily generalize the notion of a principal ideal. For α_0 any element of $\mathbf{K}$, we let $(\alpha_0) = \{\alpha_0\beta \mid \beta \text{ in } R\}$ and call (α_0) a principal fractional ideal.

Definition 5.32. Let R be an integral domain with quotient field $\mathbf{K}$. Let I be a nonzero fractional ideal of $\mathbf{K}$. Then I^{-1} is defined by

$$I^{-1} = \{\alpha \text{ in } \mathbf{K} \mid \alpha I \subseteq R\}.$$

By definition, $II^{-1} \subseteq R$. The following is a key technical lemma, which we shall not prove.

Lemma 5.33. *Let R be an integral domain with quotient field $\mathbf{K}$. Let I be a nonzero fractional ideal of $\mathbf{K}$. If R is a Dedekind domain, then I^{-1} is a fractional ideal of $\mathbf{K}$ with $II^{-1} = R$.*

We mentioned that, for ideals I, J, and K of an integral domain R, the condition that $I = JK$ is stronger than the conditions that $I \subseteq J$, but these conditions are not in general equivalent. But they *are* equivalent for Dedekind domains.

Corollary 5.34. *Let R be a Dedekind domain and let I and J be ideals of R with $I \subseteq J$. Then there is an ideal K of R with $I = JK$.*

Proof: Let $K = J^{-1}I$. Then K is a fractional ideal of $\mathbf{K}$, and, since $I \subseteq J$, $K = J^{-1}I \subseteq J^{-1}J = R$, so K is in fact an ideal of R. But then, since multiplication of fractional ideals is associative (as you can easily check), $JK = JJ^{-1}I = I$. □

Here is the main general result about Dedekind domains, and the reason why we introduced them.

Theorem 5.35. *Let R be an integral domain. The following are equivalent.*

(1) R is a Dedekind domain.

(2) Every nonzero ideal I of R can be factored essentially uniquely into a product of prime ideals, i.e., $I = P_1^{e_1} P_2^{e_2} \cdots P_k^{e_k}$ with each P_i a prime ideal, with $P_i \neq P_j$ for $i \neq j$, and with $e_1, e_2, \dots e_k$ positive integers, and if also $I = Q_1^{f_1} Q_2^{f_2} \cdots Q_\ell^{f_\ell}$ with each Q_i a prime ideal, with $Q_i \neq Q_j$ for $i \neq j$ and with $f_1, f_2, \dots f_\ell$ positive integers, then $\ell = k$, and, after possible reordering, $Q_i = P_i$ and $f_i = e_i$ for each i.

(In this case, we can also factor every nonzero fractional ideal essentially uniquely as $I = P_1^{e_1} P_2^{e_2} \cdots P_{k_1}^{e_{k_1}} P_{k_1+1}^{e_{k_1+1}} P_{k_1+2}^{e_{k_1+2}} \cdots P_{k_1+k_2}^{e_{k_1+k_2}}$ with the P_i's mutually distinct prime ideals and with $e_1, \dots, e_{k_1}$ positive integers and $e_{k_1+1}, \dots, e_{k_1+k_2}$ negative integers.)

Corollary 5.36. *Let R be a Dedekind domain. Then R is a PID if and only if R is a UFD.*

Proof: In this proof we will be very careful to distinguish between prime elements and prime ideals. We begin by recalling that an element of a UFD is a prime element if and only if it is an irreducible element.

We know that every PID is a UFD. Thus, what we must prove here is that every Dedekind domain that is a UFD is also a PID. Thus, let R be a Dedekind domain that is a UFD. We want to show that every ideal of R is principal. Since, by Theorem 5.35, every nonzero ideal is a product of prime ideals, it clearly suffices to show that every prime ideal of R is principal.

Let P be a prime ideal of R. Let α be any nonzero element of P. Then we can factor α into a product of prime elements $\alpha = \pi_1^{e_1} \dots \pi_k^{e_k}$. Then $P \supseteq (\alpha) = (\pi_1)^{e_1} \dots (\pi_k)^{e_k}$ so by the definition of a prime ideal, $P \supseteq (\pi_i)$ for some i. To simplify the notation, let $\pi = \pi_i$. We will show that (π) is a maximal ideal. Then, since $P \supseteq (\pi)$, in fact $P = (\pi)$, a principal ideal.

Thus, let Q be any ideal of R with $Q \supseteq (\pi)$. We need to show $Q = R$. Choose any element ρ of Q that is not in (π). Then $Q \supseteq (\pi, \rho) \supset (\pi)$. By Corollary 5.34 there is an ideal Q' of R with $QQ' = (\pi)$. Let σ be any element of Q'. Then $\rho\sigma$ is an element of (π), so the element $\rho\sigma$ is divisible by the element π. But π is a prime element that does not divide the element ρ, so π must divide the element σ, and thus σ is an element of the ideal (π). Thus we have

$$QQ' = (\pi) \qquad \text{and} \qquad Q' \subseteq (\pi).$$

We certainly have that $QQ' \subseteq Q'$, so we see that we must have $Q' = (\pi)$, and then $Q(\pi) = (\pi)$. This readily implies that 1 is in Q, and hence $Q = R$. □

5.4 Algebraic Number Fields and Dedekind Domains

We are about to achieve one of our major goals. We began by investigating unique factorization of elements. We saw that this property holds in $\mathbb{Z}$ but saw that it may or may not hold in rings of integers in particular quadratic fields. The appropriate generalization of unique factorization is not unique factorization of elements, but rather unique factorization of ideals, and we are about to see that this property holds in the ring of integers of *every* algebraic number field.

Theorem 5.37. *Let* $\mathbf{K}$ *be an algebraic number field and let* $R = \mathcal{O}(\mathbf{K})$ *be the ring of integers in* $\mathbf{K}$. *Then* R *is a Dedekind domain.*

Corollary 5.38 (Dedekind). *Let* $\mathbf{K}$ *be an algebraic number field and let* $R = \mathcal{O}(\mathbf{K})$ *be the ring of integers in* $\mathbf{K}$. *Then every nonzero ideal* I *of* R *can be factored essentially uniquely into a product of prime ideals.*

Proof: This is immediate from Theorem 5.35 and Theorem 5.37. □

In order to prove Theorem 5.37, we shall have to introduce the notion of the norm of an ideal of R, a notion that is important in its own right. As a matter of notation, for a set S, we let $\#(S)$ denote the cardinality of S, i.e., the number of elements of S.

Lemma 5.39. *Let* $\mathbf{K}$ *be an algebraic number field and let* $R = \mathcal{O}(\mathbf{K})$. *Let* $n = \deg_{\mathbf{K}/\mathbf{Q}}$.

(1) If $I = (m)$, *the principal ideal generated by the nonzero integer* m, *then* $\#(R/I) = |m|^n$.

(2) If I *is any ideal of* R, *then* $\#(R/I)$ *is finite.*

Proof:

(1) By Proposition 5.12, we know that R is a free $\mathbb{Z}$-module of rank n. Let $\{\alpha_1, \ldots, \alpha_n\}$ be a basis for R. Then I has basis $\{m\alpha_1, \ldots, m\alpha_n\}$ and we see that R/I is isomorphic (as a $\mathbb{Z}$-module) to $[\mathbb{Z}/m\mathbb{Z}]^n$, a set of cardinality $|m|^n$.

(2) We claim that I contains a nonzero integer m. To see this, choose any nonzero element α and consider its minimum polynomial $m_\alpha(X)$. We know that $m_\alpha(X) = X^d + a_{d-1}X^{d-1} + \ldots + a_1X + a_0$ with all the coefficients integers, and $m_\alpha(\alpha) = 0$. Note that $a_0 \neq 0$ as otherwise $m_\alpha(X)$ would have X as a factor, and $m_\alpha(X)/X$ would be a polynomial of lower degree having α as a root. But then we can solve for a_0:

$$a_0 = -\alpha^d - a_{d-1}\alpha^{d-1} - \ldots - a_1\alpha,$$

and $m = a_0$ is in I.

Let $J = (m)$. Then $I \supseteq J$, so R/I is a quotient of R/J. But by (1), R/J is a finite set, so R/I must be a finite set as well. □

Definition 5.40. Let $\mathbf{K}$ be an algebraic number field, let $R = \mathcal{O}(\mathbf{K})$, and let I be any ideal of R. The *norm* $\|I\|$ of the ideal I is $\|I\| = \#(R/I)$.

Proof of Theorem 5.37: We must verify that R satisfies the three conditions for a Dedekind domain in Definition 5.22.

(1) Here we use some general theory. In general, an ideal I of a commutative ring R is maximal if and only if R/I is a field, and is prime if and only if R/I is an integral domain. But every finite integral domain is a field. Putting these implications together, we have the following: Let I be a prime ideal of R. Then R/I is a finite integral domain, hence a field, and hence I is a maximal ideal of R.

(2) Observe that for any two ideals I and J of R with I a proper subset of J, R/J is a proper quotient of R/I, and hence $\|J\|$ divides $\|I\|$. Now let $I_1 \subset I_2 \subset I_3 \ldots$ be a sequence of ideals in R. Then $\|I_1\| > \|I_2\| > \|I_3\| > \ldots$ is a strictly decreasing sequence of positive integers and so must be finite.

(3) We proved this in Lemma 5.29. □

Lemma 5.41. *Let* $\mathbf{K}$ *be an algebraic number field. Let*

$$\mathcal{I}(\mathbf{K}) = \{\textit{nonzero fractional ideals of } \mathbf{K}\}$$

and let

$$\mathcal{I}_{\mathrm{Prin}}(\mathbf{K}) = \{\textit{nonzero principal fractional ideals of } \mathbf{K}\}.$$

Then $\mathcal{I}(\mathbf{K})$ *is an abelian group under multiplication of fractional ideals, and* $\mathcal{I}_{\mathrm{Prin}}(\mathbf{K})$ *is a subgroup of* $\mathcal{I}(\mathbf{K})$.

Proof: Clearly, multiplication of fractional ideals is commutative and associative, and $\mathcal{O}(\mathbf{K}) = (1)$ is an identity. Inverses are given by Lemma 5.33: the inverse in $\mathcal{I}(\mathbf{K})$ of the fractional ideal I is the fractional ideal I^{-1} of Definition 5.32. Clearly, the subset $\mathcal{I}_{\mathrm{Prin}}(\mathbf{K})$ is closed under multiplication and taking inverses, so is a subgroup. □

Definition 5.42. Let $\mathbf{K}$ be an algebraic number field. The *ideal class group* of $\mathbf{K}$ is the quotient group

$$\mathcal{C}(\mathbf{K}) = \mathcal{I}(\mathbf{K})/\mathcal{I}_{\mathrm{Prin}}(\mathbf{K}).$$

Here is one of the fundamental theorems of algebraic number theory.

Theorem 5.43 (Minkowski). *For any algebraic number field* $\mathbf{K}$, $\mathcal{C}(\mathbf{K})$ *is finite.*

Given this theorem, we may make the following definition.

Definition 5.44. Let $\mathbf{K}$ be an algebraic number field. The *class number* of $\mathbf{K}$ is

$$h(\mathbf{K}) = \#(\mathcal{C}(\mathbf{K})),$$

the order of the ideal class group of $\mathbf{K}$.

Corollary 5.45. *Let* $\mathbf{K}$ *be an algebraic number field. The following are equivalent:*

(1) $h(\mathbf{K}) = 1$.

(2) $\mathcal{O}(\mathbf{K})$ *is a PID.*

(3) $\mathcal{O}(\mathbf{K})$ *is a UFD.*

Proof: (1) and (2) are equivalent by definition, and (2) and (3) are equivalent by Corollary 5.36. □

We have the projection map $\pi : \mathcal{I}(\mathbf{K}) \to \mathcal{C}(\mathbf{K})$ and we let $[I] = \pi(I)$, and call $[I]$ the ideal class of the ideal I. Thus the ideal class $[I]$ of I is trivial if and only if I is a principal ideal.

Theorem 5.43 and Corollary 5.45 tell us that, for any algebraic number field $\mathbf{K}$, even if unique factorization of elements does not hold in $\mathcal{O}(\mathbf{K})$, in some sense it only misses by a finite amount.

Remark 5.46. There is an effective procedure for finding $\mathcal{C}(\mathbf{K})$ for any algebraic number field $\mathbf{K}$.

We conclude this section with a result that we record for future reference.

Lemma 5.47. *Let* $\mathbf{K}$ *be an algebraic number field and let* I *and* J *be any two ideals of* $R = \mathcal{O}(\mathbf{K})$. *Then* $\|IJ\| = \|I\| \cdot \|J\|$.

Proof: From abelian group theory, we know that R/I is isomorphic to the quotient of R/IJ by I/IJ, so $\#(R/IJ) = \#(R/I) \cdot \#(I/IJ)$. Since R is a Dedekind domain, I/IJ is isomorphic to R/J. □

5.5 Prime Ideals in $\mathcal{O}(\sqrt{D})$

In this section we determine, and describe, all the prime ideals in $R = \mathcal{O}(\sqrt{D})$. This is a long procedure, and so we proceed in stages.

There is one slight complication. In most cases, an element α of R is of the form $\alpha = a + b\sqrt{D}$, and it is clear what we mean by saying p divides a or p divides b. However, if $D \equiv 1 \pmod 4$, then a and b may be half-integers, $a = a'/2$ and $b = b'/2$ with a' and b' odd integers. In this case, for p an odd prime, we will say that p divides a or b if p divides a' or p divides b', and all the arguments go through unchanged. (This is because the denominator 2 is relatively prime to p.) But the case $D \equiv 1 \pmod 4$ and $p = 2$ will require a special argument in some places.

Lemma 5.48. *Let α_0 be any element of R of norm p, where p is a prime. If β is any element of R with norm n divisible by p, then β is divisible by α_0 or by $\overline{\alpha_0}$ in R.*

Proof: Let $\alpha_0 = a_0 + b_0\sqrt{D}$. Then $a_0^2 - b_0^2 D = \pm p$ gives $a_0^2 \equiv b_0^2 D \pmod p$. Note that $b_0 \not\equiv 0 \pmod p$. Let k_0 be a solution of $k_0^2 \equiv D \pmod p$, and note that k_0 is unique up to sign. Then $a_0 \equiv \pm k_0 b_0 \pmod p$.

Suppose that $d \not\equiv 0 \pmod p$. Then, by the same logic, $c \equiv \pm k_0 d \pmod p$. Replacing k_0 by $-k_0$ if necessary, and replacing α_0 by $\overline{\alpha_0}$, if necessary, which switches the sign of b_0, we may assume that $a_0 \equiv k_0 b_0 \pmod p$ and $c \equiv k_0 d \pmod p$. Direct calculation then shows that $\beta/\alpha_0 = \pm[e + f\sqrt{D}]/p$ where $e = ac - bdD$ and $f = bc - ad$. Now $e \equiv k_0 b k_0 d - bdD \equiv bd[k_0^2 - D] \equiv 0 \pmod p$ and $f \equiv k_0 dc - k_0 cd \equiv 0 \pmod p$ so β/α_0 is in R.

If $d \equiv 0 \pmod p$ then $c \equiv 0 \pmod p$ so $\beta = p\beta'$ with β' in R, and then certainly β is divisible by α_0 in R. □

Lemma 5.49. *Let I be any ideal in R. Then $I = (\alpha)$ for some α in R, or $I = (g, \alpha)$ for some integer g and some element α of R with $\|\alpha\|$ divisible by g.*

Proof: If $I = (0)$ we are done, and if I is a principal ideal then $I = (\alpha)$ for some α in R and we are done.

Suppose $I = (n, \alpha)$ for some integer n. Let $g = \gcd(n, \|\alpha\|)$. Then for some integers a and b, $g = na + \|\alpha\|b = na \pm \alpha\overline{\alpha}b$ is in I, and then $I = (g, \alpha)$ and we are done.

Thus, to complete the proof we must show that every nonzero ideal I in R is of the form $I = (\alpha)$ or $I = (n, \alpha)$. To see this, choose any nonzero element β_0 of I. Then I contains the nonzero integer $\beta\overline{\beta_0}$, and then I contains $\beta_0\overline{\beta_0}\sqrt{D}$ as well. Let $S_1 = \{|n'| \neq 0 \mid n' \text{ is an integer in } I\}$ and let $S_2 = \{|b'| \neq 0 \mid a' + b'\sqrt{D} \text{ is in } I\}$. Then S_1 is a nonempty set of positive integers and so has a smallest element n. Similarly, S_2 has a smallest element b (which may be a half–integer). Let $\alpha = a + b\sqrt{D}$ be in I. Now if $\beta = c + d\sqrt{D}$ is any element of I, it follows from the division algorithm that d is a multiple of b, $d = jb$ for some integer j. Then $m = \beta - j\alpha$ is an integer, and it again follows from the division algorithm that m is a multiple of n, $m = \ell n$ for some integer ℓ. Thus we see that $\beta = n\ell + \alpha j$ and so $I = (n, \alpha)$. □

Remark 5.50.

(1) It is sometimes useful to add a "redundant" generator to I. Namely, if $I = (\alpha)$, then $\|\alpha\| = \pm\alpha\overline{\alpha}$ is in I, and so we may write $I = (\|\alpha\|, \alpha)$.

(2) We see from (1) that if $I = (\alpha)$ is a principal ideal, then $I = (g, \alpha)$ is a principal ideal for $g = \pm\|\alpha\|$. On the other hand, if $g = \pm\|\alpha\|$, then $I = (g, \alpha) = (\pm\|\alpha\|, \alpha) = (\pm\alpha\overline{\alpha}, \alpha) = (\alpha)$ is a principal ideal.

(3) If $I = (g, \alpha)$ as above, then any element β of I is of the form $g\gamma + \alpha\delta$, and then $\|\beta\| = |\beta\overline{\beta}| = g\gamma\overline{\gamma} + g[\gamma\overline{\alpha}\overline{\delta} + \overline{\gamma}\alpha\delta] + \alpha\overline{\alpha}\gamma\overline{\gamma}$. So we see that $\|\beta\|$ is divisible by g for every β in I.

Proposition 5.51. *Every prime ideal P of R is of the following form:*

(1) $P = (\alpha_0)$ where $\|\alpha_0\| = p$ for some prime p; or

(2) $P = (p)$ for some prime p where R does not have an element that is not divisible by p but whose norm is divisible by p; or

(3) $P = (p, \alpha_1)$ where p is a prime, and α_1 is an element of R not divisible by p but with with $\|\alpha_1\|$ divisible by p.

Proof: We prove the theorem by ruling out every ideal that is not of one of the above forms.

Let I be an ideal of R. By Remark 5.50, we may assume I is of the form (g, α) with g an integer dividing $\|\alpha\|$.

First, suppose g has more than one prime factor. Write $g = g_1g_2$ with g_1 and g_2 relatively prime, $g_1 \neq \pm1$, $g_2 \neq \pm1$. Let $I_1 = (g_1, \alpha)$ and $I_2 = (g_2, \alpha)$. Then $I_1I_2 = (g, g_1\alpha, g_2\alpha, \alpha^2) \subseteq (g, \alpha) = I$ but $I_1 \not\subseteq I$, as I_1 has the element g_1 with $\|g_1\| = g_1^2$ not divisible by g, and I_2 has the element g_2 with $\|g_2\| = g_2^2$ not divisible by g. Thus I is not a prime ideal.

Thus, we must have $g = \pm p^j$ for some $j \geq 1$. Suppose that $j \geq 3$. Let $I_1 = (p, \alpha)$. Then $I_1^j \subseteq I$, but $I_1 \not\subseteq I$ as I_1 has the element p with $\|p\| = p^2$ not divisible by g.

Suppose that $j = 2$. If p^2 divides α in R, then $I = (p^2) = (p)(p)$ and $I \not\supseteq (p)$ so I is not prime. If p divides α, write $\alpha = p\beta$. Then $I = (p^2, p\beta) = (p)(p, \beta)$ is not prime unless $(p) = R$, which is impossible, or $(p, \beta) = R$, in which case $I = (p)$. Finally, consider the case where p does not divide α in R. Again, let $I_1 = (p, \alpha)$ and observe that $I_1^2 \subseteq I$. We claim that $I_1 \not\subseteq I$, and we prove this by showing that p is not in I. We prove this by contradiction. Suppose that p is in I. Then $p = p^2\gamma + \alpha\delta$ for some elements γ and δ of R. But then $p = p^2\overline{\gamma} + \overline{\alpha}\overline{\delta}$, and then $p\alpha = p^2\overline{\gamma}\alpha + \alpha\overline{\alpha}\overline{\delta}$. But by assumption $\|\alpha\|$ is divisible by p^2, so $\alpha\overline{\alpha} = p^2 m$ for some integer m. Substituting, we see that $p\alpha = p^2\overline{\gamma}\alpha + p^2 m\overline{\delta}$, so $\alpha = p[\overline{\gamma}\alpha + m\overline{\delta}]$ is divisible by p in R, a contradiction.

Thus, the only possibility for a prime ideal is $I = (p, \alpha)$ with $\|\alpha\|$ divisible by p. If α is divisible by p then $I = (p)$, and we are in case (2).

Thus, we are left with the possibility that $I = (p, \alpha_1)$ for some element α_1 of R with $\|\alpha_1\|$ divisible by p but with α_1 not divisible by p, in which case we are in case (3). □

Proposition 5.52. *In the situation of Proposition 5.51,*

(1) in case (1), $(p) = P\overline{P}$;

(2) in case (2), $(p^2) = P\overline{P}$;

(3) in case (3), $(p) = P\overline{P}$. *Also, in this case, if R has an element α_0 of norm p, then $P = (\alpha_0)$ or $P = (\overline{\alpha_0})$, while if R does not have an element of norm p, then P is not a principal ideal.*

Proof:

(1) and (2) are obvious.

(3) Let $\alpha_1 = a_1 + b_1\sqrt{D}$, and note that $a_1^2 - b_1^2 D \equiv 0 \pmod{p}$. Then $(p, \alpha_1)(p, \overline{\alpha_1}) = (p^2, p\alpha_1, p\overline{\alpha_1}, \alpha_1\overline{\alpha_1})$ contains $p[\alpha_1 + \overline{\alpha_1}] = 2a_1 p$ and $p[\alpha_1\sqrt{D} + \overline{\alpha_1}\sqrt{D}] = 2b_1 D p$.

Suppose p is odd. Case (a): a_1 is not divisible by p. Then $2a_1$ is not divisible by p, and so p is in $P\overline{P}$. Case (b): a_1 is divisible by p. Then b_1 is not divisible by p. If D is not divisible by p then $2b_1 D$ is not divisible by p, and so p is in $P\overline{P}$. If D is divisible by p, then $\|\alpha_1\|$ is divisible by p but not by p^2, and so p is in $P\overline{P}$.

Now suppose $p = 2$. Case (a): $D \equiv 2 \pmod 4$. Then a_1 is even, so b_1 is odd, and then $P = (2, \sqrt{D})$, in which case $P\overline{P} = (4, 2\sqrt{D}, D) = (2)$. Case (b): $D \equiv 3 \pmod 4$. Then a_1 and b_1 are both odd, $P\overline{P} = (4, 2\alpha_1, 2\overline{\alpha_1}, a_1^2 - b_1^2 D)$ and $a_1^2 - b_1^2 D \equiv 2 \pmod 4$, so $P\overline{P} = (2)$. Case (c): $D \equiv 1 \pmod 4$. Then $P\overline{P} = (4, 2a_1' + 2b_1'\sqrt{D}, 2a_1' - 2b_1'\sqrt{D}, \|\alpha_1\|) = (4, 4a_1, 2a_1' - 2b_1'\sqrt{D}, \|\alpha_1\|) = (2)$, as a_1 and b_1 are half–integers.

If R has an element α_0 of norm p, then by Lemma 5.48, α is divisible by α_0, in which case $P = (\alpha_0)$, or α is divisible by $\overline{\alpha_0}$, in which case $P = (\overline{\alpha_0})$.

Conversely, suppose that P is a principal ideal, $P = (\alpha)$ for some element α of R. Then $\overline{P} = (\overline{\alpha})$, and so $(p) = P\overline{P} = (\alpha\overline{\alpha})$. But then $p^2 = \|p\| = \|\alpha\| \cdot \|\overline{\alpha}\| = \|\alpha\|^2$ so $\|\alpha\| = p$. Thus, if R does not have an element of norm p, P cannot be a principal ideal. □

Proposition 5.53. *In the situation of Proposition 5.51, for each prime p,*

(1) in case (1), there are two ideals of this form if $P \neq \overline{P}$ and a unique ideal of this form if $P = \overline{P}$;

(2) in case (2), there is a unique ideal of this form;

(3) in case (3), there are two ideals of this form if $P \neq \overline{P}$ and a unique ideal of this form if $P = \overline{P}$.

Proof:

(1) Suppose α_0' is any element of R with $\|\alpha_0'\| = p$. Note by Lemma 5.48 that α_0' is divisible by (α_0) or by $\overline{\alpha_0}$, and that α_0 is divisible by α_0' or by $\overline{\alpha_0'}$, which readily implies that $(\alpha_0') = (\alpha_0)$ or $(\alpha_0') = (\overline{\alpha_0})$.

(2) This is obvious.

(3) Suppose α_1' is any element of R with $\|\alpha_1'\|$ divisible by p but with α_1' not divisible by p. Then, in the notation of the proof of Lemma 5.48, we must have $\alpha_1 = a_1 + b_1\sqrt{D}$ with $b_1 \not\equiv 0 \pmod p$ and $a_1 \equiv kb_1 \pmod p$, and $\alpha_1' = a_1' + b_1'\sqrt{D}$ with $b_1' \not\equiv 0 \pmod p$ and $a_1' \equiv k'b_1' \pmod p$, with $k' \equiv \pm k \equiv \pm k_0 \pmod p$. Suppose $k' \equiv k \pmod p$. Since p and b_1 are relatively prime, there is an integer m with $b_1' \equiv b_1 m \pmod p$. But then $a_1' \equiv kb_1' \equiv kb_1 m \equiv a_1 m \pmod p$, so $\alpha_1' - m\alpha_1 = c + d\sqrt{D}$ with c and d both divisible by p, i.e., $\alpha_1' - m\alpha_1 = p\gamma$ for $\gamma = [c + d\sqrt{D}]/p$ an element of R, and then $\alpha_1' = p\gamma + \alpha_1 m$ is in (p, α_1), and vice versa, so $(p, \alpha_1') = (p, \alpha_1)$. If $k' \equiv -k \pmod p$, then, replacing α_1 by $\overline{\alpha_1}$, the same argument shows $(p, \alpha_1') = (p, \overline{\alpha_1})$. □

Proposition 5.54. *In the situation of Proposition 5.51, let $e_D = 1$ if $D \equiv 1 \pmod 4$, and let $e_D = 2$ if $D \equiv 2$ or $3 \pmod 4$. For any prime p,*

(1) if p does not divide $e_D D$, then $P \neq \overline{P}$ in cases (1) and (3); and

(2) if p divides $e_D D$, case (2) does not occur, and $P = \overline{P}$ in cases (1) and (3).

Proof: First, suppose that p is an odd prime. By adding a "redundant" generator if necessary (see Remark 5.50(1)), we may assume that we are in case (3) and that $P = (p, \alpha_1)$. Let $\alpha_1 = a_1 + b_1\sqrt{D}$. Then $P = \overline{P}$ if and only if $2a_1$ is in P. Let us assume that $2a_1$ is in P.

Now every element of P has norm divisible by p, so p must divide $2a_1$, and hence p divides a_1. Then p does not divide b_1, as by assumption p does not divide α_1, but by assumption p divides $\|\alpha_1\| = |a_1^2 - b_1^2 D|$, so p divides D.

Conversely, suppose p divides D. Then, by Proposition 5.53, we may choose any suitable element α_1 and then $P = (p, \alpha_1)$. Choose $\alpha_1 = \sqrt{D}$. Then $P = (p, \sqrt{D}) = (p, -\sqrt{D}) = \overline{P}$.

Next, suppose that $p = 2$. If $D \equiv 2 \pmod 4$, then, just as above, $P = (2, \sqrt{D}) = (2, -\sqrt{D}) = \overline{P}$. If $D \equiv 3 \pmod 4$, then $P = (2, 1 + \sqrt{D}) = (2, 1 - \sqrt{D}) = \overline{P}$. If $D \equiv 1 \pmod 4$, then $\alpha_1 = a_1 + b_1\sqrt{D}$ with a_1 and b_1 half–integers, so in this case $2a_1$ is an odd integer, so, by the same argument as above, $2a_1$ is not in P, and hence in this case $P \neq \overline{P}$.

Finally, in every case where p divides $e_D D$, we have exhibited an explicit element α_1 not divisible by p but with $\|\alpha_1\|$ divisible by p, excluding case (2). □

We can now get a complete description of the prime ideals in R.

Theorem 5.55. *Let $R = \mathcal{O}(\sqrt{D})$. Let $e_D = 1$ if $D \equiv 1 \pmod 4$, and let $e_D = 2$ if $D \equiv 2$ or $3 \pmod 4$. The following is a complete listing, without duplication, of the prime ideals P in R:*

(1) For every prime p dividing $e_D D$,
(a) if R has an element α_0 of norm p, the ideal $P = (\alpha_0)$. In this case P is a principal ideal, $P = \overline{P}$, and $P^2 = (p)$;
(b) if R does not have an element of norm p, the ideal $P = (p, \alpha_1)$ where α_1 is not divisible by p but $\|\alpha_1\|$ is divisible by p. In this case P is not a principal ideal, $P = \overline{P}$, and $P^2 = (p)$.

(2) For every prime p not dividing $e_D D$,
(a) if R has an element α_1 that is not divisible by p but with $\|\alpha_1\|$

divisible by p, the ideal $P = (p, \alpha_1)$ and the ideal $\overline{P} = (p, \overline{\alpha_1})$. In this case P and $\overline{P}$ are not principal ideals, $P \neq \overline{P}$, and $P\overline{P} = (p)$;
(b) if R does not have an element that is not divisible by p but whose norm is divisible by p, the principal ideal $P = (p)$.

Proof: Assembling Propositions 5.51, 5.52, 5.53, and 5.54 gives us almost all of this theorem. There is only one thing left to do. Proposition 5.51 shows that every prime ideal must be of the form above. To complete the proof we must show that every ideal of the above form is indeed a prime ideal. Thus let P be as in the statement of the theorem, and suppose that I and J are ideals with $P \supseteq IJ$. We must show that $P \supseteq I$ or $P \supseteq J$.

To begin with, we note that every element of P has norm divisible by p. Also, by Lemma 5.49 we may assume that $I = (m, \beta)$ and $J = (n, \gamma)$ with m dividing $\|\beta\|$ and n dividing $\|\gamma\|$. Then $IJ = (mn, n\beta, m\gamma, \beta\gamma)$. Since $P \supseteq IJ$ every element of IJ must have norm divisible by p. In particular, $\|mn\| = m^2n^2$ is divisible by p. Thus at least one of m and n is divisible by p. We shall assume that m is divisible by p. (Otherwise interchange I and J.) To proceed further, we must break the proof up into several cases. We number the cases as in the statement of the theorem.

Case (2)(b): Here $P = (p)$. Now p divides m and m divides $\|\beta\|$, so p divides $\|\beta\|$ and hence p divides β. Then $P = (p) \supseteq I$.

Case (1): If p divides β then $P \supseteq (p) \supseteq I$ and we are done, so assume not. By adding a "redundant" generator if necessary, we may assume that $P = (p, \alpha)$ with α not divisible by p but with $\|\alpha\|$ divisible by p. Write $\alpha = a + b\sqrt{D}$. Note that $b \not\equiv 0 \pmod{p}$ and define k by $a \equiv kb \pmod{p}$. Then, as in the proof of Proposition 5.53, every element of P is of the form $a' + b'\sqrt{D}$ with $a' \equiv kb' \pmod{p}$. Similarly, $\beta = c + d\sqrt{D}$ with $d \not\equiv 0 \pmod{p}$ and $c \equiv kd \pmod{p}$ or $c \equiv -kd \pmod{p}$. In the former case, again as in the proof of Proposition 5.53, β is in P and so $P \supseteq I$. In the latter case, by the same argument, β is in $\overline{P}$ and so $\overline{P} \supseteq I$. But here $P = \overline{P}$, so $P \supseteq I$.

Case (2)(a): In the event that $D \equiv 1 \pmod{4}$, we have a preliminary step. In this event p must be odd, so we may replace β and γ by 2β and 2γ, if necessary, to ensure that $\beta = a + b\sqrt{D}$ and $\gamma = c + d\sqrt{D}$ with a, b, c, and d integers.

By the argument for Case (1), $P \supseteq I$ except possibly in the situation that β is in $\overline{P}$, so suppose we are in that situation. Then $n\beta$ is in P, as $P \supseteq IJ$. If n is not divisible by p, that gives β in P, and then $P = \overline{P}$, which is impossible. Thus, we must have n divisible by p. Now consider the ideal $J = (n, \gamma)$. By the same argument as above, $P \supseteq J$ except possibly in the situation that γ is in $\overline{P}$. Thus, the final situation to consider is where β and γ are both in $\overline{P}$. In this situation $\overline{\beta}$ and $\overline{\gamma}$ are in P. Then, on the one hand, $\overline{\beta\gamma}$ is in P, as P is an ideal, and, on the other hand, $\beta\gamma$ is in P as $P \supseteq IJ$. Thus $\delta = \beta\gamma - \overline{\beta\gamma}$ is in P. But direct computation shows that $\delta = e + f\sqrt{D}$ with $e \equiv 0 \pmod{p}$ and $f \equiv 4kbd \pmod{p}$. Thus $0 \equiv k[4kbd] = [2k]^2bd \pmod{p}$. Since $b \not\equiv 0 \pmod{p}$ and $d \not\equiv 0 \pmod{p}$ we must have $k \equiv 0 \pmod{p}$ or $2 \equiv 0 \pmod{p}$. But in either of these cases $P = \overline{P}$, which we have ruled out. □

Remark 5.56. In the situation of Theorem 5.55, let p be an odd prime not dividing D. From the proof of Lemma 5.48, we can see that if D is a quadratic residue $\pmod{p}$ then case (2)(a) occurs, while if D is a quadratic nonresidue $\pmod{p}$ then case (2)(b) occurs.

Lemma 5.57. *In the situation of Theorem 5.55, let* $D \equiv 1 \pmod{4}$ *and let* $p = 2$. *If* $D \equiv 1 \pmod{8}$ *then case (2)(a) occurs, while if* $D \equiv 5 \pmod{8}$ *then case (2)(b) occurs.*

Proof: If $\alpha = a + b\sqrt{D}$ with a and b integers and with $\|\alpha\|$ even, then either a and b are both even or they are both odd. In either event, α is divisible by 2. Thus, the only possibility for α_1 is $\alpha_1 = a_1 + b_1\sqrt{D}$ with a_1 and b_1 both half–integers, i.e., $\alpha = [a' + b'\sqrt{D}]/2$ with a' and b' odd integers. Using the fact that $c^2 \equiv 1 \pmod{8}$ for any odd integer c, we see that in this case $\|\alpha\| \equiv [1 - D]/4 \pmod{2}$. Thus, if $D \equiv 1 \pmod{8}$, we may choose α_1 to be any element of R of this form, and we are in case 2(a), while if $D \equiv 5 \pmod{8}$, there is no possible choice for α_1 and we are in case 2(b).□

Remark 5.58. Here is a summary of all the possibilities, together with an expression of each of the prime ideals in terms of congruence conditions.

(1) $D \equiv 2$ or $3 \pmod{4}$ and

 (a) p is an odd prime dividing D:

$$P = (p, \sqrt{D}) = \{a + b\sqrt{D} \mid a \equiv 0 \pmod{p}\}.$$

 (b) p is an odd prime not dividing D and D is a quadratic residue $\pmod{p}$:

$$P = (p, k_0 + \sqrt{D}) = \{a + b\sqrt{D} \mid a \equiv k_0 b \pmod{p}\},$$
$$P = (p, -k_0 + \sqrt{D}) = \{a + b\sqrt{D} \mid a \equiv -k_0 b \pmod{p}\},$$

 where $k_0^2 \equiv D \pmod{p}$.

(c) p is an odd prime not dividing D and D is a quadratic nonresidue $(\operatorname{mod} p)$:

$$P = (p) = \{a + b\sqrt{D} \mid a \equiv b \equiv 0 \pmod{p}\}.$$

(d) $D \equiv 2 \pmod{4}$ and $p = 2$:

$$P = (2, \sqrt{D}) = \{a + b\sqrt{D} \mid a \equiv 0 \pmod{2}\}.$$

(e) $D \equiv 3 \pmod{4}$ and $p = 2$:

$$P = (p, 1 + \sqrt{D}) = \{a + b\sqrt{D} \mid a \equiv b \pmod{2}\}.$$

(2) $D \equiv 1 \pmod{4}$ and

(a) p is an odd prime dividing D:

$$P = (p, \sqrt{D}) = \{[a'+b'\sqrt{D}]/2 \mid a' \equiv b' \pmod{2} \text{ and } a' \equiv 0 \pmod{p}\}.$$

(b) p is an odd prime not dividing D and D is a quadratic residue $(\operatorname{mod} p)$:

$$\begin{aligned} P &= (p, [k_0 + \sqrt{D}]/2) \\ &= \{[a' + b'\sqrt{D}]/2 \mid a' \equiv b' \pmod{2} \text{ and } a' \equiv k_0 b' \pmod{p}\}, \\ P &= (p, [-k_0 + \sqrt{D}]/2) \\ &= \{[a' + b'\sqrt{D}]/2 \mid a' \equiv b' \pmod{2} \text{ and } a' \equiv -k_0 b' \pmod{p}\}, \end{aligned}$$

where $k_0^2 \equiv D \pmod{p}$.

(c) p is an odd prime not dividing D and D is a quadratic nonresidue $(\operatorname{mod} p)$:

$$P = (p) = \{[a'+b'\sqrt{D}]/2 \mid a' \equiv b' \pmod{2} \text{ and } a \equiv b \equiv 0 \pmod{p}\}.$$

(d) $D \equiv 1 \pmod{8}$ and $p = 2$:

$$\begin{aligned} P &= (2, [1 + \sqrt{D}]/2) = \{[a' + b'\sqrt{D}]/2 \mid a' \equiv b' \pmod{4}\}, \\ P &= (2, [-1 + \sqrt{D}]/2) = \{[a' + b'\sqrt{D}]/2 \mid a' \equiv -b' \pmod{4}\}. \end{aligned}$$

(e) $D \equiv 5 \pmod{8}$ and $p = 2$:

$$P = (2) = \{[a'+b'\sqrt{D}]/2 \mid a' \equiv b' \equiv 0 \pmod{2} \text{ and } a' \equiv b' \pmod{4}\}.$$

5.6 Examples of Ideals in $\mathcal{O}(\sqrt{D})$

In this section, we will give concrete examples of nonprincipal ideals in $\mathcal{O}(\sqrt{D})$. We will start off by giving classes of examples, and we will finish off by giving examples for particular values of D. We adopt the standard notation that $h(D)$ denotes the class number of the field $h(\mathbf{Q}(\sqrt{D}))$.

Proposition 5.59. *Let $R = \mathcal{O}(\sqrt{D})$.*

(1) If $D \equiv 1 \pmod 8$, let $I = (2, [1+\sqrt{D}]/2)$.

(2) If $D \equiv 2 \pmod 4$, let $I = (2, \sqrt{D})$.

(3) If $D \equiv 3 \pmod 4$, let $I = (2, 1+\sqrt{D})$.

If there is no element α of R with $\|\alpha\| = 2$, then I is not a principal ideal.

(4) In cases (2) and (3), $I^2 = (2)$, a principal ideal.

(5) In case (1), suppose that R has an element β, which is not divisible by 2, with $\|\beta\| = 4$. Then $I^2 = (\beta)$, a principal ideal.

Proof: We have proved (1), (2), (3), and (4) in Theorem 5.55 and Lemma 5.57.

As for (5), in this case we must have $\beta = [a + b\sqrt{D}]/2$ with a and b odd integers. Replacing β by $\overline{\beta}$, if necessary, we may assume that $a \equiv b \pmod 4$. Computation then shows that $a \equiv b \pmod 8$ if $D \equiv 1 \pmod{16}$ and that $a \equiv b + 4 \pmod 8$ if $D \equiv 9 \pmod{16}$.

Let $\alpha_1 = [1+\sqrt{D}]/2$. Then $I^2 = (4, 2\alpha_1, \alpha_1^2) = (4, 2\alpha_1 - 2\alpha_1^2, \alpha_1^2) = (4, [1-D]/2, \alpha_1^2) = (4, \alpha_1^2)$. Certainly β divides 4, and a long but routine computation shows that β divides α_1^2. Thus $I^2 \subseteq (\beta)$. Another computation shows that $b\beta - \alpha_1^2 = 4c$ for some integer c, so $b\beta$ is in I^2. But certainly 4β is in I^2, so β is in I^2. Thus $I^2 \supseteq (\beta)$. Hence $I^2 = (\beta)$. □

Corollary 5.60. *Let $D \equiv 1$, 2, 3, 6, or 7 $\pmod 8$ and let $R = \mathcal{O}(\sqrt{D})$. Let I be the ideal of Proposition 5.59.*

(1) If $D < 0$, $D \neq -1$, -2, or -7, then I is a nonprincipal ideal of R. Consequently, $h(D) > 1$.

(2) If D is divisible by a prime congruent to $5 \pmod 8$, or if D is divisible by a prime congruent to $3 \pmod 8$ and by a prime congruent to $5 \pmod 7$, then I is a nonprincipal ideal of R. Consequently, $h(D) > 1$.

(3) If $D \equiv 2$, 3, 6, or 7 $\pmod 8$ and D is as in (1) or (2), then $h(D)$ is even. If $D \equiv 1 \pmod 8$, D is as in (1) or (2), and R has an element β that is not divisible by 2, with $\|\beta\| = 4$, then $h(D)$ is even.

Proof: It is immediate from Lemma 2.67, Lemma 2.77, Theorem 5.55, and Proposition 5.59 that I is a nonprincipal ideal of $\mathcal{O}(\sqrt{D})$, and that, if D is as in (3), I^2 is a principal ideal. Thus, in this case, $[I]$ is an element of order 2 of $\mathcal{C}(\mathbf{Q}(\sqrt{D}))$, and so this group has even order. □

Example 5.61. If $D = -1$, then $\alpha = 1 + \sqrt{D}$ is an element of $\mathcal{O}(\sqrt{D})$ with $\|\alpha\| = 2$. If $D = -2$, then $\alpha = \sqrt{D}$ is an element of $\mathcal{O}(\sqrt{D})$ with $\|\alpha\| = 2$. If $D = -7$, then $\alpha = [1 + \sqrt{D}]/2$ is an element of $\mathcal{O}(\sqrt{D})$ with $\|\alpha\| = 2$.

Here are examples of real quadratic fields $\mathcal{O}(\sqrt{D})$ having elements α of $\mathcal{O}(\sqrt{D})$ with $\|\alpha\| = 2$:

$$\begin{array}{ll}
D = 2: & \alpha = \sqrt{2}, \\
D = 3: & \alpha = 1 + \sqrt{3}, \\
D = 6: & \alpha = 2 + \sqrt{6}, \\
D = 7: & \alpha = 3 + \sqrt{7}, \\
D = 11: & \alpha = 3 + \sqrt{11}, \\
D = 14: & \alpha = 4 + \sqrt{14}, \\
D = 17: & \alpha = [5 + \sqrt{17}]/2, \\
D = 19: & \alpha = 13 + 3\sqrt{19}, \\
D = 22: & \alpha = 14 + 3\sqrt{22}, \\
D = 23: & \alpha = 5 + \sqrt{23}, \\
D = 31: & \alpha = 39 + 7\sqrt{31}, \\
D = 33: & \alpha = [5 + \sqrt{33}]/2, \\
D = 34: & \alpha = 6 + \sqrt{34}, \\
D = 38: & \alpha = 6 + \sqrt{38}, \\
D = 41: & \alpha = [7 + \sqrt{41}]/2, \\
D = 43: & \alpha = 59 + 9\sqrt{43}, \\
D = 46: & \alpha = 156 + 23\sqrt{46}, \\
D = 47: & \alpha = 7 + \sqrt{47}, \\
D = 51: & \alpha = 7 + \sqrt{51}, \\
D = 57: & \alpha = [7 + \sqrt{57}]/2, \\
D = 59: & \alpha = 23 + 3\sqrt{59}, \\
D = 62: & \alpha = 8 + \sqrt{62}, \\
D = 66: & \alpha = 8 + \sqrt{66}, \\
D = 67: & \alpha = 221 + 27\sqrt{67}, \\
D = 71: & \alpha = 59 + 7\sqrt{71}, \\
D = 73: & \alpha = [9 + \sqrt{73}]/2, \\
D = 79: & \alpha = 9 + \sqrt{79}, \\
D = 86: & \alpha = 102 + 11\sqrt{86}, \\
D = 89: & \alpha = [9 + \sqrt{89}]/2, \\
D = 94: & \alpha = 1464 + 151\sqrt{94}, \\
D = 97: & \alpha = [69 + 7\sqrt{97}]/2.
\end{array}$$

Thus the values of D in this example are *not* covered by Corollary 5.60.

Example 5.62. Here are examples of elements β as in Proposition 5.59. Thus, for these values of D, $h(D)$ is even:

$$\begin{array}{lll} D = 65: & \beta = [7-\sqrt{65}]/2, & \beta\overline{\beta} = -4; \\ D = 105: & \beta = [11-\sqrt{105}]/2, & \beta\overline{\beta} = 4; \\ D = 185: & \beta = [13+\sqrt{185}]/2, & \beta\overline{\beta} = -4; \\ D = 265: & \beta = [49-3\sqrt{265}]/2, & \beta\overline{\beta} = 4; \\ D = 273: & \beta = [17+\sqrt{273}]/2, & \beta\overline{\beta} = 4; \\ D = 305: & \beta = [17+\sqrt{305}]/2, & \beta\overline{\beta} = -4; \\ D = 345: & \beta = [19-\sqrt{345}]/2, & \beta\overline{\beta} = 4; \\ D = 385: & \beta = [59+3\sqrt{385}]/2, & \beta\overline{\beta} = 4; \\ D = 465: & \beta = [151+7\sqrt{465}]/2, & \beta\overline{\beta} = 4. \end{array}$$

Proposition 5.63.

(1) *Let p_1 and p_2 be odd primes. Suppose that $p_2 \equiv 1 \pmod 4$ and that p_1 is a quadratic nonresidue $\pmod{p_2}$.*

  (a) *If D is divisible by p_2 and D is a quadratic residue $\pmod{p_1}$, then $h(D) > 1$.*

  (b) *If D is divisible by p_1p_2, then $h(D)$ is even.*

(2) *Let p be a prime, $p \equiv 3 \pmod 4$. If D is divisible by p and $D \equiv 2 \pmod 8$, then $h(D)$ is even.*

(3) *Let p_1, p_2, and p_3 be primes, all congruent to $3 \pmod 4$. If D is divisible by $p_1p_2p_3$, then $h(D)$ is even.*

Proof: We have proved (1)(a) in Theorem 2.82. We recapitulate: since $p_2 \equiv 1 \pmod 4$, $-p_1$ is also a quadratic nonresidue $\pmod{p_2}$. Thus, by congruence considerations $\pmod{p_2}$, $R = \mathcal{O}(\sqrt{D})$ does not have an element of norm p_1. On the other hand, since D is a quadratic residue $\pmod{p_1}$, $\alpha_1 = k + \sqrt{D}$ is an element of R of norm divisible by p_1, where $k^2 \equiv D \pmod p$. Thus, by Theorem 5.55, $I = (p_1, \alpha_1)$ is a nonprincipal ideal of R. The proof of (1)(b) is almost identical, but here we choose $\alpha_1 = \sqrt{D}$ and now I is a nonprincipal ideal with I^2 a principal ideal, by Theorem 5.55.

(2) follows similarly from the proof of Theorem 2.87.

As for (3), let (a/p) be the quadratic residue symbol: $(a/p) = 1$ if a is a quadratic residue $\pmod p$ and $(a/p) = -1$ if a is a quadratic nonresidue $\pmod p$. We use two facts. First, for $p \equiv 3 \pmod 4$, $(-a/p) = -(a/p)$, and second, by the Law of Quadratic Reciprocity, for $p \equiv q \equiv 3 \pmod 4$, $(p/q) = -(q/p)$. Here p and q are primes.

We wish to show that there is some prime $p = p_1$, p_2, or p_3 with R not having an element of norm p. Then $I = (p, \sqrt{D})$ is a nonprincipal ideal with I^2 a principal ideal, by Theorem 5.55 again. The argument for this breaks up into several cases, and we do two of them. (The others, which are similar, we leave for the reader.) (a) Suppose that $(p_1/p_2) = 1$ and that $(p_1/p_3) = -1$. Then $(-p_1/p_2) = -1$, and by congruence considerations $(\bmod\, p_2)$ and $(\bmod\, p_3)$, R does not have an element of norm p_1. (b) Suppose that $(p_1/p_2) = 1$ and that $(p_1/p_3) = 1$. If $(p_2/p_3) = 1$, then $(p_2/p_1) = -1$ and $(-p_2/p_3) = -1$, and by congruence considerations $(\bmod\, p_1)$ and $(\bmod\, p_3)$, R does not have an element of norm p_2. If $(p_2/p_3) = -1$, then $(p_3/p_2) = -1$ and $(-p_3/p_1) = -1$, and by congruence considerations $(\bmod\, p_1)$ and $(\bmod\, p_2)$, R does not have an element of norm p_3. □

As we shall now see, for D negative we can easily get stronger information on the ideal class group, and hence the class number, of the field $\mathbf{Q}(\sqrt{D})$.

Theorem 5.64. *Let $p_1, \ldots, p_k$ be distinct primes and set $D = -p_1 \cdots p_k$. Then $\mathcal{C}(\mathbf{Q}(\sqrt{D}))$ has a subgroup isomorphic to $(\mathbb{Z}/2\mathbb{Z})^{k-1}$. Consequently, $h(D)$ is divisible by 2^{k-1}. If $D \equiv 3 \pmod 4$, then $\mathcal{C}(\mathbf{Q}(\sqrt{D}))$ has a subgroup isomorphic to $(\mathbb{Z}/2\mathbb{Z})^k$. Consequently, $h(D)$ is divisible by 2^k.*

Proof: Let $S = \{p_1, \ldots, p_k\}$. The subsets of S form a group $\mathcal{G}$ of order 2^k with the group operation $*$ being symmetric difference, i.e., if T_1 and T_2 are two subsets of S, then $T_1 * T_2 = \{p \text{ in } S \mid p \text{ in } T_1 \text{ or in } T_2 \text{ but not in both}\}$. Indeed, $\mathcal{G}$ is isomorphic to $(\mathbb{Z}/2\mathbb{Z})^k$. Let $\mathcal{G}_0$ be the subgroup of order 2 of $\mathcal{G}$ given by $\mathcal{G}_0 = \{\emptyset, S\}$. (Here $\emptyset$ denotes the empty set, as usual.)

For a subset T of S, let d_T be the product of the elements of T (with $d_T = 1$ if $T = \emptyset$), and let I_T be the ideal $I_T = (d_T, \sqrt{-D})$. We have a map $\pi : \mathcal{G} \to \mathcal{C}(\mathbf{Q}(\sqrt{D}))$ given by $\pi(T) = [I_T]$. Note that π is a homomorphism as $[I_{T_1}][I_{T_2}] = [I_{T_1 * T_2}]$.

It is straightforward to check that I_T is a principal ideal if and only if $T = \emptyset$ or $T = S$. This gives that $\mathrm{Ker}(\pi) = \mathcal{G}_0$, so $\mathrm{Im}(\pi)$, which is a subgroup of $\mathcal{C}(\mathbf{Q}(\sqrt{D}))$, is isomorphic to $\mathcal{G}/\mathcal{G}_0$, which is isomorphic to $(\mathbb{Z}/2\mathbb{Z})^{k-1}$.

In case $D \equiv 3 \pmod 4$, we have the ideal $J = (2, 1 + \sqrt{D})$. $[J]$ is an element of order 2 in $\mathcal{C}(\mathbf{Q}(\sqrt{D}))$. $[J][I_T]$ is not a principal ideal for any subset T of S, so $[J]$ is not in $\mathrm{Im}(\pi)$. Thus, in this case $\mathcal{C}(\mathbf{Q}(\sqrt{D}))$ has a subgroup isomorphic to $(\mathbb{Z}/2\mathbb{Z})^k$. □

We proved Theorem 5.64 by elementary means. This is just the easy half of the following much deeper theorem of Gauss:

Theorem 5.65 (Gauss). *Let $D = \pm p_1 \cdots p_k$ for distinct primes $p_1, \ldots, p_k$.*

(1) If $D < 0$ and $D \equiv 1$ or $2 \pmod 4$, then $h(D)$ is divisible by 2^{k-1}.

(2) If $D < 0$ and $D \equiv 3 \pmod 4$, then $h(D)$ is divisible by 2^k.

(3) If $D > 0$ and $D \equiv 1$ or $2 \pmod 4$, then $h(D)$ is divisible by 2^{k-2}.

(4) If $D > 0$ and $D \equiv 3 \pmod 4$, then $h(D)$ is divisible by 2^{k-1}.

Theorem 5.64 showed how to find quadratic fields $\mathbf{Q}(\sqrt{D})$ with $h(D)$ arbitrarily large. But in these cases $\mathcal{C}(\mathbf{Q}(\sqrt{D}))$ consisted entirely of elements of order 2 (plus the identity). We will now show how to find quadratic fields $\mathbf{Q}(\sqrt{D})$ where $\mathcal{C}(\mathbf{Q}(\sqrt{D}))$ contains elements of order n for any n (and in particular for n odd).

Theorem 5.66. *Let n be an arbitrary integer and let $q > 1$ be odd. Let a be a positive integer with a and q relatively prime. Let D be the unique square-free integer defined by $b^2 D = a^2 - q^n$, where b is an integer. Suppose that q^j and $a + b\sqrt{D}$ do not have a common nonunit factor in $\mathcal{O}(\sqrt{D})$ for any proper factor j of n. Let I be the ideal*

$$I = (q, a + b\sqrt{D}).$$

Then $I^n = (a + b\sqrt{D})$, a principal ideal. Furthermore, $[I]$ is an element of $\mathcal{C}(\mathbf{Q}(\sqrt{D}))$ of order n.

Proof: We claim that $I^k = (q^k, a + b\sqrt{D})$ for $k = 1, \ldots, n$. We prove this by induction. It is certainly true for $k = 1$. Now assume it is true for some value k. Then

$$\begin{aligned} I^{k+1} = II^k &= (q, a + b\sqrt{D})(q^k, a + b\sqrt{D}) \\ &= (q^{k+1}, q(a + b\sqrt{D}), q^k(a + b\sqrt{D}), (a + b\sqrt{D})^2) \\ &= (q^{k+1}, q(a + b\sqrt{D}), (a + b\sqrt{D})^2). \end{aligned}$$

We claim $I^{k+1} = J = (q^{k+1}, a + b\sqrt{D})$. Clearly $I^{k+1} \subseteq J$ as each of the generators of I^{k+1} is in J. To show the reverse inclusion we must show that each of the generators of J is in I^{k+1}. This is evident for q^{k+1}, so it remains to consider $a + b\sqrt{D}$. Now $[a + b\sqrt{D}]^2 = a^2 + b^2 D + 2ab\sqrt{D} = [a^2 - b^2 D] + 2b^2 D + 2ab\sqrt{D} = q^n + 2b\sqrt{D}[a + b\sqrt{D}]$ is in I^{k+1}, so $2b\sqrt{D}[a + b\sqrt{D}]$ is in I^{k+1}, as is $[\sqrt{D}][2b\sqrt{D}][a + b\sqrt{D}] = 2bD[a + b\sqrt{D}]$. Thus I^{k+1} contains both $q[a + b\sqrt{D}]$ and $2bD[a + b\sqrt{D}]$. Since q and $2bD$ are relatively prime integers, there are integers x and y with $qx + 2bDy = 1$, which implies that $(a + b\sqrt{D})$ is in I^{k+1}, as required.

Now $I^n = (q^n, a + b\sqrt{D}) = (a + b\sqrt{D})$ as $q^n = \pm[a + b\sqrt{D}][a - b\sqrt{D}]$, so I^n is a principal ideal. Thus $[I]$ has order j in $\mathcal{C}(\mathbf{Q}(\sqrt{D}))$ for some j dividing n. We want to show $j = n$.

Suppose $j < n$, in which case j is a proper factor of n. If I^j is principal, then $I^j = (\gamma_j)$ for some element γ_j of $\mathcal{O}(\sqrt{D})$. Then γ_j divides both p^j and $a + b\sqrt{D}$, so by hypothesis γ_j is a unit, and then $I^j = \mathcal{O}(\sqrt{D})$. But then $I^n = (I^j)^{n/j} = \mathcal{O}(\sqrt{D})$, contradicting $I^n = (a + b\sqrt{D})$. □

Suppose that $a^2 < q^n$ in Theorem 5.66. Then $D < 0$. Note then that in any particular case we can check the hypothesis in Theorem 5.66 that q^j and $a + b\sqrt{D}$ do not have a common nonunit factor in $\mathcal{O}(\sqrt{D})$ for any proper factor j of n. For any such common factor must have norm dividing q^n, and, for D negative, $\mathcal{O}(\sqrt{D})$ has only finitely many elements of norm N for any integer N. But, more interestingly, we can give general conditions that ensure that this hypothesis holds.

Corollary 5.67. *In the situation of Theorem 5.66, assume that $a^2 < q^n$. Let p be the smallest prime factor of n, and set $m = n/p$. (If n is prime, note that $m = 1$.) Suppose that $-D > q^m - 1$ if $D \equiv 2$ or $3 \pmod 4$ and that $-D > 4q^m - 1$ if $D \equiv 1 \pmod 4$. Then [I] is an element of $\mathcal{C}(\mathbf{Q}(\sqrt{D}))$ of order n.*

Proof: Let us set $\gamma_n = a + b\sqrt{D}$, so $I^n = (\gamma_n)$. Note that γ_n has norm $|\gamma_n\overline{\gamma_n}| = |[a + b\sqrt{D}][a - b\sqrt{D}]| = q^n$.

We prove the corollary by contradiction. Suppose that $[I]$ has order j in $\mathcal{C}(\mathbf{Q}(\sqrt{D}))$ for some proper factor j of n. Then I^j is a principal ideal, i.e., $I^j = (\gamma_j)$ for some element γ_j of $\mathcal{O}(\sqrt{D})$. Then $I^n = (I^j)^{n/j} = (\gamma_j^{n/j})$. Thus γ_n and $\gamma_j^{n/j}$ are associates, so they have the same norm. But the norm is multiplicative, and that implies that γ_j has norm q^j.

Let $\gamma_j = c + d\sqrt{D}$. Since a, b, and q are relatively prime, d cannot be 0, and then c cannot be 0 either. Note that $|d| \geq 1$ if $D \equiv 2$ or $3 \pmod 4$ and $|d| \geq 1/2$ if $D \equiv 1 \pmod 4$, and similarly for $|c|$. Now γ_j has norm $q^j = c^2 - d^2D = c^2 + d^2(-D)$, so we see that $q^j \geq 1 - D$ if $D \equiv 2$ or $3 \pmod 4$ and that $q^j \geq (1 - D)/4$ if $D \equiv 1 \pmod 4$. But, as a little algebra shows, this contradicts our hypothesis on q and D. □

Example 5.68. Here are some instances of this corollary:

(1) $1^2+26 = 3^3$ so if $I = (3, 1+\sqrt{-26})$, then $[I]$ has order 3 in $\mathcal{C}(\mathbf{Q}(\sqrt{-26}))$, and hence $h(-26)$ is divisible by 3.

(2) $2^2+23 = 3^3$ so if $I = (3, 2+\sqrt{-23})$, then $[I]$ has order 3 in $\mathcal{C}(\mathbf{Q}(\sqrt{-23}))$, and hence $h(-23)$ is divisible by 3.

(3) $5^2 + 2^2 \cdot 14 = 3^4$ so if $I = (3, 5 + 2\sqrt{-14})$, then $[I]$ has order 4 in $\mathcal{C}(\mathbf{Q}(\sqrt{-14}))$, and hence $h(-14)$ is divisible by 4.

(4) $14^2 + 47 = 3^5$ so if $I = (3, 14 + \sqrt{-47})$, then $[I]$ has order 5 in $\mathcal{C}(\mathbf{Q}(\sqrt{-47}))$, and hence $h(-47)$ is divisible by 5.

(5) $116^2 + 3^2 \cdot 241 = 5^6$ so if $I = (5, 116 + 3\sqrt{-241})$, then $[I]$ has order 6 in $\mathcal{C}(\mathbf{Q}(\sqrt{-241}))$, and hence $h(-241)$ is divisible by 6.

(6) $46^2 + 71 = 3^7$ so if $I = (3, 46 + \sqrt{-71})$, then $[I]$ has order 7 in $\mathcal{C}(\mathbf{Q}(\sqrt{-71}))$, and hence $h(-71)$ is divisible by 7.

(7) $80^2 + 161 = 3^8$ so if $I = (3, 80 + \sqrt{-161})$, then $[I]$ has order 8 in $\mathcal{C}(\mathbf{Q}(\sqrt{-161}))$, and hence $h(-161)$ is divisible by 8.

(8) $139^2 + 362 = 3^9$ so if $I = (3, 139 + \sqrt{-362})$, then $[I]$ has order 9 in $\mathcal{C}(\mathbf{Q}(\sqrt{-362}))$, and hence $h(-362)$ is divisible by 9.

Remark 5.69.

(1) We saw in Theorem 5.55 that $I\overline{I}$ is a principal ideal for every prime ideal I of $\mathcal{O}(\sqrt{D})$, and hence this holds for any ideal of $\mathcal{O}(\sqrt{D})$. Thus, for any ideal I, $[\overline{I}] = [I]^{-1}$ in $\mathcal{C}(\mathbf{Q}(\sqrt{D}))$.

(2) More precisely, let $I = (g, \alpha)$ with $\|\alpha\|$ divisible by g but with g and α having no common integer divisor other than ± 1. We saw in Theorem 5.55 that if $g = p$ is a prime, then $I\overline{I} = (p)$, and hence $I\overline{I} = (g)$ for an arbitrary integer g.

Now we use the work we have done to examine the situation in $\mathcal{O}(\sqrt{D})$ for specific values of D.

Example 5.70. Throughout this example $R = \mathcal{O}(\sqrt{D})$ and $\mathbf{K} = \mathbf{Q}(\sqrt{D})$. The assertions here can all be verified by direct computation and we shall omit the details.

(1) $D = -5$: Here we return to our original example of nonunique factorization. Recall the factorizations into irreducible elements:

$$6 = 2 \cdot 3 = [1 + \sqrt{-5}][1 - \sqrt{-5}].$$

This corresponds to the factorizations into principal ideals:

$$(6) = (2)(3) = (1 + \sqrt{-5})(1 - \sqrt{-5}).$$

But this is *not* a counterexample to unique factorization into prime ideals as the ideals (2), (3), $(1 + \sqrt{-5})$, and $(1 - \sqrt{-5})$ are *not* prime ideals in R. Let $I_1 = (2, 1 + \sqrt{-5})$ and $I_2 = (3, 1 + \sqrt{-5})$. Note I_1 and I_2 are not principal ideals as R does not have any elements of norm 2 or 3. Then,

$$(2) = I_1^2, \qquad (3) = I_2\overline{I_2}, \qquad (1 + \sqrt{-5}) = I_1 I_2, \qquad (1 - \sqrt{-5}) = I_1\overline{I_2},$$

and this gives the following factorization of the ideal (6) into a product of prime ideals:

$$(6) = I_1^2 I_2 \overline{I_2}.$$

It is known that $h(-5) = 2$ and so $[I_1]$ (or $[I_2]$ or $[\overline{I_2}]$) is the nontrivial element of $\mathcal{C}(\mathbf{K})$.

(2) $D = 6$: We have the following factorizations in R:

$$6 = 2 \cdot 3 = [\sqrt{6}]^2.$$

This corresponds to the factorization of principal ideals:

$$(6) = (2)(3) = (\sqrt{6})^2.$$

While this superficially looks a lot like the example we have just given for $D = -5$, in fact it could not be more different. The elements 2, 3, and $\sqrt{6}$ of R are not irreducible. To be precise, we have the following factorizations into irreducible elements:

$$2 = -[2+\sqrt{6}][2-\sqrt{6}], \quad 3 = [3+\sqrt{6}][3-\sqrt{6}], \quad \sqrt{6} = -[2-\sqrt{6}][3+\sqrt{6}].$$

Then the element 6 has the factorization into irreducible elements:

$$6 = [2 - \sqrt{6}]^2[3 + \sqrt{6}]^2.$$

If we let $I_1 = (2 + \sqrt{6}) = \overline{I_1}$ and $I_2 = (3 + \sqrt{6}) = \overline{I_2}$, then $(2) = I_1^2$ and $(3) = I_2^2$. We thus obtain the following factorization of the ideal (6) into a product of prime ideals:

$$(6) = I_1^2 I_2^2.$$

But in this case R is a UFD by Corollary 2.53, and so the above factorization of 6 into irreducible elements is essentially unique. (The elements $2+\sqrt{6}$ and $2-\sqrt{6}$ are associates, as are the elements $3+\sqrt{6}$ and $3-\sqrt{6}$.) The factorization of (6) into prime ideals just reflects this factorization of elements. Finally, as R is a UFD, $h(6)=1$.

(3) $D=-23$: We have the following factorizations:

$$27 = 3^3 = [2+\sqrt{-23}][2-\sqrt{-23}].$$

This suggests that we should consider $I=(3, 2+\sqrt{-23})$, a prime ideal. Note that I is a nonprincipal ideal as R has no elements of norm 3. Then, by Theorem 5.66 (and Remark 5.69),

$$I\overline{I} = (3), \qquad I^3 = (2+\sqrt{-23}), \qquad \overline{I}^3 = (2-\sqrt{-23}).$$

Thus, we see that we have the following factorization of the ideal (27) into a product of prime ideals:

$$(27) = I^3\overline{I}^3.$$

It is known that $h(-23)=3$ and, by Example 5.68(2), $[I]$ is an element of $\mathcal{C}(\mathbf{K})$ of order 3, so $[I]$ is a generator of this group.

(4) $D=-26$: We have the following factorizations:

$$27 = 3^3 = [1+\sqrt{-26}][1-\sqrt{-26}].$$

This suggests that we should consider $I=(3, 1+\sqrt{-26})$, a prime ideal. Note that I is a nonprincipal ideal as R has no elements of norm 3. Then, by Theorem 5.66 (and Remark 5.69),

$$I\overline{I} = (3), \qquad I^3 = (1+\sqrt{-26}), \qquad \overline{I}^3 = (1-\sqrt{-26}).$$

Thus, we see that we have the following factorization of the ideal (27) into a product of prime ideals:

$$(27) = I^3\overline{I}^3.$$

By Example 5.68(1), $[I]$ is an element of $\mathcal{C}(\mathbf{K})$ of order 3. Let $J = (2, \sqrt{-26})$. Then J is not a principal ideal but $J^2=(2)$ is a principal ideal, by Theorem 5.55, so $[J]$ is an element of $\mathcal{C}(\mathbf{K})$ of order 2. Let $K = IJ = (6, 2+\sqrt{-26})$. Then K is an element of $\mathcal{C}(\mathbf{K})$ of order 6. It is known that $h(-26)=6$, so $[K]$ is a generator of this group.

(5) $D = -47$: We have the following factorizations:

$$243 = 3^5 = [14 + \sqrt{-47}][14 - \sqrt{-47}].$$

This suggests that we should consider $I = (3, 14 + \sqrt{-47})$, a non-principal prime ideal. Then $I\overline{I} = (3)$ and $I^5 = (14 + \sqrt{-47})$, so $(243) = (3)^5 = I^5\overline{I}^5$. By Example 5.68(6), $[I]$ is an element of $\mathcal{C}(\mathbf{K})$ of order 5. It is known that $h(-47) = 5$, so $[I]$ is a generator of this group.

We also have the factorizations

$$16807 = 7^5 = [128 + 3\sqrt{-47}][128 - 3\sqrt{-47}],$$

leading to the ideal $J = (7, 128 + 3\sqrt{-47})$, and a similar argument shows that $[J]$ is an element of $\mathcal{C}(\mathbf{K})$ of order 5. Thus, $[J]$ must be a power of $[I]$. Calculation shows that $I^2J = (-4 + \sqrt{-47})$, a principal ideal, so $[J] = [I]^{-2} = [I]^3$ in $\mathcal{C}(\mathbf{K})$.

(6) $D = -239$: We have the following factorizations:

$$243 = 3^5 = [2 + \sqrt{-239}][2 - \sqrt{-239}],$$

leading to the ideal $I = (3, 1 + \sqrt{-235})$, with $[I]$ an element of order 5 in $\mathcal{C}(\mathbf{K})$.

We also have the factorizations

$$4913 = 17^3 = [33 + 4\sqrt{-239}][33 - 4\sqrt{-239}],$$

leading to the ideal $J = (17, 33 + 4\sqrt{-235})$, with $[J]$ an element of order 3 in $\mathcal{C}(\mathbf{K})$.

Then $K = IJ = (51, 4 + \sqrt{-239})$ is an element of order 15 in $\mathcal{C}(\mathbf{K})$. It is known that $h(-239) = 15$, so $[K]$ is a generator of this group.

(7) $D = 79$: We have the following factorizations:

$$-15 = -3 \cdot 5 = [8 - \sqrt{79}][8 + \sqrt{79}],$$

leading to the ideals $I = (3, 8 - \sqrt{235})$ and $J = (5, 8 - \sqrt{235})$. It can be shown that R has no elements of norm 3 or 5, from which it follows that I and J are not principal ideals. $I^3 = (17 + 2\sqrt{79})$, a principal ideal, and $IJ = (-8 + \sqrt{79})$, a principal ideal. It is known that $h(79) = 3$, so $[I]$ is a generator of this group, and $[J] = [I]^{-1}$.

(8) $D = 235$: We have the following factorizations:

$$234 = 2 \cdot 3^2 \cdot 13 = -[1+\sqrt{235}][1-\sqrt{235}],$$

leading to the ideals $I_1 = (3, 1+\sqrt{235})$, $I_2 = (13, 1+\sqrt{235})$, and $J = (2, 1+\sqrt{235})$. Note that I_1, I_2, and J are not principal ideals, as, since ± 3, ± 13, and ± 2 are quadratic nonresidues $(\bmod 5)$, R does not have elements of norm 3, 13, or 2.

Then $I_1\overline{I_1} = (3)$, $I_2\overline{I_2} = (13)$, and $J^2 = (2)$. (Recall that $\overline{J} = J$.) More interestingly, $I_1^2 I_2 J = (1+\sqrt{235})$ and so $\overline{I_1}^2\overline{I_2}J = (1-\sqrt{235})$. In any event, we have the factorization into prime ideals:

$$(234) = I_1^2\overline{I_1}^2 I_2\overline{I_2}J^2.$$

Let $K = I_1 J = (6, 1+\sqrt{235})$. Then $K^3 = (34 - 2\sqrt{235})$, and $\|34 - 2\sqrt{235}\| = 216 = 6^3$. It can be shown that R does not have an element of norm 6, from which it follows that K is not a principal ideal, and hence $[K]$ has order 3 in $\mathcal{C}(\mathbf{K})$. $J^2 = (2)$, so $[J]$ has order 2 in $\mathcal{C}(\mathbf{K})$. This implies that $[I_1]$ has order 6 in $\mathcal{C}(\mathbf{K})$ (and calculation shows that $I_1^6 = (67 + 4\sqrt{235})$). It is known that $h(235) = 6$, so $[I_1]$ is a generator of this group.

(9) $D = 401$: We have the following factorizations:

$$400 = 2^4 \cdot 5^2 = -[1+\sqrt{401}][1-\sqrt{401}],$$

leading to the ideal $I = (5, 1+\sqrt{401})$. It can be shown that R has no element of norm 5, from which it follows that I is not a principal ideal. $I^5 = (22 - 3\sqrt{401})$, a principal ideal. It is known that $h(401) = 5$, so $[I]$ is a generator of this group.

(10) $D = 483$: We have the following factorizations:

$$483 = 3 \cdot 7 \cdot 23 = [\sqrt{483}]^2,$$

leading to the ideals $I = (3, \sqrt{483})$ and $J = (2, 1+\sqrt{483})$. Then $K = IJ = (6, 3+\sqrt{483})$. Congruence considerations $(\bmod 483)$ show that R does not have any elements of norm 2, 3, or 6, so I, J, and K are all nonprincipal ideals. On the other hand, $I^2 = (3)$, $J^2 = (2)$, and $K^2 = (6)$ are all principal ideals. Thus, $\{(1), [I], [J], [K]\}$ is a subgroup of $\mathcal{C}(\mathbf{K})$ isomorphic to $[\mathbb{Z}/2\mathbb{Z}]^2$. It is known that $h(235) = 4$, so this subgroup is in fact all of $\mathcal{C}(\mathbf{K})$.

Remark 5.71. As we have observed, for D negative, it is a finite process to check whether $\mathcal{O}(\sqrt{D})$ has an element of norm n for any particular value of n. In case $D > 0$, we sometimes have been able to rule out this possibility through congruence considerations. The theory of continued fractions, which we do not develop here—it is long, though elementary—enables us to decide this for any n with $|n|$ relatively small compared to D. (We alluded to this in parts (7), (8), and (9) of Example 5.70.)

Remark 5.72. For any particular value of D, $h(D)$ can be computed from *Dirichlet's class number formula*, which we shall not present here. The nature of this formula, however, is not such that we can draw general conclusions about the behavior or properties of $h(D)$. As you might imagine, the formula is more complicated for positive values of D than for negative values of D. The class number formula evidently yields an integer for $D < 0$, but it is not a priori evident that this number is positive. The class number formula evidently yields a real number for $D > 0$, but it is not even a priori evident that this real number is an integer!

Remark 5.73. Tables of the values of $h(D)$ for all D, positive and negative, with $|D| < 500$, can be found in the book *Number Theory* by Z. I. Borevich and I. R. Shafarevich, Academic Press, New York, 1966. (Note the following error in the tables: the correct value for $h(-485)$ is 20.)

Remark 5.74. We have the following three conjectures of Gauss:

Conjecture. *There are exactly nine imaginary quadratic fields* $\mathbf{Q}(\sqrt{D})$ *with* $h(D) = 1$. *They are* $\mathbf{Q}(\sqrt{D})$ *for* $D = -1, -2, -3, -7, -11, -19, -43, -67,$ *and* -163.

Conjecture. *For any* N, *there are only finitely many imaginary quadratic fields* $\mathbf{Q}(\sqrt{D})$ *with* $h(D) = N$.

Conjecture. *There are infinitely many real quadratic fields* $\mathbf{Q}(\sqrt{D})$ *with* $h(D) = 1$.

The first two of these conjectures are major twentieth-century theorems. The history of their proof is an interesting one, and we refer the interested reader to the article "Gauss' Class Number Problem for Imaginary Quadratic Fields," by D. Goldfeld, *Bull. Amer. Math. Soc.* 13 (1985), 23–37.

The last of these conjectures is still wide open.

The historically minded reader may wonder about the attribution of various theorems and conjectures in this section to Gauss. Gauss's work appeared in his monumental book *Disquisitiones Arithmeticae*, published in 1801, while algebraic number theory had its beginnings in the mid and late nineteenth century. The explanation is that Gauss investigated binary quadratic forms. With the advent of algebraic number theory, it was realized that his work could be reinterpreted in the context of the ideal theory of quadratic fields.

5.7 Behavior of Ideals in Algebraic Number Fields

In this section, we consider "what happens" to prime ideals in $\mathbb{Z}$ when we consider them as ideals in $\mathcal{O}(\mathbf{K})$ for some algebraic number field $\mathbf{K}$. We will make this rather vague statement clear below, but first we need some preliminaries. We let $R = \mathcal{O}(\mathbf{K})$. Also, we let $n = \deg_{\mathbf{K}/\mathbf{Q}}$.

It is easy to check that if I is any fractional ideal of $\mathbf{Q}$, then $RI = \{\sum \alpha_j i_j \mid \alpha_j \text{ in } R, i_j \text{ in } I\}$ is a fractional ideal of $\mathbf{K}$. In this situation we write $I_{\mathbf{K}}$ for RI.

Lemma 5.75. *Let P be any prime ideal of R. Then P divides $(p)_{\mathbf{K}}$ for exactly one prime number p.*

Proof: It is easy to check that for any two integers m_1 and m_2, $(m_1 m_2)_{\mathbf{K}} = (m_1)_{\mathbf{K}}(m_2)_{\mathbf{K}}$. Now consider P. By the proof of Lemma 5.39, we know that P contains an integer m. Let m have prime factorization $m = p_1^{a_1} p_2^{a_2} \cdots p_k^{a_k}$. Then P divides $(m)_{\mathbf{K}} = ((p_1)_{\mathbf{K}})^{a_1}((p_2)_{\mathbf{K}})^{a_2} \cdots ((p_k)_{\mathbf{K}})^{a_k}$. By the definition of a prime ideal, this implies that P divides $(p_i)_{\mathbf{K}}$ for some i. Suppose that P divides $(p_j)_{\mathbf{K}}$ as well, for some $p_j \neq p_i$. Then p_i is in P and p_j is in P, so $1 = p_i x + p_j y$ for some integers x and y is in P, and then $P = R$, a contradiction. □

In the situation of Lemma 5.75, we say that P lies over p. Thus we see that every prime ideal of R lies over some prime number p. Our object in this section is to investigate the prime ideals lying over an arbitrary prime number p. To this end, we consider any prime number p. By unique factorization of ideals in R, we have

$$(p)_{\mathbf{K}} = P_1^{e_1} P_2^{e_2} \cdots P_g^{e_g},$$

with each P_i a prime ideal. Note that $\|(p)_{\mathbf{K}}\| = p^n$ and $\|P_i\|$ divides $\|(p)_{\mathbf{K}}\|$, so we must have $\|P_i\| = p^{f_i}$ for some f_i. The integer e_i is called the *ramification index* of P_i and the integer f_i is called the *residue class field degree* of P_i. (Note that R/P_i is a field as P_i is a prime ideal and hence a maximal ideal of the Dedekind domain R.) Here is our basic result.

Theorem 5.76. *In the above situation,*

$$n = \sum_{i=1}^{g} e_i f_i.$$

Furthermore, if $\mathbf{K}$ *is a Galois extension of* $\mathbf{Q}$*, then all of the* e_i*'s are equal, and all of the* f_i*'s are equal. Let* e *be the common value of the* e_i*'s and* f *be the common value of the* f_i*'s. Thus, in the case of a Galois extension* $\mathbf{K}$ *of* $\mathbf{Q}$*,*

$$n = efg.$$

Proof: By Lemma 5.39,

$$\#((p)_{\mathbf{K}}) = p^n$$

and by the multiplicativity of the norm (Lemma 5.47),

$$\#(P_1^{e_1} P_2^{e_2} \cdots P_g^{e_g}) = \#(P_1)^{e_1} \cdot \#(P_2)^{e^2} \cdot \ldots \#(P_g)^{e_g} = p^{f_1 e_1} p^{f_1 e_1} \ldots p^{f_g e_g}.$$

Comparing the exponents of p yields the first claim of the theorem.

If $\mathbf{K}$ is a Galois extension of $\mathbf{Q}$, the Galois group leaves $(p)_{\mathbf{K}}$ invariant and permutes the P_i's transitively, so the e_i's and the f_i's are all equal, yielding the second claim of the theorem. □

Example 5.77. Let us examine the possible behavior for a quadratic field $\mathbf{K}$. Here $n = 2$ so we have the following possibilities:

$$\begin{array}{llll} (1) & g = 1, & e_1 = 2, & f_1 = 1, \\ (2a) & g = 2, & e_1 = e_2 = 1, & f_1 = f_2 = 1, \\ (2b) & g = 1, & e_1 = 1, & f_1 = 2. \end{array}$$

We have numbered the cases as above as they correspond to the cases with the same number in Theorem 5.55. (In Theorem 5.55 we divided case (1) into cases (1a) and (1b) depending on whether P was a principal ideal, but that distinction is not relevant here.) In case (1), p is said to *ramify* in $\mathbf{K}$. In case (2a), p is said to *split* in $\mathbf{K}$. In case (2b), p is said to be *inert* in $\mathbf{K}$.

Note that in Theorem 5.55, we had a simple numerical criterion for deciding whether or not we were in case (1). (The distinction between cases (2a) and (2b) is much more subtle, as we see from Remark 5.58.) This is in fact true in general. To any algebraic number field $\mathbf{K}$ we can associate an integer $\Delta_{\mathbf{K}}$, the *discriminant* of $\mathbf{K}$. $\Delta_{\mathbf{K}}$ can be computed directly from a knowledge of $\mathcal{O}(\mathbf{K})$.

Theorem 5.78 (Dedekind). *Let* $\mathbf{K}$ *be an algebraic number field and let* p *be a prime number. Let*

$$(p)_{\mathbf{K}} = P_1^{e_1} P_2^{e_2} \cdots P_g^{e_g}$$

as above. Then $e_i > 1$ *for some* i *if and only if* p *divides* $\Delta_{\mathbf{K}}$. *(In this case, we say that* P_i *is ramified.)*

Example 5.79. In the case of a quadratic field $\mathbf{K} = \mathbf{Q}(\sqrt{D})$,

$$\begin{aligned} \Delta_{\mathbf{K}} &= D \quad \text{if } D \equiv 1 \pmod 4, \\ \Delta_{\mathbf{K}} &= 4D \quad \text{if } D \equiv 2 \text{ or } 3 \pmod 4. \end{aligned}$$

We see that the general theory specializes to the result we had in Theorem 5.55, as, in the notation of that theorem, p divides $e_D D$ if and only if p divides $\Delta_{\mathbf{K}}$.

In fact, even for quadratic fields the situation can be very intricate. We close this section by citing the following theorem:

Theorem 5.80. *Let* m *be any integer and let* S_i, S_r, *and* S_s *be any three finite disjoint sets of primes. Then there are infinitely many positive values of* D, *and infinitely many negative values of* D, *for which, setting* $\mathbf{K} = \mathbf{Q}(\sqrt{D})$,

(1) $h(\mathbf{K})$ *is divisible by* m;

(2) all the primes in S_i *are inert in* $\mathbf{K}$;

(3) all the primes in S_r *ramify in* $\mathbf{K}$;

(4) all the primes in S_s *split in* $\mathbf{K}$.

5.8 Ideal Elements

One last question about ideals faces us—a linguistic one. Where does the name "ideal" come from? The answer to this question itself involves some interesting mathematics. The theory of ideals was first developed by Kummer in his studies of algebraic number fields. Kummer discovered that any ideal I in an algebraic number field $\mathbf{K}$ becomes a principal ideal upon passing to a larger algebraic number field $\mathbf{K}'$, and he called a generator of this principal ideal an *ideal element* of $\mathbf{K}$.

Theorem 5.81 (Kummer). *Let* $\mathbf{K}$ *be an algebraic number field and let* I *be an ideal in* $R = \mathcal{O}(\mathbf{K})$. *Then there is an algebraic number field* $\mathbf{K}' \supseteq \mathbf{K}$ *and an element* α' *of* $R' = \mathcal{O}(\mathbf{K}')$ *such that*

$$I = (\alpha') \cap \mathbf{K}.$$

(In other words, the original ideal I *consists of those elements of the ideal* (α') *that are in* $\mathbf{K}$.*)*

(So α' is *not* in general an element of $\mathbf{K}$, but the ideal I consists of the multiples of α' that *are* in $\mathbf{K}$. Thus the name "ideal element" of $\mathbf{K}$ is a good one for α'.)

Proof: We adopt a similar notation to that in the previous section. If J is any ideal of R, we denote by $J_{\mathbf{K}'}$ the ideal $R'J$ of R'.

By Theorem 5.43, $\mathcal{C}(\mathbf{K})$ is a finite group. Thus $[I]$ has finite order n, so $I^n = (\beta)$ for some β in $\mathbf{K}$. We set $\mathbf{K}' = \mathbf{K}(\alpha')$ where α' is a root of the polynomial $X^n - \beta$. Then, as ideals in $\mathbf{K}'$,

$$(\alpha')^n = (\beta)_{\mathbf{K}'} = (I_{\mathbf{K}'})^n,$$

so, by unique factorization of ideals in $\mathbf{K}'$, we have $I_{\mathbf{K}'} = (\alpha')$. But then $I = I_{\mathbf{K}'} \cap \mathbf{K} = (\alpha') \cap \mathbf{K}$. □

Later on, mathematicians realized that the notion of an ideal was an important one, and have focused on that notion rather than on the notion of an ideal element. (Indeed, our proof of Theorem 5.81 is certainly not Kummer's proof, as our proof uses Minkowski's result Theorem 5.43, a later development in algebraic number theory.) But let us see what the ideal elements of one of the ideals we have considered above are.

Example 5.82. We refer to Example 5.70(1), with $\mathbf{K} = \mathcal{O}(\sqrt{-5})$. We have the ideals $I_1 = (2, 1+\sqrt{-5}) = (1+\sqrt{-5}, 1-\sqrt{-5})$, $I_2 = (3, 1+\sqrt{-5})$, and $\overline{I_2} = (3, 1-\sqrt{-5})$.

First consider the ideal I_1. Let $\mathbf{K}'$ be the field $\mathbf{K}' = \mathbf{Q}(\sqrt{2}, \sqrt{-5})$. Then we have the following factorizations of elements in $\mathcal{O}(\mathbf{K}')$:

$$1+\sqrt{-5} = \sqrt{2}\frac{\sqrt{2}+\sqrt{-10}}{2},$$
$$1-\sqrt{-5} = \sqrt{2}\frac{\sqrt{2}-\sqrt{-10}}{2}.$$

It may seem obvious that these are factorizations, but there is something we need to check: we need to check that all the factors are algebraic integers.

(Recall that an algebraic integer is the root of a monic polynomial with integer coefficients.) Now $\sqrt{2}$ is certainly an algebraic integer, as it is a root of the monic polynomial $X^2 - 2$. But $(\sqrt{2}+\sqrt{-10})/2$ and $(\sqrt{2}-\sqrt{-10})/2$ are also algebraic integers as they are both roots of the monic polynomial $X^4 + 4X^2 + 9$. Since $1 + \sqrt{-5}$ and $1 - \sqrt{-5}$ are both multiples of $\sqrt{2}$, it then readily follows that $\sqrt{2}$ is an ideal generator of I_1.

Next consider the ideals I_2 and $\overline{I_2}$. Now let $\mathbf{K}'$ be the field $\mathbf{K}' = \mathbf{Q}(\sqrt{3}, \sqrt{-5})$. Then we have the following factorizations of elements in $\mathcal{O}(\mathbf{K}')$:

$$3 = [\sqrt{3}]^2$$

$$1 + \sqrt{-5} = \sqrt{3}\frac{\sqrt{3}+\sqrt{-15}}{3},$$

$$1 - \sqrt{-5} = \sqrt{3}\frac{\sqrt{3}-\sqrt{-15}}{3}.$$

Again we need to check that all the factors are algebraic integers. Now $\sqrt{3}$ is an algebraic integer, as it is a root of the monic polynomial $X^2 - 3$, and $(\sqrt{3}+\sqrt{-15})/3$ and $(\sqrt{3}-\sqrt{-15})/3$ are also algebraic integers, as they are both roots of the monic polynomial $X^4 + 8X^2 + 4$. It then readily follows that $\sqrt{3}$ is an ideal generator of both I_2 and $\overline{I_2}$.

Remark 5.83. Let us begin with an arbitrary algebraic number field $\mathbf{K}$. Since $\mathcal{C}(\mathbf{K})$ is a finite group, it easily follows that there is a field in which *every* ideal of $\mathcal{O}(\mathbf{K})$ becomes principal, and in fact a smallest such field $\mathbf{K}'$. But this does not imply that every ideal of $\mathcal{O}(\mathbf{K}')$ is principal, as there may be new ideals of $\mathcal{O}(\mathbf{K}')$ that we have to consider. So we may have to pass to another smallest field $\mathbf{K}''$ in order that all the ideals of $\mathcal{O}(\mathbf{K}')$ become principal, etc. It is natural to ask whether this procedure must always eventually stop, and the answer is no. It is known that this procedure may go on forever, i.e., that we may obtain an infinite sequence of fields $\mathbf{K} \subset \mathbf{K}' \subset \mathbf{K}'' \ldots$ related in this way.

5.9 Dirichlet's Unit Theorem

To explain *Dirichlet's Unit Theorem* we need to take a more abstract view of algebraic number fields. We will briefly recapitulate a bit of field theory. First we define an (abstract) algebraic number field (compare Definition 5.3).

Definition 5.84. A field $\mathbf{K}$ is an algebraic number field if it is a finite extension of $\mathbf{Q}$, i.e., $\mathbf{K}$ is an extension of $\mathbf{Q}$ with degree $\deg_{\mathbf{K}/\mathbf{Q}}$, the dimension of $\mathbf{K}$ as a vector space over $\mathbf{Q}$, finite.

Lemma 5.85. *Let $\mathbf{K}$ be an algebraic number field. Then every element of $\mathbf{K}$ is algebraic, i.e., is a root of a polynomial in $\mathbf{Q}[X]$.*

Proof: Exactly the same as the proof of Lemma 5.4. □

In the following lemma, $(f(X))$ denotes the ideal of $\mathbf{F}[X]$ generated by $f(X)$.

Lemma 5.86. *Let $\mathbf{F}$ be a field and let $f(X)$ be an irreducible polynomial of degree n in $\mathbf{F}[X]$. Then $\mathbf{E} = \mathbf{F}[X]/(f(X))$ is a field and is an extension of $\mathbf{F}$ of degree n.*

Proof: Although part of this theorem follows from general considerations, we will give a concrete proof of all of it. Let $f(X) = a_nX^n + \ldots + a_0$ and let $\pi : \mathbf{F}[X] \to \mathbf{F}[X]/(f(X)) = \mathbf{E}$ be the projection. Set $x = \pi(X)$. Then $f(x) = a_nx^n + \ldots + a_0 = \pi(a_nX^n + \ldots + a_0) = \pi(f(X)) = 0$ in $\mathbf{E} = \mathbf{F}[X]/(f(X))$. Thus x is a root of $f(X)$ in $\mathbf{E}$.

We claim that $S = \{x^{n-1}, x^{n-2}, \ldots, x, 1\}$ is a basis for $\mathbf{E}$ as a vector space over $\mathbf{F}$. First, let us see that S spans $\mathbf{E}$: Consider an arbitrary element y of $\mathbf{E}$. Then $y = \pi(g(X))$ for some polynomial $g(X)$. By the division algorithm for polynomials we may write $g(X)$ uniquely as $g(X) = f(X)q(X) + r(X)$ where either $r(X) = 0$ or $r(X)$ is a polynomial of degree at most $n - 1$, $r(X) = b_{n-1}X^{n-1} + \ldots + b_0$. But then $y = \pi(g(X)) = g(x) = f(x)q(x) + r(x) = 0 \cdot q(x) + r(x) = r(x) = b_{n-1}x^{n-1} + \ldots + b_0$ is in the span of S. Next, let us see that S is linearly independent: Suppose that some nontrivial linear combination $c_{n-1}x^{n-1} + \ldots + c_0 \cdot 1$ of elements of S is 0. Then x is a root of $h(X) = c_{n-1}X^{n-1} + \ldots + c_0$. Then (recalling that $\mathbf{F}[X]$ is a Euclidean domain), x is a root of the nonzero polynomial $k(X) = \gcd(f(X), h(X))$. Thus, $k(X)$ cannot be a constant polynomial. But $k(X)$ divides $f(X)$, contradicting the hypothesis that $f(X)$ is irreducible.

Now we claim that $\mathbf{E}$ is a field. Let y be an arbitrary nonzero element of $\mathbf{E}$. Then, as above, $y = g(x)$ for some nonzero polynomial $g(X)$ that is not divisible by $f(X)$. Since $f(X)$ is assumed irreducible, $1 = \gcd(f(X), g(X))$, and then $1 = f(X)s(X) + g(X)t(X)$ for some polynomials $s(X)$ and $t(X)$. But then $1 = f(x)s(x) + g(x)t(x) = 0 \cdot s(x) + yt(x) = yt(x)$, so y is invertible in $\mathbf{E}$. □

Corollary 5.87. *Every algebraic number field $\mathbf{K}$ can be obtained by a finite succession of extensions as in Lemma 5.86, beginning with $\mathbf{Q}$.*

Proof: Let $\{\alpha_1, \ldots \alpha_n\}$ be a basis for $\mathbf{K}$ as a vector space over $\mathbf{Q}$. Set $\mathbf{K}_0 = \mathbf{Q}$. Now α_1 is a root of an irreducible polynomial $f_1(X)$ in $\mathbf{K}_0[X]$. Set $\mathbf{K}_1 = \mathbf{K}_0[X]/(f_1[X])$. Now α_2 is a root of an irreducible polynomial $f_2(X)$ in $\mathbf{K}_1[X]$. Set $\mathbf{K}_2 = \mathbf{K}_1[X]/(f_2[X])$. Continue, to obtain $\mathbf{K}_n = \mathbf{K}$.□

We now consider embeddings of algebraic number fields in $\mathbf{C}$. An *embedding* of $\mathbf{K}$ in $\mathbf{C}$ is an isomorphism φ from $\mathbf{K}$ to a subfield of $\mathbf{C}$. We say that φ is a real embedding if its image is a subfield of $\mathbf{R}$ and a complex embedding otherwise.

Theorem 5.88. *Let* $\mathbf{K}$ *be an algebraic number field with* $\deg_{\mathbf{K}/\mathbf{Q}} = n$. *Let* r *be the number of real embeddings of* $\mathbf{K}$ *in* $\mathbf{C}$ *and let* s *be the number of pairs of conjugate complex embeddings of* $\mathbf{K}$ *in* $\mathbf{C}$. *Then*

$$r + 2s = n.$$

Proof: It is a theorem that every finite extension of $\mathbf{Q}$ can be obtained by a single application of Lemma 5.86, i.e., $\mathbf{K} = \mathbf{Q}[X]/(f(X))$ for an irreducible polynomial $f(X)$ in $\mathbf{Q}[X]$ of degree n. Let x in $\mathbf{K}$ be as in the proof of Lemma 5.86. Then $\mathbf{K}$ is spanned by the powers of x, so any embedding φ is determined by $\varphi(x)$. Furthermore, $f(x) = 0$. Then, for any embedding φ of $\mathbf{K}$, we must have $0 = \varphi(0) = \varphi(f(x)) = f(\varphi(x))$. In other words, $\varphi(x)$ must be a root of $f(X)$ in $\mathbf{C}$. It is easy to check that we may obtain an embedding by choosing $\varphi(x)$ to be any root of $f(X)$ in $\mathbf{C}$.

It is a theorem that every irreducible polynomial in $\mathbf{Q}[X]$ has distinct roots. The Fundamental Theorem of Algebra tells us that every polynomial of degree n in $\mathbf{C}[X]$, and hence every polynomial of degree n in $\mathbf{Q}[X]$, has n roots in $\mathbf{C}$. Also, as is well known, the complex (i.e., nonreal) roots of every polynomial in $\mathbf{R}[X]$, and hence of every polynomial in $\mathbf{Q}[X]$, occur in conjugate pairs. Assembling these facts, we see that $f(X)$ factors in $\mathbf{C}[X]$ as

$$f(X) = (X - \rho_1) \ldots (X - \rho_r)(X - \sigma_1)(X - \overline{\sigma_1}) \ldots (X - \sigma_s)(X - \overline{\sigma_s})$$

for distinct real numbers $\rho_1, \ldots, \rho_r$ and distinct complex (i.e., nonreal) numbers $\sigma_1, \overline{\sigma_1}, \ldots, \sigma_s, \overline{\sigma_s}$ with $r + 2s = n$. As we have observed, we obtain an embedding φ of $\mathbf{K}$ from each choice $\varphi(x) = \rho_i$, σ_i, or $\overline{\sigma_i}$, and these are all the embeddings. □

Theorem 5.88 says that, given an "abstract" algebraic number field of degree n, as we have defined it here, there are n ways to identify it with a "concrete" algebraic number field, as we have previously defined it.

Example 5.89.

(1) Let D be a square-free integer, $D \neq 1$, and let $\mathbf{K} = \mathbf{Q}[X]/(X^2 - D)$. Then $X^2 - D = (X - \sqrt{D})(X + \sqrt{D})$, and so we obtain embeddings of $\mathbf{K}$ in $\mathbf{C}$ by $\varphi(x) = \sqrt{D}$ and $\varphi(x) = -\sqrt{D}$. We thus see:

for $\mathbf{K}$ a real quadratic field, i.e., $D > 0$, $r = 2$ and $s = 0$;

for $\mathbf{K}$ an imaginary quadratic field, i.e., $D < 0$, $r = 0$ and $s = 1$.

(2) Let $\mathbf{K}$ be obtained in two stages: $\mathbf{K}_1 = \mathbf{Q}[X]/(X^2 - 2)$ and $\mathbf{K} = \mathbf{K}_1[X]/(X^2 - 3)$. In fact $\mathbf{K}$ can be obtained in one stage as $\mathbf{K} = \mathbf{Q}[X]/(X^4 - 10X^2 + 1)$. Then $X^4 - 10X^2 + 1 = (X - (\sqrt{2} + \sqrt{3}))(X - (\sqrt{2} - \sqrt{3}))(X - (-\sqrt{2} + \sqrt{3}))(X - (-\sqrt{2} - \sqrt{3}))$, and so we obtain embeddings of $\mathbf{K}$ in $\mathbf{C}$ by $\varphi(x) = \sqrt{2} + \sqrt{3}$, $\sqrt{2} - \sqrt{3}$, $-\sqrt{2} + \sqrt{3}$, or $-\sqrt{2} - \sqrt{3}$. In this case $r = 4$ and $s = 0$.

(3) Let $\mathbf{K} = \mathbf{Q}[X]/(X^3 - 2)$. Then $X^3 - 2 = (X - \sqrt[3]{2})(X - \omega\sqrt[3]{2})(X - \omega^2\sqrt[3]{2})$ where $\omega = (-1 + i\sqrt{3})/2$ is a primitive cube root of 1, and so we obtain embeddings of $\mathbf{K}$ in $\mathbf{C}$ by $\varphi(x) = \sqrt[3]{2}$, $\omega\sqrt[3]{2}$, or $\omega^2\sqrt[3]{2}$. In this case $r = 1$ and $s = 1$.

Observe that the units of an algebraic number field $\mathbf{K}$ form a group under multiplication. Here is the result to which we have been heading.

Theorem 5.90 (Dirichlet). *Let $\mathbf{K}$ be an algebraic number field of degree n, with r real embeddings and s pairs of conjugate complex embeddings. Let $U_{\mathbf{K}}$ be the group of units in $\mathbf{K}$. Then $U_{\mathbf{K}}$ is isomorphic to $R_{\mathbf{K}} \times F_{\mathbf{K}}$, where $R_{\mathbf{K}}$ is the finite cyclic group of roots of 1 in $\mathbf{K}$, and $F_{\mathbf{K}}$ is a free abelian group of rank $r + s - 1$.*

We call a set of generators for $F_{\mathbf{K}}$ a *fundamental system of units* for $\mathbf{K}$.

Example 5.91.

(1) Let $\mathbf{K} = \mathbf{Q}(\sqrt{D})$ for $D < 0$ be an imaginary quadratic field. From Example 5.89(1) we see that $r + s - 1 = 0$, so that $U_{\mathbf{K}}$ is isomorphic to the finite group $R_{\mathbf{K}}$. For $D = -1$, $R_{\mathbf{K}} = \{\pm 1, \pm i\}$ is a group of order 4. For $D = -3$, $R_{\mathbf{K}} = \{\pm 1, \pm\omega, \pm\omega^2\}$ is a group of order 6. Otherwise, $R_{\mathbf{K}} = \{\pm 1\}$ is a group of order 2.

(2) Let $\mathbf{K} = \mathbf{Q}(\sqrt{D})$ for $D > 0$ be a real quadratic field. Then $R_{\mathbf{K}} = \{\pm 1\}$ is a group of order 2. From Example 5.89(1) we see that $r + s - 1 = 1$, so that $F_{\mathbf{K}}$ is isomorphic to $\mathbb{Z}$. Let ε_0 be a fundamental unit of $\mathbf{K}$.

If $\varepsilon_0 = a + b\sqrt{D}$, then a and b are integers, or perhaps half–integers if $D \equiv 1 \pmod 4$, with $a^2 - b^2D = \pm 1$, so do not immediately give a solution of Pell's equation. But some power $\varepsilon_{\text{Pell}}$ of ε_0 does, and so we recover our previous description of the structure of solutions of Pell's equation (and in particular the fact that Pell's equation always has infinitely many solutions) as a special case of this theorem.

(3) Let $\mathbf{K} = \mathbf{Q}(\sqrt{2}, \sqrt{3})$. Then $R_{\mathbf{K}} = \{\pm 1\}$ is a group of order 2. From Example 5.89(2) we see that $r + s - 1 = 3$, so that $F_{\mathbf{K}}$ is isomorphic to $\mathbb{Z}^3$. We can easily obtain three elements of $F_{\mathbf{K}}$ as follows: $\mathbf{K} \supset \mathbf{Q}(\sqrt{2})$ with fundamental unit $\varepsilon_1 = 1+\sqrt{2}$, $\mathbf{K} \supset \mathbf{Q}(\sqrt{3})$ with fundamental unit $\varepsilon_2 = 2 + \sqrt{3}$, and $\mathbf{K} \supset \mathbf{Q}(\sqrt{6})$ with fundamental unit $\varepsilon_3 = 5 + 2\sqrt{6}$. (Note that ε_2 and ε_3 give solutions to Pell's equation. ε_1 does not, but ε_1^2 does.) These elements generate a free subgroup F' of $F_{\mathbf{K}}$ of rank 3, so F' must be a subgroup of $F_{\mathbf{K}}$ of finite index, but a priori need not be all of $F_{\mathbf{K}}$. In fact, it is not. It is known that $\{1+\sqrt{2}, \sqrt{2}+\sqrt{3}, (\sqrt{2}+\sqrt{6})/2\}$ is a fundamental system of units for $\mathbf{K}$, and it is then easy to check that F' is a subgroup of $F_{\mathbf{K}}$ of index 4.

(4) Let $\mathbf{K} = \mathbf{Q}(\sqrt[3]{2})$. Then $R_{\mathbf{K}} = \{\pm 1\}$ is a group of order 2. From Example 5.89(3) we see that $r + s - 1 = 1$, so that $F_{\mathbf{K}}$ is isomorphic to $\mathbb{Z}$. $1 + \sqrt[3]{2} + (\sqrt[3]{2})^2$ is a fundamental unit of $\mathbf{K}$ (and $(1 + \sqrt[3]{2} + (\sqrt[3]{2})^2)(-1 + \sqrt[3]{2}) = 1$). Similarly, $\mathbf{K} = \mathbf{Q}(\omega\sqrt[3]{2})$ has fundamental unit $1 + \omega\sqrt[3]{2} + (\omega\sqrt[3]{2})^2$ and $\mathbf{K} = \mathbf{Q}(\omega^2\sqrt[3]{2})$ has fundamental unit $1 + \omega^2\sqrt[3]{2} + (\omega^2\sqrt[3]{2})^2$.

5.10 Exercises

Exercise 5.1. Let $R = \mathcal{O}(\sqrt{D})$.

(a) Let $I = (\alpha_1, \ldots, \alpha_k)$ with $\gcd(\|\alpha_1\|, \ldots, \|\alpha_k\|)$ relatively prime. Show that $I = (1)$. (Of course, $(1) = R$.)

(b) More generally, let $I = (\alpha_1, \ldots, \alpha_k)$ and suppose there is an element α of R with α dividing each α_i and with $\|\alpha\| = \gcd(\|\alpha_1\|, \ldots, \|\alpha_k\|)$. Show that $I = (\alpha)$.

Exercise 5.2. Let $R = \mathcal{O}(\sqrt{D})$. Let m_1 and m_2 be relatively prime integers and let α be any element of R with $\|\alpha\|$ dividing m_1m_2. Show that $(m_1, \alpha)(m_2, \alpha) = (\alpha)$.

Exercise 5.3. In each case, it is known that $\mathcal{C}(\mathbf{Q}(\sqrt{D}))$ is a cyclic group of the given order $h(D)$. Find an ideal I of $\mathcal{O}(\sqrt{D})$ with $[I]$ a generator of $\mathcal{C}(\mathbf{Q}(\sqrt{D}))$.

(a1)	$D = -103$,	$h(D) = 5$	(a2)	$D = -127$,	$h(D) = 5$
(b1)	$D = -71$,	$h(D) = 7$	(b2)	$D = -151$,	$h(D) = 7$
(c1)	$D = -199$,	$h(D) = 9$	(c2)	$D = -367$,	$h(D) = 9$
(d1)	$D = -74$,	$h(D) = 10$	(d2)	$D = -86$,	$h(D) = 10$
(e1)	$D = -167$,	$h(D) = 11$	(e2)	$D = -271$,	$h(D) = 11$
(f1)	$D = -191$,	$h(D) = 13$	(f2)	$D = -263$,	$h(D) = 13$
(g1)	$D = -101$,	$h(D) = 14$	(g2)	$D = -134$,	$h(D) = 14$
(h1)	$D = -439$,	$h(D) = 15$	(h2)	$D = -751$,	$h(D) = 15$
(i1)	$D = -293$,	$h(D) = 18$	(i2)	$D = -335$,	$h(D) = 18$
(j1)	$D = -743$,	$h(D) = 21$	(j2)	$D = -1931$,	$h(D) = 21$
(k1)	$D = -461$,	$h(D) = 30$	(k2)	$D = -509$,	$h(D) = 30$
(l1)	$D = -1031$,	$h(D) = 35$	(l2)	$D = -2087$,	$h(D) = 35$
(m1)	$D = -794$,	$h(D) = 42$	(m2)	$D = -1046$,	$h(D) = 42$

Exercise 5.4. In a Dedekind domain R, we can define the gcd and lcm of ideals similarly to our definition of the gcd and lcm of elements in Chapter 2:

> If I and J are ideals of R, then $G = \gcd(I, J)$ if G is an ideal of R dividing both I and J, and if any other ideal dividing I and J divides G. Similarly, if I and J are ideals of R, then $L = \operatorname{lcm}(I, J)$ if L is an ideal of R divisible by both I and J, and if any other ideal divisible by I and J is divisible by L. Also, two ideals I and J of R are relatively prime if $\gcd(I, J) = (1)$.

(a) Show that $G = \gcd(I, J)$ and $L = \operatorname{lcm}(I, J)$ always exist.

(b) Show that $G = I + J = \{i + j \mid i \text{ in } I,\ j \text{ in } J\}$ and that $L = I \cap J$.

(c) Show that I and J are relatively prime if and only if they have no common prime ideal factor. In this case, show that $L = IJ$.

(d) Express G and L in terms of the factorizations of I and J into prime ideals.

(e) Show that $GL = IJ$.

Exercise 5.5. For I an ideal of R, define $a \equiv b \pmod{I}$ if $a - b$ is an element of I. Show that the Chinese Remainder Theorem holds for ideals in a Dedekind domain R: if I and J are relatively prime ideals, then the simultaneous congruences $x \equiv a \pmod{I}$ and $x \equiv b \pmod{J}$ have a solution for any a and b, and this solution is unique $\pmod{IJ}$.

Exercise 5.6. Let I and J be ideals in a Dedekind domain R. Show that R/J is isomorphic to I/IJ.

Exercise 5.7. Use the descriptions of the ideals P in Remark 5.58 to show that these ideals are maximal and hence prime, thus providing another proof of Theorem 5.55.

Exercise 5.8. Let $\mathbf{K}$ be an algebraic number field with $\deg_{\mathbf{K}/\mathbf{Q}} = n$. Let α be an algebraic integer in $\mathbf{K}$. It follows from field theory that the degree d of the polynomial $m_\alpha(X)$ divides n. Let this polynomial have constant term a. The norm $\|\alpha\|$ of the element α is defined to be $|a^{n/d}|$. We have defined the norm of an ideal of $\mathcal{O}(\mathbf{K})$ in general in Definition 5.40, and it is a theorem that for a principal ideal (α), $\|(\alpha)\| = \|\alpha\|$. Verify that with this definition of the norm of an element, this theorem is true when $\mathbf{K} = \mathbf{Q}(\sqrt{D})$.

Exercise 5.9. Let $\mathbf{K}'$ be an algebraic number field that is an extension of the algebraic number field $\mathbf{K}$. Let $R = \mathcal{O}(\mathbf{K})$ and let $R' = \mathcal{O}(\mathbf{K}')$.

(a) Let I_0 be a fractional ideal of $\mathbf{K}$ such that $R'I_0 = R'$. Show that $I_0 = R$.

(b) Let I be any ideal of R. Show that $R'I \cap R = I$.

Exercise 5.10. Let $R = \mathbb{Z}[X]$, the ring of polynomials in the variable "X" with integer coefficients. R is known to be a UFD. Note that parts (a) and (b) below show that R is *not* a Dedekind domain.

(a) Show that (2) and (X) are prime ideals of R.

(b) Show that $(2, X)$ is a maximal ideal of R.

(c) Find a pair of ideals I and J such that $J \supset I$ but such that there is no ideal K with $I = JK$.

(d) Find an ideal of R that *cannot* be written as a product of prime ideals. (This is true even though every element of R *can* be written as a product of prime elements.)

(e) Let $I = (2, X)$. Show that, for any positive integer n, the minimum number of elements in a generating set for I^n is $n + 1$. In particular, no power of I is principal.

Exercise 5.11. Let $R' = \{a + b\sqrt{-3} \mid a \text{ and } b \text{ are integers}\}$. We observed in Example 5.30(2) that R' is not integrally closed in its quotient field. Observe that we have the factorizations $4 = [1+\sqrt{-3}][1-\sqrt{-3}] = 2 \cdot 2$ in R'.

(a) Show that $(1+\sqrt{-3})$, $(1-\sqrt{-3})$, and (2) are not prime ideals in R'.

(b) Show that the ideal $(2, 1+\sqrt{-3})$ is not invertible.

(c) Find an example of ideals I_1, I_2, and J in R' with $I_1 \neq I_2$ but with $I_1 J = I_2 J$.

Exercise 5.12. Let R be the ring of "polynomials in positive fractional powers of X." That is, R consists of expressions such as $2+3X^{1/4}+5X^{2/3}+7X^3$ with addition and multiplication given by the usual rules of exponents.

(a) Show that R does not satisfy the ascending chain condition (ACC).

(b) Find a nonzero proper ideal I of R with $I^2 = I$.

Appendix A

Mathematical Induction

Mathematical induction is an extremely powerful and useful proof technique. In this appendix we describe it and its variants, and draw some of its particularly useful consequences.

A.1 Mathematical Induction and Its Equivalents

Mathematical induction is nothing other than a formalized version of dominoes.

Suppose we have an infinite collection of dominoes, numbered 1, 2, 3, ... Suppose that the first domino falls. Also, suppose that the dominoes are arranged so that if each domino falls, the next one will also fall. What will be the result? Clearly, it is that they will all fall.

To formalize this, let $F(n)$ be the proposition

$F(n)$: Domino n falls.

Then we are assuming that

(1) $F(1)$ and

(2) if $F(n)$, then $F(n+1)$;

and we derive from this the conclusion

$F(n)$ for all positive integers n.

Stated in this way, there is clearly nothing special about $F(n)$, and we may substitute any proposition. This leads us to the principle of mathematical induction.

Axiom A.1 (The Principle of Mathematical Induction). *Let $P(n)$ be any proposition about positive integers. Suppose that*

(1) $P(1)$ is true;

(2) for each positive integer n, if $P(n)$ is true, then $P(n+1)$ is true.

Then $P(n)$ is true for every positive integer n.

Let us emphasize that the point of (2) is that we do not need to prove $P(n)$, but rather that we can assume $P(n)$ and use it to prove $P(n+1)$. (In a typical proof, at the point we involve the truth of $P(n)$, we often state "By the inductive hypothesis ....")

As a practical matter, when using mathematical induction to prove a proposition, it is usually the case that verifying (1) is easy but verifying (2) takes work. Occasionally it is the case that verifying (2) is easy but verifying (1) takes work. Rarely it is the case that verifying both (1) and (2) take work. (And it is virtually never the case that verifying both (1) and (2) is easy—that would be getting something for nothing.)

There is a variant of mathematical induction called complete induction. Again, we will first introduce it in the context of dominoes, so again let us suppose we have an infinite number of dominoes numbered 1, 2, 3, ... Again, let us suppose that the first domino falls. But now let us suppose that the dominoes are arranged a bit differently, so that it is not necessarily the case that if domino n falls, then domino $n+1$ falls, for every n. Suppose it is instead the case that if dominoes 1 through n all fall, the domino $n+1$ falls, for every n. What will happen? Again, all the dominoes will fall.

There is another possibility. Assume the first domino falls, and the arrangement of the dominoes is that domino $n+1$ is not necessarily knocked down by domino n, but by domino k for some k between 1 and n, for every n. What will happen? Again, all the dominoes will fall. But in this third situation, we generally do not know which domino k will knock down domino $n+1$. So we handle that lack of information by simply assuming that we are back in the second case, that is, if dominoes 1 through n all fall (and hence the mysterious domino k falls, whatever k may happen to be), domino $n+1$ will also fall, for every n.

There are some other possibilities as well, where the first domino falls, and where domino $n+1$ falls if some (perhaps known, perhaps mysterious) combination of the preceding n dominoes falls. But again, we can handle this by simply assuming that we are back in the second case, that is, if dominoes 1 through n all fall, domino $n+1$ will also fall, for every n.

This second case, translating from dominoes falling to general propositions, gives us complete induction.

Axiom A.2 (The Principle of Complete Induction). *Let $P(n)$ be any proposition about positive integers. Suppose that*

(1) $P(1)$ is true;

(2) for each positive integer n, if $P(k)$ is true for all integers k between 1 and n, then $P(n+1)$ is true.

Then $P(n)$ is true for every positive integer n.

These two variants of induction are logically equivalent. However, in order to prove this it is convenient to introduce a third variant, well-ordering, which is important in its own right.

On the face of it, well-ordering looks very different than induction, but it turns out to be equivalent to it.

First, the domino version: Suppose we have an infinite number of dominoes numbered 1, 2, 3, ..., not all of which have fallen down. Then among the dominoes that are still standing, there is one with the smallest number. Stated slightly more formally, in this case among the set S consisting of the dominoes that are still standing, there is a domino that has the smallest number. Translated into general terms, this gives the statement of well-ordering.

Axiom A.3 (The Well-Ordering Principle). *Any nonempty subset S of the set of positive integers has a smallest element.*

We now show that all three of these variants of induction are logically equivalent. The proof of this is rather subtle and tricky.

Theorem A.4. *The following are equivalent:*

(1) the Principle of Mathematical Induction,

(2) the Principle of Complete Induction, and

(3) the Well-Ordering Principle.

Proof: As is common in this situation, we produce a "round-robin" proof, showing that (1)⇒(2)⇒(3)⇒(1).

(1)⇒(2): We need to be clear about what we must prove. We are supposing that the hypotheses of complete induction are satisfied, and we need to show that we can use the method of mathematical induction to obtain the conclusion of complete induction.

So suppose the hypotheses of complete induction are satisfied for a proposition $P(n)$:

(1) $P(1)$ is true.

(2) If $P(k)$ is true for every integer k with $1 \leq k \leq n$, then $P(n+1)$ is true, for every n.

Let us consider a new proposition $Q(n)$. $Q(n)$ is the proposition

$$Q(n) : P(k) \text{ is true for every integer } k \text{ with } 1 \leq k \leq n.$$

Note that $P(1)$ and $Q(1)$ say the same thing, so our assumption that $P(1)$ is true gives

(1′) $Q(1)$ is true.

Also the hypothesis of (2) becomes $Q(n)$ is true, and the conclusion of (2) is that $P(n+1)$ is true. But we are assuming $Q(n)$ is true, so we know that $P(k)$ is true for every k with $1 \leq k \leq n$, and that, together with the fact that $P(n+1)$ is true, shows that $P(k)$ is true for every k with $1 \leq k \leq n+1$. But that assertion is $Q(n+1)$. Thus we see that our assumption (2) gives

(2′) For each integer n, $Q(n)$ implies $Q(n+1)$.

But now we may apply mathematical induction to (1′) and (2′) to conclude that $Q(n)$ holds for every positive integer n. So for every positive integer n,

$$P(k) \text{ is true for every integer } k \text{ with } 1 \leq k \leq n.$$

In particular, choosing $k = n$, we see that

$$P(n) \text{ is true for every integer } n,$$

and that is what we wanted to prove.

(2)⇒(3): Here we are supposing that the hypotheses of well-ordering are satisfied, and we need to show that we can use the method of complete induction to obtain the conclusion of well-ordering.

Actually, we will put the precise statement of well-ordering aside for the moment and deal with a related statement instead. Let $S(n)$ be the proposition

$S(n)$: Every subset T of the set of positive integers
that contains the integer n has a smallest element.

First we observe that

(1) $S(1)$ is true, for if the set T contains the positive integer 1, then 1 is certainly the smallest positive integer in T (as 1 is the smallest positive integer, period).

Next we claim that

(2) if $S(k)$ is true for every integer k with $1 \leq k \leq n$, then $S(n+1)$ is true. Suppose that T contains the positive integer $n+1$. There are two possibilities: either $n+1$ is the smallest positive integer in T, or it is not. In the first case, T certainly has a smallest element, namely $n+1$. In the second case, T contains some positive integer k with $k < n+1$, i.e., with $1 \leq k \leq n$. But then, since $S(k)$ is assumed to be true, we can again conclude that T has a smallest element.

Thus, both hypotheses of complete induction are satisfied and we can apply complete induction to conclude

$S(n)$ is true for every positive integer n.

Now let us return to well-ordering per se. We are given a nonempty subset S of the set of positive integers. To say that S is nonempty means that it contains some positive integer n_0. We have just concluded that $S(n)$ is true for *every* positive integer n, so in particular it is true for the integer $n = n_0$. But $S(n_0)$ states that *every* set T that is a subset of the set of positive integers and that contains n_0 has a smallest element. But our set S is such a set, so we conclude that S has a smallest element, and that is what we wanted to prove.

(3) $\Rightarrow$(1): Here we are supposing that the hypotheses of mathematical induction are satisfied, and we need to show that we can use the well-ordering principle to obtain the conclusion of mathematical induction.

So suppose that $P(n)$ is a proposition about the positive integer n, and we have that

(1) $P(1)$ is true, and

(2) if $P(n)$ is true, then $P(n+1)$ is true, for every n.

We want to conclude that $P(n)$ is true for every n.

Our proof in this case is a proof by contradiction. Suppose it is not the case that $P(n)$ is true for every n. Then $P(n)$ is false for at least one n.

Thus if we define the set S by

$$S = \{n \mid P(n) \text{ is false}\},$$

then S is a nonempty set. We now apply the well-ordering principle to conclude that S has a least element n_0. There are two cases: (1′) $n_0 = 1$ or (2′) $n_0 > 1$.

Case (1′): $n_0 = 1$. Since n_0 is in S, by the definition of S we have that $P(1)$ is false. But this directly contradicts hypothesis (1) of mathematical induction, which states that $P(1)$ is true.

Case (2′): $n_0 > 1$. Since n_0 is in S, by the definition of S we have that $P(n_0)$ is false. But in this case $n_0 - 1$ is a positive integer and, since n_0 is the smallest positive integer in S, we see that $n_0 - 1$ is not in S. Then, by the definition of S, we have that $P(n_0 - 1)$ is true. But now we can apply hypothesis (2) of mathematical induction: since $P(n_0 - 1)$ is true, so is $P((n_0 - 1) + 1)$, i.e., $P(n_0)$ is true, which is again a contradiction.

Thus, our assumption that the conclusion of induction is false leads (in any case) to a contradiction, so we conclude that the conclusion of induction is true, that $P(n)$ is true for every positive integer n, and that is what we wanted to prove. □

Remark A.5. We should observe that we have not proved mathematical induction. In fact, it is one of the basic properties that define the structure of the positive integers, and, as we have indicated, we take it as an axiom.

A.2 Consequences of Mathematical Induction

In this section, we draw two of the most often used consequences of mathematical induction. Here is the first one:

Theorem A.6. *Let $a_1 > a_2 > a_3 > \ldots$ be a strictly decreasing sequence of positive integers. Then this sequence is finite.*

Proof: Let $S = \{a_1, a_2, a_3, \ldots\}$ be the set of all the elements of the sequence. Then, by the Well-Ordering Principle, S has a smallest element, which is a_k for some k. We claim the sequence stops at a_k.

We prove this by contradiction. Suppose not. Then S contains the next term a_{k+1}. But $a_k > a_{k+1}$ since the sequence is strictly decreasing. This contradicts a_k being the smallest element of S. Hence this supposition is false and the sequence stops at a_k.

Thus, the sequence has a finite number (to be precise, k, but we don't know what k is) of terms. □

The second consequence is the Pigeonhole Principle. The Pigeonhole Principle is an intuitively obvious principle that gets its name from an old (but not obsolete) technology. Here it is:

> Suppose a postal employee is sorting letters into pigeonholes, and there are more letters than pigeonholes. Then some pigeonhole must contain more than one letter.

We have said that the pigeonhole principle is intuitively obvious. But of course that does not mean that it does not require proof. We state it precisely and prove it now, as a consequence of mathematical induction.

Theorem A.7 (The Pigeonhole Principle). *If m objects are sorted into n categories, and $m > n$, then at least one category contains more than one object.*

Proof: We prove this by induction on n. Let $P(n)$ be the statement of the theorem for a given value of n. We verify the hypotheses of induction:

(1) $P(1)$ is true: If there is only one category, then all $m > 1$ objects are in that category.

(2) $P(n)$ implies $P(n+1)$: Suppose $m > n+1$ objects are sorted into n categories. Pick a category. If that category has $k > 1$ objects, we are done. So suppose not. Then it has $k = 0$ or 1 objects. Consider the remaining n categories and the remaining $m - k$ objects. If $k = 0$, $m-k = m > n+1 > n$, while if $k = 1$, $m-k = m-1 > (n+1)-1 = n$, so in either event $m - k > n$ and by the inductive hypothesis at least one of the remaining categories contains more than one object. So in any case $P(n+1)$ is true.

Then, by induction, we conclude that $P(n)$ is true for every positive integer n. □

The pigeonhole principle has a couple of variants, which are proved similarly.

Theorem A.8. *If m objects are sorted into n categories and $m < n$, then at least one category is empty.*

Proof: We prove this by complete induction on m. Let $P(m)$ be the statement of the theorem for a given value of m. We verify the hypotheses of complete induction:

(1) $P(1)$ is true: Pick a category. If that category is empty, we are done. Otherwise, the single object is in that category. In that case, every other category is empty.

(2) $P(m)$ implies $P(m+1)$: Suppose $m+1 < n$ objects are sorted into n categories. Pick a category. If that category is empty, we are done. So suppose not. Then it contains $j \geq 1$ objects. Consider the remaining $n-1$ categories and the remaining $m' = m+1-j$ objects. Since $m+1 < n$, $m' = m+1-j < n-j \leq n-1$. Then, by the inductive hypothesis, at least one of the remaining categories is empty. So in any case $P(m+1)$ is true.

Thus, by complete induction, we conclude that $P(m)$ is true for every positive integer m. □

Theorem A.9. *Suppose that n objects are sorted into n categories. The following are equivalent.*

(1) Every category contains at most one object.

(2) Every category contains at least one object.

(3) Every category contains exactly one object.

Proof: If $n = 1$, then the single object is in the single category, and all three statements are true, and so are equivalent.

Assume henceforth that $n \geq 2$. We begin by observing that (3) is logically equivalent to (1) and (2). Hence (3) implies (1) and (3) implies (2). If we show that (1) implies (2), that will show that (1) implies (3), and if we show that (2) implies (1), that will show that (2) implies (3). So we must prove these two implications.

(1) implies (2): We prove the contrapositive: not-(2) implies not-(1). Suppose (2) is false and some category is empty. Then the remaining $n-1$ categories contain n objects, so by the original pigeonhole principle (Theorem A.7) some category must contain more than one object, and (1) is false.

(2) implies (1): We prove the contrapositive: not-(1) implies not-(2). Suppose (1) is false and some category contains $j > 1$ objects. Then the remaining $n-1$ categories contain $m = n - j < n-1$ objects, so by the above variant on the pigeonhole principle (Theorem A.8), some category must be empty, and (1) is false. (We may have $m = 0$, but then all of the other categories are empty and (1) is certainly false.) □

A.3 Exercises

Exercise A.1. A *geometric progression* with first term a and ratio r is a sequence of the form $a,\ ar,\ ar^2,\ ar^3,\ \ldots$. Show that, for $r \neq 1$, the sum of the first n terms of this progression is $a(r^n - 1)/(r-1)$, i.e., show that

$$\sum_{i=0}^{n-1} ar^i = a\frac{r^n - 1}{r-1}.$$

(Of course, if $r = 1$, then the progression is constant and the sum of its first n terms is an.)

Exercise A.2. A decomposition of a positive integer n is a way of writing n as a sum of positive integers *in order*. Let $d(n)$ be the number of decompositions of n. For example, $d(4) = 8$ as 4 has the decompositions $4,\ 3+1,\ 1+3,\ 2+2,\ 2+1+1,\ 1+2+1,\ 1+1+2,\ 1+1+1+1$. Show that, for every positive integer n, $d(n) = 2^{n-1}$. (Hint: show that $d(n) = 1 + d(1) + d(2) + \ldots + d(n-1)$ for $n \geq 2$.)

Exercise A.3. Let $S = \{x_0, x_1, x_2, \ldots\}$ be a set of positive integers with $x_0 = 1$ and $x_{i-1} < x_i \leq 2x_{i-1}$ for $i \geq 1$.

(a) Show that every positive integer n can be written as a sum of distinct elements of S.

(b) We might call this a "sub-binary" expansion of n as if $x_i = 2^i$ for all i (in which case $S = \{1, 2, 4, \ldots\}$), this is just the binary expansion of n. Show that in this case the expansion of n is unique for every positive integer n, but that in every other case there are positive integers n whose expansion is not unique.

Exercise A.4.

(a) Let $x = m_1 + \sqrt{m_1^2 - 1}$ for some positive integer m_1. Show that, for every $n \geq 1$, $x^n = m_n + \sqrt{m_n^2 - 1}$ for some positive integer m_n. (For example, if $x = 2 + \sqrt{3}$, then $x^2 = 7 + 4\sqrt{3} = 7 + \sqrt{48}$, $x^3 = 26 + 15\sqrt{3} = 26 + \sqrt{675}, \ldots$.)

(b) More generally, let $x = m_1 + \sqrt{m_1^2 - N}$ for some positive integer m_1 and some integer N. Show that, for every $n \geq 1$, $x^n = m_n + \sqrt{m_n^2 - N^n}$ for some positive integer m_n. (For example, if $x = 2 + \sqrt{6}$, in which case $N = -2$, then $x^2 = 10 + 4\sqrt{6} = 10 + \sqrt{96}$, $x^3 = 44 + 18\sqrt{6} = 44 + \sqrt{1944}$, $\ldots$.)

Exercise A.5. A *characteristic* of degree d is defined to be a sequence of $2d$ integers $(m_1, \ldots, m_d, n_1, \ldots, n_d)$ with each m_i and each n_i either 0 or 1. A characteristic is *even* or *odd* according as $m_1n_1 + \ldots + m_dn_d$ is even or odd. Let $e(d)$ be the number of even characteristics of degree d and let $o(d)$ be the number of odd characteristics of degree d. (For example, $e(1) = 3$ as there are 3 even characteristics of degree 1, namely $(0,0)$, $(0,1)$, and $(1,0)$; while $o(1) = 1$ as there is 1 odd characteristic of degree 1, namely $(1,1)$.)

(a) Show that $e(d+1) = 3e(d) + o(d)$ and that $o(d+1) = 3o(d) + e(d)$ for every positive integer d. (Hint: think about extending a characteristic of degree $d-1$ to a characteristic of degree d.)

(b) Show that $e(d) = 2^{d-1}(2^d+1)$ and $o(d) = 2^{d-1}(2^d-1)$ for every positive integer d.

Exercise A.6. Fix positive integers r, s, and t and define a sequence by

$$a_1 = r, \qquad a_{n+1} = (s+1)a_n + t \text{ for } n \geq 1.$$

(a) Suppose $r = s = t = 1$, so the sequence is defined by $a_1 = 1$, $a_{n+1} = 2a_n + 1$. Show that $a_n = 2^n - 1$ for every n.

(b) Suppose $r = s = 1$, so the sequence is defined by $a_1 = 1$, $a_{n+1} = 2a_n + t$. Show that $a_n = (t+1)2^{n-1} - t$ for every n.

(c) Suppose $r = t = 1$, so the sequence is defined by $a_1 = 1$, $a_{n+1} = (s+1)a_n + 1$. Show that $a_n = ((s+1)^n - 1)/s$ for every n.

(d) Suppose $s = t = 1$, so the sequence is defined by $a_1 = r$, $a_{n+1} = 2a_n + 1$. Show that $a_n = (r+1)2^{n-1} - 1$ for every n.

(e) Experiment with arbitrary values of r, s, and t and come up with, and prove, a formula for a_n in general.

Exercise A.7. The *Fibonacci numbers* are defined by

$$f_1 = 1, \quad f_2 = 1, \quad f_{n+2} = f_n + f_{n+1} \text{ for } n \geq 1.$$

The first few Fibonacci numbers are given by

n	1	2	3	4	5	6	7	8	9	10
f_n	1	1	2	3	5	8	13	21	34	55

(a) Show that $f_1 + f_2 + \ldots + f_n = f_{n+2} - 1$.

(b) Show that $f_1 + f_3 + \ldots + f_{2n-1} = f_{2n}$.

(c) Show that $f_2 + f_4 + \ldots + f_{2n} = f_{2n+1} - 1$.

(d) Show that $(3/2)^{n-2} < f_n < 2^{n-2}$ for $n \geq 4$.

Exercise A.8. The Fibonacci numbers can be generalized as follows. Choose a and b nonzero and let

$$g_1 = a, \quad g_2 = b, \quad g_{n+2} = g_n + g_{n+1} \text{ for } n \geq 1.$$

(a) Show that $g_1 + g_2 + \ldots + g_n = g_{n+2} - b$.

(b) Show that $g_1 + g_3 + \ldots + g_{2n-1} = g_{2n} - (b - a)$.

(c) Show that $g_2 + g_4 + \ldots + g_{2n} = g_{2n+1} - a$.

Exercise A.9. The generalized Fibonacci numbers with $a = 1$ and $b = 3$ are known as the *Lucas numbers*. The first few Lucas numbers are given by:

n	1	2	3	4	5	6	7	8	9	10
ℓ_n	1	3	4	7	11	18	29	47	76	123

Observe that $\ell_1 = f_1$. Show that $\ell_n = 2f_{n-1} + f_n$ and that $f_n = (2/5)\ell_{n-1} + (1/5)\ell_n$ for every $n \geq 2$.

Exercise A.10. Define sequences $\{p_0, p_1, p_2, \ldots\}$ and $\{q_0, q_1, q_2, \ldots\}$ by

$$p_0 = 1, \quad p_n = p_{n-1} + 2q_{n-1} \text{ for } n \geq 1,$$
$$q_0 = 1, \quad q_n = 2p_{n-1} + q_{n-1} \text{ for } n \geq 1.$$

The first few of these are given by

n	0	1	2	3	4	5	6
p_n	1	1	3	7	17	41	99
q_n	0	1	2	5	12	29	70

(a) Observe that $p_0 = 1$, $p_1 = 1$, $q_0 = 0$, and $q_1 = 1$. Show that $p_n = p_{n-2} + 2p_{n-1}$ for $n \geq 2$, and that $q_n = q_{n-2} + 2q_{n-1}$ for $n \geq 2$.

(b) Show that $p_n = q_{n-1} + q_n$ and that $q_n = (1/2)p_{n-1} + (1/2)p_n$.

(c) Show that $p_n^2 - 2q_n^2 = (-1)^n$.

Exercise A.11. Define sequences $\{p_0, p_1, p_2, \ldots\}$ and $\{q_0, q_1, q_2, \ldots\}$ by

$$p_0 = 1, \quad p_n = 2p_{n-1} + 3q_{n-1} \text{ for } n \geq 1,$$
$$q_0 = 0, \quad q_n = p_{n-1} + 2q_{n-1} \text{ for } n \geq 1.$$

The first few of these are given by

n	0	1	2	3	4	5	6
p_n	1	2	7	26	97	362	1351
q_n	0	1	4	15	56	209	780

(a) Observe that $p_0 = 1$, $p_1 = 2$, $q_0 = 0$, and $q_1 = 1$. Show that $p_n = -p_{n-2} + 4p_{n-1}$ for $n \geq 2$, and that $q_n = -q_{n-2} + 4q_{n-1}$ for $n \geq 2$.

(b) Show that $p_n = -q_{n-1} + 2q_n$ and that $q_n = -(1/3)p_{n-1} + (2/3)p_n$.

(c) Show that $p_n^2 - 3q_n^2 = 1$.

Exercise A.12. Define sequences $\{p_0, p_1, p_2, \ldots\}$ and $\{q_0, q_1, q_2, \ldots\}$ by

$$p_0 = 1, \quad p_n = 2p_{n-1} + 5q_{n-1} \text{ for } n \geq 1,$$
$$q_0 = 1, \quad q_n = p_{n-1} + 2q_{n-1} \text{ for } n \geq 1.$$

The first few of these are given by

n	0	1	2	3	4	5	6
p_n	1	2	9	38	161	682	2889
q_n	0	1	4	17	72	305	1292

(a) Observe that $p_0 = 1$, $p_1 = 2$, $q_0 = 0$, and $q_1 = 1$. Show that $p_n = p_{n-2} + 4p_{n-1}$ for $n \geq 2$, and that $q_n = q_{n-2} + 4q_{n-1}$ for $n \geq 2$.

(b) Show that $p_n = q_{n-1} + 2q_n$ and that $q_n = (1/5)p_{n-1} + (2/5)p_n$.

(c) Show that $p_n^2 - 5q_n^2 = (-1)^n$.

Exercise A.13. Define sequences $\{p_0, p_1, p_2, \ldots\}$ and $\{q_0, q_1, q_2, \ldots\}$ by

$$p_0 = 1, \quad p_n = 5p_{n-1} + 12q_{n-1} \text{ for } n \geq 1,$$
$$q_0 = 1, \quad q_n = 2p_{n-1} + 5q_{n-1} \text{ for } n \geq 1.$$

The first few of these are given by

n	0	1	2	3	4	5	6
p_n	1	5	49	485	4801	47525	470449
q_n	0	2	20	198	1960	19402	192060

(a) Observe that $p_0 = 1$, $p_1 = 5$, $q_0 = 0$, and $q_1 = 2$. Show that $p_n = -p_{n-2} + 10p_{n-1}$ for $n \geq 2$, and that $q_n = -q_{n-2} + 10q_{n-1}$ for $n \geq 2$.

(b) Show that $p_n = -(1/2)q_{n-1} + (5/2)q_n$ and that $q_n = -(1/12)p_{n-1} + (5/12)p_n$.

(c) Show that $p_n^2 - 6q_n^2 = 1$.

Exercise A.14. Fix a rational number D that is not a perfect square. Choose nonzero rational numbers a and b and set $N = a^2 - b^2D$. Define sequences $\{p_0, p_1, p_2, \ldots\}$ and $\{q_0, q_1, q_2, \ldots\}$ by

$$\begin{aligned} p_0 &= 1, \quad p_n = ap_{n-1} + bDq_{n-1} \text{ for } n \geq 1, \\ q_0 &= 0, \quad q_n = bp_{n-1} + aq_{n-1} \text{ for } n \geq 1. \end{aligned}$$

(a) Observe that $p_0 = 1$, $p_1 = a$, $q_0 = 0$, and $q_1 = b$. Show that $p_n = -Np_{n-2} + 2ap_{n-1}$ for $n \geq 2$, and that $q_n = -Nq_{n-2} + 2aq_{n-1}$ for $n \geq 2$.

(b) Show that $p_n = (-Nq_{n-1} + aq_n)/b$ and that $q_n = (-Np_{n-1} + ap_n)/(bD)$.

(c) Show that $p_n^2 - Dq_n^2 = N^n$.

Exercise A.15. Let $D = 5$, $a = 1/2$, $b = 1/2$, and define sequences $\{p_0, p_1, p_2, \ldots\}$ and $\{q_0, q_1, q_2, \ldots\}$ as in Exercise A.14. Show that $\ell_n = 2p_n$ and $f_n = 2q_n$ for every $n \geq 1$, where ℓ_n is the n-th Lucas number and f_n is the n-th Fibonacci number.

Exercise A.16. An *Egyptian fraction decomposition* is a way of writing a fraction a/b as a sum of fractions with numerator 1 and distinct denominators. For example, $3/5$ has an Egyptian fraction decomposition $3/5 = 1/2 + 1/10$ and $3/7$ has an Egyptian fraction decomposition $3/7 = 1/3 + 1/11 + 1/231$. Show that every fraction a/b with a and b positive integers and $a < b$ has an Egyptian fraction decomposition. (Hint: if $a = 1$ we are done. Otherwise, let c be the largest integer with $1/c < a/b$ and let $a'/b' = a/b - 1/c$. Show that $a'/b' < 1/c$ and that $a' < a$. Then argue by complete induction on a.) Note that an Egyptian fraction decomposition can never be unique. For we have the algebraic identity $\frac{1}{x} = \frac{1}{x+1} + \frac{1}{x(x+1)}$ and we may apply that here to get, for example, $1/2 = 1/2 = 1/3 + 1/6 = 1/3 + 1/7 + 1/42 = 1/3 + 1/7 + 1/43 + 1/1806 = \ldots$.

Exercise A.17. You may be familiar with magic squares. A magic square is a square array with the sums of its rows, columns, and diagonals all the same. Let us define an *anti-magic square* to be a square array with the sums of its rows, columns, and diagonals all different. Show that there does not exist an n-by-n antimagic square, for any n, all of whose entries are -1, 0, or 1.

Exercise A.18. Let n be an arbitrary positive integer. Let $S = \{a_1, \ldots, a_n\}$ be any set of n distinct integers between 1 and $2n - 2$. Show that S has two elements a_i and a_j with $a_i + a_j = 2n - 1$.

Exercise A.19. Let n be an arbitrary positive integer. Let $S = \{a_1, \ldots, a_n\}$ be any set of $n + 1$ distinct integers between 1 and $2n$. Show that S has two elements a_i and a_j that are consecutive integers.

Appendix B

Congruences

B.1 The Notion of Congruence

At its core, congruence is just shorthand for a simple relationship between numbers. But it is so pervasive, and so useful, that it has metamorphosed into a point of view. Here is the basic definition.

Definition B.1. Let n be a nonzero integer. Two integers x and a are *congruent* modulo n, written

$$x \equiv a \pmod{n}$$

if their difference $x - a$ is divisible by n.

Thus, for example, $40 \equiv 12 \pmod{7}$ as $40 - 12 = 28$ is divisible by 7, $65 \equiv 0 \pmod{13}$ as $65 - 0 = 65$ is divisible by 13, $9 \equiv -2 \pmod{11}$ as $9 - (-2) = 11$ is divisible by 11, and $123456789 \equiv 987654321 \pmod{9}$ as $123456789 - 987654321 = -864197532 = 9(-96021948)$ is divisible by 9.

Let us draw an immediate consequence of this definition.

Lemma B.2. *For a fixed a, the integers x satisfying the congruence*

$$x \equiv a \pmod{n}$$

are the integers $x = a + nk$ for any integer k.

Proof: On the one hand, if x is of the form $x = a + nk$, then $x - a = nk$, which is certainly divisible by n, so $x \equiv a \pmod{n}$.

On the other hand, if $x \equiv a \pmod{n}$, then $x - a$ is divisible by n, so $x - a = nk$ for some integer k, and hence $x = a + nk$. □

Thus, for example, the integers x with $x \equiv 0 \pmod{2}$ are the integers of the form $x = 2k$, i.e., the even integers, and the integers x with $x \equiv 1 \pmod{2}$ are the integers of the form $x = 1 + 2k$, i.e., the odd integers.

We think of two integers that are congruent modulo n as being equivalent in a certain way, or, technically speaking, that congruence modulo n is an *equivalence relation*. That is the content of the next proposition.

Proposition B.3.

(1) For any integer a, $a \equiv a \pmod{n}$.

(2) For any two integers a and b, if $a \equiv b \pmod{n}$, then $b \equiv a \pmod{n}$.

(3) For any three integers a, b, and c, if $a \equiv b \pmod{n}$ and $b \equiv c \pmod{n}$, then $a \equiv c \pmod{n}$.

Proof:

(1) $a - a = 0$ is divisible by n, so $a \equiv a \pmod{n}$.

(2) If $a - b \equiv 0 \pmod{n}$, then $a - b$ is divisible by n, so $a - b = nk$ for some k. But then $b - a = -nk = n(-k)$, so $b - a$ is divisible by n, and then $b \equiv a \pmod{n}$.

(3) If $a \equiv b \pmod{n}$, then $a - b = nk_1$ for some k_1. If $b \equiv c \pmod{n}$, then $b - c = nk_2$ for some k_2. But then

$$a - c = (a - b) + (b - c) = nk_1 + nk_2 = n(k_1 + k_2),$$

so $a - c$ is divisible by n, and then $a \equiv c \pmod{n}$. □

Next we shall see that congruence is compatible with three of the four basic arithmetic operations—addition, subtraction, and multiplication. The situation with division is more complicated, and we defer it to the next section. (See Lemma B.10 and Lemma B.13.)

Proposition B.4. *Suppose that $a_1 \equiv b_1 \pmod{n}$ and $a_2 \equiv b_2 \pmod{n}$. Then*

$$\begin{aligned} a_1 + a_2 &\equiv b_1 + b_2 \pmod{n}, \\ a_1 - a_2 &\equiv b_1 - b_2 \pmod{n}, \\ a_1 a_2 &\equiv b_1 b_2 \pmod{n}. \end{aligned}$$

Proof: Since $a_1 \equiv b_1 \pmod{n}$, $a_1 = b_1 + nk_1$ for some k, by Lemma B.2. Since $a_2 \equiv b_2 \pmod{n}, a_2 = b_2 + nk_2$ for some k_2, also by Lemma B.2. Then

$$a_1 + a_2 = (b_1 + nk_1) + (b_2 + nk_2) = b_1 + b_2 + n(k_1 + k_2),$$

so, again by Lemma B.2,

$$a_1 + a_2 \equiv b_1 + b_2 \pmod{n},$$

and also

$$a_1 - a_2 = (b_1 + nk_1) - (b_2 + nk_2) = b_1 - b_2 + n(k_1 - k_2),$$

so, again by Lemma B.2,

$$a_1 - a_2 \equiv b_1 - b_2 \pmod{n},$$

and, finally,

$$a_1a_2 = (b_1 + nk_1)(b_2 + nk_2) = b_1b_2 + n(k_1b_2 + k_2b_1 + nk_1k_2),$$

so, once again by Lemma B.2,

$$a_1a_2 \equiv b_1b_2 \pmod{n}.$$

□

Next we have the following result, which states, in fancier language, that for a positive integer n, the integers are a *complete set of representatives* of the congruence classes modulo n. But we state the result in a much more down-to-earth way. Note, however, that in order to prove this result we need to use our work in Section 2.1.

But even before we state it, we observe that this encapsulates the usual result of division. When we divide the integer x by the positive integer n, we get a quotient (which we do not care about here) and a remainder. We can always get a remainder between 0 and $n-1$, and when we impose this restriction on the remainder, it is unique.

Theorem B.5. *Let n be a positive integer. For any integer x, the congruence*

$$x \equiv a \pmod{n}$$

is valid for exactly one of the integers $0, 1, 2, \ldots, n-1$.

Proof: First we will show that $x \equiv a \pmod{n}$ for some such integer a between 0 and $n-1$, and then we will show that only one such integer a works.

We begin by applying Lemma 2.7, which states that the integers are a Euclidean domain, and we refer to Definition 2.5 to see what that means. Doing so, we see that, given x, there are integers k_0 and a_0 with

$$x = a_0 + nk_0 \quad \text{and} \quad |a_0| < |n|.$$

Now there are two possibilities:

(1) $a_0 \geq 0$. If this is true then we have $0 \leq a_0 < n$, so a_0 is one of the integers $0, 1, 2, \ldots, n-1$. We set $a = a_0$ and $x \equiv a \pmod{n}$ as required.

(2) $a_0 < 0$. If this is true then we have $-n < a_0 < 0$, so a_0 is one of the integers $-(n-1), -(n-2), \ldots, -1$. In this case, $x = a_0 + nk_0 = (a_0+n)+n(k_0-1)$, so $x \equiv a_0 = n \pmod{n}$. We set $a = a_0+n$, and we see that $x \equiv a \pmod{n}$ and furthermore that a is one of the integers 1, 2, ... , $n-1$, so again a is as required.

Thus, we have that the congruence is true for some a between 0 and $n-1$. Now we must show that there is only one such a. We do this by assuming there is another value a' and showing in fact it must just be a.

So suppose

$$x \equiv a \pmod{n} \quad \text{and} \quad x \equiv a' \pmod{n}$$

with both a and a' among the integers $0, 1, \ldots, n-1$. On the one hand, $-(n-1) \leq a' - a$, as $a' - a$ is smallest when a' is as small as possible and a is as big as possible, i.e., when $a' = 0$ and $a = n-1$. On the other hand, $-(n-1) \leq a' - a$, as $a' - a$ is largest when a' is as big as possible and a is as small as possible, i.e., when $a' = n-1$ and $a = 0$. Putting these together, we see that $-(n-1) \leq a' - a \leq n-1$.

But, by Proposition B.3, the two congruences $x \equiv a \pmod{n}$ and $x \equiv a' \pmod{n}$ imply $a' \equiv a \pmod{n}$, i.e., that $a' - a$ is a multiple of n. But the only multiple of n between $-(n-1)$ and $(n-1)$ is 0, so we see that $a' - a = 0$, i.e., that $a' = a$, as required. □

Actually, the restriction that n be positive was purely for convenience. Here is the more general result.

Corollary B.6. *Let n be a nonzero integer. For any integer x, the congruence*

$$x \equiv a \pmod{n}$$

is valid for a exactly one of the integers $0, 1, 2, \ldots, |n| - 1$.

Proof: If $n > 0$ this is just Theorem B.5.

Suppose $n < 0$, and let $n' = -n$. Then $n' > 0$ and we can apply Theorem B.5 to conclude $x \equiv a \pmod{n'}$ for a exactly one of $0, 1, \ldots, n' - 1$, i.e., for a exactly one of $0, 1, \ldots, |n| - 1$. But $x \equiv a \pmod{n'}$ means $x - a = n'k$ for some k, so $x - a = n'k = -n(k) = n(-k)$ and so $x \equiv a \pmod{n}$, and vice versa. □

With this in hand we can explore the relationships between congruences to different bases.

Proposition B.7. *Let n_1 be any integer dividing n, and set $d = n/n_1$.*

(1) If $x \equiv a \pmod{n}$, then $x \equiv a \pmod{n_1}$.

(2) If $x \equiv a \pmod{n_1}$, then $x \equiv a + n_1 b \pmod{n}$ for b exactly one of the integers $0, 1, 2, \ldots, |d| - 1$.

Proof:

(1) Since $x \equiv a \pmod{n}$, we have that, for some k, $x - a = nk = (n_1 d)k = n_1(dk)$, so $x \equiv a \pmod{n_1}$.

(2) By Lemma B.2, $x \equiv a \pmod{n_1}$ means that x is of the form $x = a + n_1 k$ for some k, and vice versa.

Then certainly $x \equiv a + n_1 k \pmod{n}$, so if we simply claimed $x \equiv a + n_1 b \pmod{n}$ for some b, we would be done. But we are further claiming that we may choose b to be one of the integers $0, 1, \ldots, |d| - 1$, and furthermore that this choice is unique. Again, we will prove this in two stages.

First, by Corollary B.6, we have that $k \equiv b \pmod{d}$ for some value of b between 0 and $|d| - 1$. Then $k = b + dj$ for some j, so

$$x = a + n_1 k = a + n_1(b + dj) = a + n_1 b + n_1 dj = a + n_1 b + nj$$

and hence

$$x \equiv a + n_1 b \pmod{n}.$$

Second, suppose $x \equiv a + n_1 b \pmod{n}$ and $x \equiv a + n_1 b' \pmod{n}$ with both b and b' between 0 and $|d| - 1$. Then, by Proposition B.3, $a + n_1 b' \equiv a + n_1 b \pmod{n}$, and further, by Proposition B.4, $n_1 b' \equiv n_1 b \pmod{n}$. Thus $n_1 b' - n_1 b = nk = n_1 dk$ for some k, so $b' - b = dk$. Thus we see that $b' \equiv b \pmod{d}$. But from here on the proof is the same as the proof of uniqueness in Theorem B.5: if $b' \equiv b \pmod{d}$ with both b and b' between 0 and $|d| - 1$, then $b' = b$. □

Proposition B.7 tells us what the solutions to these two congruences actually are. But it is worthwhile to simply count the number of solutions.

Corollary B.8. *Let n_1 be any integer dividing n, and set $d = n/n_1$.*

(1) The congruence $x \equiv a \pmod{n}$ has a unique solution $\pmod{n_1}$.

(2) The congruence $x \equiv a \pmod{n_1}$ has $|d|$ solutions $\pmod{n}$.

Proof: Part (1) of this corollary follows immediately from part (1) of Proposition B.7, and part (2) of this corollary follows immediately from part (2) of Proposition B.7. □

We should be precise here about what we are counting. For example, if $x \equiv 0 \pmod{10}$, then $x \equiv 0 \pmod{2}$, so we have a unique solution $\pmod{2}$, but there are an infinite number of integers that are solutions, i.e., $x = 0, \pm 10, \pm 20, \pm 30, \ldots$. Similarly, if $x \equiv 0 \pmod{2}$ then $x \equiv 0$, 2, 4, 6, or 8 $\pmod{10}$, so we have five solutions $\pmod{10}$, but again an infinite number of integer solutions, i.e., $x = 0, \pm 2, \pm 4, \pm 6, \pm 8, \pm 10, \ldots$.

Here is another consequence of Proposition B.7.

Corollary B.9. *Let n_1 be any integer dividing n and consider the system of simultaneous congruences*

$$x \equiv a_1 \pmod{n_1},$$
$$x \equiv a \pmod{n}.$$

If $a_1 \equiv a \pmod{n_1}$, this system has a solution, and this solution is unique $\pmod{n}$. If $a_1 \not\equiv a \pmod{n_1}$, this system does not have a solution.

Proof: The second congruence forces the solution to be $x \equiv a \pmod{n}$ so by Proposition B.7 it has a unique solution $\pmod{n}$, and we need only see whether this is also a solution $\pmod{n_1}$.

By Proposition B.7(1), if $x \equiv a \pmod{n}$ then $x \equiv a \pmod{n_1}$, so if $a_1 \equiv a \pmod{n_1}$, then by Proposition B.3 we also have $x \equiv a_1 \pmod{n_1}$, and both congruences are satisfied.

Conversely, suppose both congruences are satisfied. Then $x \equiv a_1 \pmod{n_1}$ and $x \equiv a \pmod{n}$, and again by Proposition B.7(1), the second of these congruences implies $x \equiv a \pmod{n_1}$, so by Proposition B.3 again we must have $a_1 \equiv a \pmod{n_1}$. Then taking the contrapositive gives the second claim. □

In general, we can ask about congruences with respect to two different bases n_1 and n_2. Corollary B.9 is one extreme, where one of the bases divides the other. The other extreme is where the two bases are *relatively prime*, i.e., have no common factor, a notion that is defined precisely in Chapter 2. (See Definition 2.15.) Here we have the Chinese Remainder Theorem, which we state and prove in the next section. Suffice it to say now that in this case, the two congruences are completely independent of each other. More precisely, if n_1 and n_2 are relatively prime, the simultaneous congruences $x \equiv a_1 \pmod{n_1}$ and $x \equiv a_2 \pmod{n_2}$ *always* have a solution, regardless of the values of a_1 and a_2, and furthermore this solution is unique $(\mathrm{mod}\, n_1 n_2)$. (See Lemma B.17, where this is stated for a pair of congruences, and the Chinese Remainder Theorem itself, Theorem B.18, where this is stated for any number of congruences.)

B.2 Linear Congruences

In this section we will consider linear congruences, i.e., congruences of the form

$$ax + b_0 \equiv b_1 \pmod{n}.$$

We note immediately that this congruence is equivalent to

$$ax \equiv b \pmod{n}$$

where $b = b_1 - b_0$, so we will just consider congruences of this form. We will proceed in three stages. First, we will consider congruences of this form with a and n relatively prime. Second, we will consider congruences of this form in general. Third, we will consider systems of congruences, where we will derive the Chinese Remainder Theorem.

We begin with a key lemma.

Lemma B.10. *Suppose that a and n are relatively prime. If*

$$ax_1 \equiv ax_2 \pmod{n},$$

then

$$x_1 \equiv x_2 \pmod{n}.$$

Proof: By definition, $ax_1 \equiv ax_2 \pmod{n}$ means that n divides $ax_1 - ax_2 = a(x_1 - x_2)$. By assumption, a and n are relatively prime, so we may apply Euclid's Lemma (Lemma 2.41) to conclude that n divides $x_1 - x_2$, which means that $x_1 \equiv x_2 \pmod{n}$. □

Theorem B.11. *Suppose that a and n are relatively prime. Then for any b, the congruence*

$$ax \equiv b \pmod{n}$$

has a solution, and this solution is unique $(\bmod\, n)$.

First Proof: For simplicity, let us first assume that n is positive.

We prove this by contradiction. Assume that there is a value of b for which the congruence $ax \equiv b \pmod{n}$ does not have a solution. By Theorem B.5, there is an integer b_0 between 0 and $n-1$ with $b_0 \equiv b \pmod{n}$, and then the congruence $ax' \equiv b_0 \pmod{n}$ also does not have a solution. (For any solution of this latter congruence would also be a solution of our original congruence, by Proposition B.3.)

For each $i = 0, 1, \ldots, n-1$, let c_i be given by

$$c_i \equiv ai \pmod{n} \quad \text{and } 0 \leq c_i \leq n-1.$$

(Note that this indeed uniquely defines c_i, by Corollary B.6.)

Let

$$\mathcal{C} = \{c_0,\ c_1,\ \ldots,\ c_{n-1}\}.$$

On the one hand, $\mathcal{C}$ has a total of n elements. On the other hand, let us ask how many possible values there are for each c_i. A priori, there are n possible values, namely $0, 1, \ldots, n-1$. But by assumption, no c_i can be equal to b_0. So (at least) one of these values is excluded, and so in fact there are fewer than n possible values for each c_i. Hence, by the Pigeonhole Principle, the different c_i's cannot all be distinct. Thus, for some $i_1 \neq i_2$, with both between 0 and $n-1$, $c_{i_1} = c_{i_2}$. But this gives

$$ai_1 \equiv ai_2 \pmod{n} \quad \text{with } 0 \leq i_1,\ i_2 \leq n-1.$$

But by Lemma B.10, since a and n are relatively prime by hypothesis, this gives

$$i_1 \equiv i_2 \pmod{n} \quad \text{with } 0 \leq i_1,\ i_2 \leq n-1,$$

which is impossible by (the last part of the proof of) Theorem B.5.

Hence the congruence $ax \equiv b \pmod{n}$ has a solution (and that solution is the value of i for which $c_i = b_0$).

If n is negative, apply the above argument with n replaced by $|n| = -n$ to conclude that $ax \equiv b \pmod{-n}$ has a solution. But then $ax \equiv b \pmod{n}$ as well (compare Corollary B.6).

Now we must show that the solution is unique $(\operatorname{mod} n)$. Suppose we have solutions $ax' \equiv b \pmod{n}$ and $ax \equiv b \pmod{n}$. Then $ax' \equiv ax \pmod{n}$ by Proposition B.4, and then $x' \equiv x \pmod{n}$ by Lemma B.10. □

Second Proof: This proof is more involved than our first proof but has the virtue of applying to more general situations (to any PID, in the language of Chapter 2).

We are assuming that a and n are relatively prime, i.e., have a gcd of 1, so from Lemma 2.7, Definition 2.18, and Theorem 2.20, we know that there are integers a' and n' with

$$aa' + nn' = 1,$$

so $aa' - 1 = -nn' = n(-n')$ is divisible by n, and so

$$aa' \equiv 1 \pmod{n}.$$

Set $x = a'b$. Then

$$ax = a(a'b) = (aa')b \equiv 1(b) = b \pmod{n},$$

so $x = a'b$ is a solution of our congruence, and indeed so is any $x \equiv a'b \pmod{n}$, by Proposition B.4.

The proof of uniqueness of the solution $(\operatorname{mod} n)$ is the same as in the first proof. □

Let us rephrase this theorem from another point of view.

Corollary B.12. *Suppose that a and n are relatively prime. Then for any b, the congruence*

$$ax \equiv b \pmod{n}$$

is equivalent to the congruence

$$x \equiv c \pmod{n}$$

for some value of c.

Furthermore, $c \equiv a'b \pmod{n}$ where a' is such that $aa' \equiv 1 \pmod{n}$.

Proof: By Theorem B.11, the congruence $ax \equiv b \pmod{n}$ has a solution $x \equiv c \pmod{n}$ for a unique value of $c \pmod{n}$. Thus, the congruence $ax \equiv b \pmod{n}$ holds if and only if the congruence $x \equiv c \pmod{n}$ holds, i.e., these two congruences are equivalent.

Setting $b = 1$ in Theorem B.11, we see that the congruence $ax \equiv 1 \pmod{n}$ has a unique solution $x \equiv a' \pmod{n}$. Setting $c = a'b$, we have

$$ac \equiv a(a'b) \equiv (aa')b \equiv 1b \equiv b \pmod{n},$$

i.e., if $c = a'b$ then $ac \equiv b \pmod{n}$, as claimed. □

Now let us ask how to go about solving a congruence $ax \equiv b \pmod{n}$, with a and n relatively prime, in practice.

First, let us suppose n is relatively small. Then we can directly apply Corollary B.12. The congruence $ax \equiv b \pmod{n}$ is equivalent to $x \equiv c \pmod{n}$, and so there are n possibilities for the congruence class of c, namely integers $0, 1, 2, \ldots, n-1$, and we may simply proceed by trial and error. For example, suppose we wish to solve the congruence $7x \equiv 2 \pmod{10}$. Then we need only try $x = 0, 1, 2, \ldots, 9$. We see, in order, that $7 \cdot 0 = 0 \equiv 0 \pmod{10}$; $7 \cdot 1 = 7 \equiv 7 \pmod{10}$; $7 \cdot 2 = 14 \equiv 4 \pmod{10}$; $7 \cdot 3 = 21 \equiv 1 \pmod{10}$; $7 \cdot 4 = 28 \equiv 8 \pmod{10}$; $7 \cdot 5 = 35 \equiv 5 \pmod{10}$; $7 \cdot 6 = 42 \equiv 2 \pmod{10}$. Thus the solution to our congruence is $x \equiv 6 \pmod{10}$.

This is fine if we want to solve a single congruence $ax \equiv b \pmod{n}$. But if we want to be able to solve multiple congruences with the same a and n (i.e., $ax \equiv b_1 \pmod{n}$, $ax \equiv b_2 \pmod{n}$, $ax \equiv b_3 \pmod{n}$, etc.) there is a better way to proceed. In the notation of Corollary B.12, we first find the solution $x = a'$ of the congruence $ax \equiv 1 \pmod{n}$, and then the solution of the congruence $ax \equiv 1 \pmod{n}$ is given by $x \equiv a'b \pmod{n}$. For example, suppose we want to solve the congruences $5x \equiv b \pmod{14}$ for different values of b. We first solve $5x \equiv 1 \pmod{14}$ by trial and error, letting $x = 0, 1, 2, \ldots, 13$. We see, in order, that $5 \cdot 0 = 0 \equiv 0 \pmod{14}$; $5 \cdot 1 = 5 \equiv 5 \pmod{14}$; $5 \cdot 2 = 10 \equiv 10 \pmod{14}$; $5 \cdot 3 = 15 \equiv 1 \pmod{14}$. Thus $a' = 3$. Then the solution to $5x \equiv 2 \pmod{14}$ is $x = 3 \cdot 2 = 6 \pmod{14}$; the solution to $5x \equiv 3 \pmod{14}$ is $x = 3 \cdot 3 = 9 \pmod{14}$; the solution to $5x \equiv 4 \pmod{14}$ is $x = 3 \cdot 4 = 12 \pmod{14}$; the solution to $5x \equiv 5 \pmod{14}$ is $x = 3 \cdot 5 = 15 \equiv 1 \pmod{14}$ (this one is obvious); the solution to $5x \equiv 6 \pmod{14}$ is $x = 3 \cdot 6 = 18 \equiv 4 \pmod{14}$; etc.

Obviously this method is practical only if n is small. But in fact we have already derived a method for finding a', which works effectively for any n. This method is Euclid's algorithm, developed in Section 2.2. For example, consider the congruence

$$37x \equiv 93 \pmod{143}.$$

In Example 2.25(3) we applied Euclid's algorithm to obtain

$$1 = 37(58) + 143(-15),$$

so if $n = 143$ and $a = 37$, then $a' = 58$, and our congruence has the solution

$$\begin{aligned} x = 58 \cdot 93 = 5394 &= 143 \cdot 37 + 103 \\ &\equiv 103 \pmod{143}. \end{aligned}$$

Of course, we live in an age where 143 is a very small number for a computer. But for n large, Euclid's algorithm is much more efficient than trial and error, and for n very large it is the only practical method, even for a computer. Here is an example with larger numbers. Consider the congruence

$$876543210x \equiv 555555556 \pmod{1123456789}.$$

In Example 2.25(5) we applied Euclid's algorithm to obtain

$$1 = 1123456789(356396689) + 876543210(-456790122),$$

so if $n = 1123456789$ and $a = 876543210$, then $a' \equiv -456790122 \pmod{1123456789}$. We could simply choose $a' = -456790122$, (and this would work perfectly well), but we would like to find a value of a' between 1 and 1123456789, so we choose $a' = -456790122 + 1123456789 = 666666667$. Then our congruence has the solution

$$\begin{aligned} x = 666666667 \cdot 555555556 &= 370370370851851852 \\ &= 1123456789 \cdot 392670330 + 481481482 \\ &\equiv 481481482 \pmod{1123456789}. \end{aligned}$$

Now we consider congruences $ax \equiv b \pmod{n}$ with a and n not relatively prime. We begin by observing that in this case the conclusion of Lemma B.10 is false. For example, $6 \cdot 4 \equiv 6 \cdot 9 \pmod{10}$ but $4 \not\equiv 9 \pmod{10}$. Similarly, Theorem B.11 (and hence Corollary B.12) is false. For example, $6x \equiv 7 \pmod{10}$ does not have a solution (as, substituting $x = 0, 1, \ldots, 9$, we see that none of these work) while the congruence $6x \equiv 8 \pmod{10}$ has more than one solution $(\bmod 10)$ (as, substituting $x = 0, 1, \ldots, 9$, we see that $x = 3$ and $x = 8$ both work).

We will proceed in parallel with our previous work. In fact, setting $d = 1$ in Lemma B.13, Theorem B.14, and Corollary B.15 will recover Lemma B.10, Theorem B.11, and Corollary B.12. But, actually, we will use our previous work to prove these new results.

Lemma B.13. *Suppose that a and n have* $\gcd d$, *and set $n_1 = n/d$. If*

$$ax_1 \equiv ax_2 \pmod{n},$$

then

$$x_1 \equiv x_2 \pmod{n_1}.$$

Proof: Set $a_1 = a/d$. By definition, $ax_1 \equiv ax_2 \pmod{n}$ means that n divides $ax_1 - ax_2 = a(x_1 - x_2)$, i.e., that dn_1 divides $da_1(x_1 - x_2)$. But then n_1 divides $a_1(x_1 - x_2) = a_1x_1 - a_1x_2$. In other words, $a_1x_1 \equiv a_1x_2 \pmod{n_1}$. But, by Lemma 2.16, a_1 and n_1 are relatively prime, so we may apply Lemma B.10 to conclude that $x_1 \equiv x_2 \pmod{n_1}$. □

Theorem B.14. *Let* $d = \gcd(a, n)$ *and set* $n_1 = n/d$. *Consider the congruence*

$$ax \equiv b \pmod{n}.$$

(1) *If b is divisible by d then this congruence has a solution, and this solution is unique* $(\bmod n_1)$.

(2) *If b is not divisible by d then this congruence does not have a solution.*

Proof: Let us write $a = da_1$ and $n = dn_1$. We know that a_1 and n_1 are relatively prime by Lemma 2.16.

(1) Suppose that b is divisible by d, and write $b = db_1$. Then our original congruence is

$$da_1x \equiv db_1 \pmod{dn_1},$$

i.e., $da_1x - db_1$ is divisible by dn_1, so $da_1x - db_1 = (dn_1)k$ for some k. But then $d(a_1x - b_1) = da_1x - db_1 = (dn_1)k = d(n_1k)$, so $a_1x - b_1 = n_1k$, and hence

$$a_1x_1 \equiv b_1 \pmod{n_1}.$$

But a_1 and n_1 are relatively prime, so we may apply Theorem B.11 to conclude this congruence has a unique solution $(\bmod n_1)$.

To be precise, we have shown that any solution of the congruence $ax = b \pmod{n}$ must be a solution of the congruence $a_1x \equiv b_1 \pmod{n_1}$. But we must also show that any solution of $a_1x \equiv b_1 \pmod{n_1}$ is indeed a solution of $ax \equiv b \pmod{n}$. So suppose that $a_1x \equiv b_1 \pmod{n_1}$. Then $a_1x - b_1 = n_1k$ for some k. But then $ax - b = (da_1)x - db_1 = d(a_1x - b_1) = d(n_1k) = (dn_1)k = nk$, and so $ax \equiv b \pmod{n}$.

(2) We prove this by proving the contrapositive: if the congruence $ax \equiv b \pmod{n}$ has a solution, then b is divisible by d. So suppose this congruence has a solution, and let x_0 be any solution. Then $ax_0 \equiv b \pmod{n}$, so $ax_0 - b = nk$ for some k. Then $b = ax - nk$. Now a is divisible by d and n is divisible by d, so each term on the right-hand side is divisible by d. Hence b is divisible by d as well. □

Corollary B.15. *Let $d = \gcd(a, n)$ and consider the congruence*

$$ax \equiv b \pmod{n}.$$

Suppose that b is divisible by d. Write $a = da_1$, $n = dn_1$, and $b = db_1$. Then this congruence is equivalent to the congruence

$$a_1x \equiv b_1 \pmod{n_1},$$

which is equivalent to the congruence

$$x \equiv c_1 \pmod{n_1}$$

for some value of c_1.

Furthermore, $c_1 \equiv a_1'b_1 \pmod{n_1}$ where a_1' is such that $a_1a_1' \equiv 1 \pmod{n_1}$.

Proof: The proof of Theorem B.14 shows that $ax \equiv b \pmod{n}$ is equivalent to $a_1x \equiv b_1 \pmod{n_1}$, and then the rest of the corollary is a restatement of Corollary B.12 in this case. □

Remark B.16. Note that (in the notation of Corollary B.12 and Corollary B.15) if $aa' + nn' = d$, then, dividing by d, we have $a_1a' + n_1n' = 1$, so $a_1a' \equiv 1 \pmod{n_1}$ and so we may choose $a_1' = a'$.

Let us see how to apply this theorem and corollary. For example, consider the congruence

$$360x \equiv 324 \pmod{2268}.$$

In Example 2.25(1), we found that $\gcd(2268, 360) = 36$. Since 324 is divisible by 36 (as $324 = 36 \cdot 9$), this congruence has a solution and it is unique (mod 63) (as $63 = 2268/36$). Indeed, this congruence is equivalent to the congruence

$$10x \equiv 9 \pmod{63}.$$

We also found there that $36 = 360(19) + 2268(-3)$, from which we see that $1 = 10(19) + 63(-3)$.

Thus, we find that our original congruence has the solution

$$x = 19 \cdot 9 = 171 \equiv 45 \pmod{63}.$$

If we want to solve for $x \pmod{2268}$, we then find that x can be congruent to any one of the following 36 values (mod 2268):

$$\begin{aligned}
x &= 45 + 63 \cdot 0 &&\equiv 45 &&\pmod{2268},\\
x &= 45 + 63 \cdot 1 &&\equiv 108 &&\pmod{2268},\\
x &= 45 + 63 \cdot 2 &&\equiv 171 &&\pmod{2268},\\
&\vdots\\
x &= 45 + 63 \cdot 35 &&\equiv 2250 &&\pmod{2268}.
\end{aligned}$$

Our final topic is the Chinese Remainder Theorem, which deals with the solution of simultaneous linear congruences. The case of two simultaneous congruences is the crucial one, and we shall do that one separately first to pave the way for the general situation.

Lemma B.17. *Let n_1 and n_2 be relatively prime, and let b_1 and b_2 be arbitrary. Then the system of simultaneous congruences*

$$x \equiv b_1 \pmod{n_1},$$
$$x \equiv b_2 \pmod{n_2}$$

has a solution, and this solution is unique $\pmod{n_1 n_2}$.

Proof: Observe that for *any* value of y, $x = b_1 + n_1 y$ is a solution of the first congruence $x \equiv b_1 \pmod{n_1}$. We wish to show that we can choose y so that x is a solution of the second congruence as well. Now this congruence becomes

$$b_1 + n_1 y \equiv b_2 \pmod{n_2}$$

or, equivalently,

$$n_1 y \equiv b_2 - b_1 \pmod{n_2}.$$

But n_1 and n_2 are relatively prime, by hypothesis, so we can apply Theorem B.11 to conclude that this congruence indeed has a solution y_0. Thus, $x_0 = b_1 + n_1 y_0$ is a solution of both congruences.

Our pair of simultaneous congruences has a solution x_0, and we claim that any $x_0' \equiv x_0 \pmod{n_1 n_2}$ is also a solution. To see this, note that $x_0' \equiv x_0 \pmod{n_1 n_2}$ means $x_0' = x_0 + n_1 n_2 k$ for some k, so $x_0' - x_0 = n_1 n_2 k$. Thus we see that $x_0' - x_0$ is divisible by n_1, i.e., that $x_0' \equiv x_0 \equiv b_1 \pmod{n_1}$, and also that $x_0' - x_0$ is divisible by n_2, i.e., that $x_0' \equiv x_0 \equiv b_2 \pmod{n_2}$.

Now we must show that our solution x_0 is unique $\pmod{n_1 n_2}$. That is, we must show that if x_0' is any solution of these simultaneous congruences, then in fact we must have $x_0' \equiv x_0 \pmod{n_1 n_2}$. To see this, observe that, since x_0 and x_0' are both solutions of the system of congruences, they are both solutions of each of the individual congruences in the system. Looking at the first congruence, we see that $x_0 \equiv b_1 \pmod{n_1}$ and $x_0' \equiv b_1 \pmod{n_1}$, and so we conclude that $x_0' \equiv x_0 \pmod{n_1}$. In other words, $x_0' - x_0$ is divisible by n_1. Looking at the second congruence, and applying exactly the same logic, we conclude that $x_0' - x_0$ is divisible by n_2. Thus $x_0' - x_0$ is divisible by both n_1 and n_2. But, by hypothesis, n_1 and n_2 are relatively prime. Then we may apply Corollary 2.42 to conclude that $x_0' - x_0$ is divisible by $n_1 n_2$, i.e., that $x_0' \equiv x_0 \pmod{n_1 n_2}$. □

Recall now that integers $n_1, n_2, \ldots, n_k$ are *pairwise relatively prime* if every pair of them is relatively prime. With this definition in hand we can state our theorem.

Theorem B.18 (Chinese Remainder Theorem). *Let $n_1, n_2, \ldots, n_k$ be pairwise relatively prime and let $b_1, b_2, \ldots, b_k$ be arbitrary. Then the system of simultaneous congruences*

$$\begin{aligned} x &\equiv b_1 \pmod{n_1}, \\ x &\equiv b_2 \pmod{n_2}, \\ &\vdots \\ x &\equiv b_k \pmod{n_k} \end{aligned}$$

has a solution, and this solution is unique $\pmod{n_1 n_2 \cdots n_k}$.

Proof: We prove this by induction on k. For $k = 1$ there is nothing to prove, and for $k = 2$ we have proved this in Lemma B.17.

Now suppose the theorem is true for $k - 1$ simultaneous congruences and consider a set of k simultaneous congruences as above.

Ignore the first $k - 2$ congruences for the moment and focus on the last two:

$$\begin{aligned} x &\equiv b_{k-1} \pmod{n_{k-1}}, \\ x &\equiv b_k \pmod{n_k}. \end{aligned}$$

We can apply Lemma B.17 to conclude that this pair of congruences is equivalent to the single congruence

$$x \equiv c \pmod{n_{k-1}n_k}$$

for some c. Then, replacing these two original congruences by this new one, we obtain the equivalent system

$$\begin{aligned} x &\equiv b_1 && \pmod{n_1}, \\ x &\equiv b_2 && \pmod{n_2}, \\ &\vdots \\ x &\equiv b_{k-2} && \pmod{n_{k-2}}, \\ x &\equiv c && \pmod{n_{k-1}n_k}. \end{aligned}$$

But this is a system of $k-1$ simultaneous congruences, so by the inductive hypothesis we know it has a solution, and this solution is unique $(\bmod\, n_1 n_2 \ldots n_{k-2}(n_{k-1}n_k))$, i.e., $(\bmod\, n_1 n_2 \cdots n_k)$, and by induction we are done. □

Let us see how to apply this theorem. If the numbers are relatively small, trial and error works again. For example, consider the pair of simultaneous congruences

$$\begin{aligned} x &\equiv 8 && \pmod{9}, \\ x &\equiv 5 && \pmod{7}. \end{aligned}$$

Then we set $x = 8 + 9y$ and we try $y = 0, 1, \ldots, 6$ until we find a value that works: $y = 0$, $x = 8 \equiv 1 \pmod{7}$; $y = 1$, $x = 17 \equiv 3 \pmod{7}$; $y = 2$, $x = 26 \equiv 5 \pmod{7}$. Thus this system has the solution

$$x \equiv 26 \pmod{63}.$$

Here is another example. Consider the triple of simultaneous congruences:

$$\begin{aligned} x &\equiv 2 && \pmod{5}, \\ x &\equiv 8 && \pmod{9}, \\ x &\equiv 5 && \pmod{7}. \end{aligned}$$

We solve the last two congruences first. (Actually, it does not matter what order we do it in.) We just did that, so we use our answer and note that the original system is equivalent to

$$\begin{aligned} x &\equiv 2 && \pmod{5}, \\ x &\equiv 26 && \pmod{63}. \end{aligned}$$

We set $x = 26 + 63y$ and try $y = 0, \ldots, 4$. (Here we are reversing the order of the search to keep the number of possibilities small.) We see that $y = 0,\ x = 26 \equiv 1 \pmod 5$; $y = 1,\ x = 89 \equiv 4 \pmod 5$; $y = 2,\ x = 152 \equiv 2 \pmod 5$. Thus our original system has the solution

$$x \equiv 152 \pmod{315}.$$

If the numbers get larger, we once again have to use Euclid's algorithm. We will first give a formula for the solution of two simultaneous congruences, and then deal with the general case.

Recall that, for n_1 and n_2 relatively prime, we used Euclid's algorithm to find n_1' and n_2' with $n_1 n_1' + n_2 n_2' = 1$. Then $n_1 n_1' \equiv 1 \pmod{n_2}$ and $n_2 n_2' \equiv 1 \pmod{n_1}$.

Lemma B.19. *Let n_1 and n_2 be relatively prime and let b_1 and b_2 be arbitrary. Then the pair of simultaneous congruences*

$$\begin{aligned} x &\equiv b_1 \pmod{n_1}, \\ x &\equiv b_2 \pmod{n_2} \end{aligned}$$

has the solution

$$x \equiv n_2 n_2' b_1 + n_1 n_1' b_2 \pmod{n_1 n_2},$$

where $n_1 n_1' \equiv 1 \pmod{n_2}$ and $n_2 n_2' \equiv 1 \pmod{n_1}$.

Proof: We simply have to check that the given value of x satisfies both congruences. We see that

$$n_2 n_2' b_1 + n_1 n_1' b_2 \equiv (n_2 n_2') b_1 \equiv 1(b_1) \equiv b_1 \pmod{n_1}$$

and

$$n_2 n_2' b_1 + n_1 n_1' b_2 \equiv (n_1 n_1') b_2 \equiv 1(b_2) \equiv b_2 \pmod{n_2}.$$

Thus this value of $x \pmod{n_1 n_2}$ is a solution, and any $x' \equiv x \pmod{n_1 n_2}$ is also a solution. Furthermore, this is the only solution $\pmod{n_1 n_2}$ as we have already shown that the solution is unique $\pmod{n_1 n_2}$. □

Let us revisit one of our examples using this method.

Consider the pair of simultaneous congruences

$$\begin{aligned} x &\equiv 8 \pmod 9, \\ x &\equiv 5 \pmod 7. \end{aligned}$$

We apply Euclid's algorithm to 9 and 7,

$$\begin{aligned}9 &= 7 \cdot 1 + 2\\ 7 &= 2 \cdot 3 + 1\\ 2 &= 1 \cdot 2,\end{aligned}$$

and then solve for the gcd 1:

$$\begin{aligned}1 &= 7 + 2(-3)\\ &= 7 + (9 + 7(-1))(-3)\\ &= 9 \cdot (-3) + 7 \cdot 4.\end{aligned}$$

Then we have $n_1 = 9$, $n_1' = -3$, $n_2 = 7$, $n_2' = 4$, $b_1 = 8$, and $b_2 = 5$, so

$$x \equiv 7 \cdot 4 \cdot 8 + 9 \cdot (-3) \cdot 5 = 89 \equiv 26 \pmod{63}$$

as before.

Now let us try a larger example where the use of Euclid's algorithm is a necessity. Consider the pair of simultaneous congruences

$$\begin{aligned}x &\equiv 19 \pmod{37},\\ x &\equiv 91 \pmod{143}.\end{aligned}$$

In Example 2.25(3) we applied Euclid's algorithm to obtain

$$1 = 37(58) + 143(-15).$$

Thus, we have $n_1 = 37$, $n_1' = 58$, $n_2 = 143$, $n_2' = -15$, $b_1 = 19$, and $b_2 = 91$, so

$$x \equiv 143(-15)(19) + 37(58)(91) = 154531 \equiv 1092 \pmod{5291}$$

is the solution.

Here is the formula in the general case. Note that we will have to use Euclid's algorithm repeatedly to find the numbers $m_1', \ldots, m_k'$. Also note that if $k = 2$, comparing Theorem B.20 to Lemma B.19, we have $m_1 = n_2$ and $m_1' = n_2'$, and also $m_2 = n_1$ and $m_2' = n_1'$ (so the subscripts are reversed).

Theorem B.20 (Chinese Remainder Formula). *Let n_1, n_2, ..., n_k be pairwise relatively prime and let b_1, b_2, ..., b_k be arbitrary. Consider the system of simultaneous congruences*

$$\begin{aligned}x &\equiv b_1 \pmod{n_1},\\ x &\equiv b_2 \pmod{n_2},\\ &\vdots\\ x &\equiv b_k \pmod{n_k}.\end{aligned}$$

Set $N = n_1 n_2 \cdots n_k$. *For each* i, *let* $m_i = N/n_i$ *and let* m_i' *be such that* $m_i m_i' \equiv 1 \pmod{n_i}$. *Then this system has the solution*

$$x \equiv m_1 m_1' b_1 + m_2 m_2' b_2 + \cdots + m_k m_k' b_k \pmod{N}.$$

Proof: We must check that this value of x satisfies all of the congruences. We will simply check that it satisfies the first one. The others are the same except for the subscripts.

The key thing to note is that n_1 divides $m_2 = N/n_2 = n_1 n_3 \cdots n_k$ and similarly that n_1 divides $m_3, \ldots, m_k$. Thus each term $m_2 m_2' b_2$, $m_3 m_3' b_3$, $\ldots$, $m_k m_k' b_k$ is divisible by n_1, i.e., is congruent to 0 $\pmod{n_1}$. Then we have

$$\begin{aligned} m_1 m_1' b_1 + m_2 m_2' b_2 + \cdots + m_k m_k' b_k &\equiv m_1 m_1' b_1 + 0 + \cdots + 0 \pmod{n_1} \\ &\equiv m_1 m_1' b_1 \pmod{n_1} \\ &\equiv (m_1 m_1') b_1 \pmod{n_1} \\ &\equiv 1(b_1) \pmod{n_1} \\ &\equiv b_1 \pmod{n_1} \end{aligned}$$

as $m_1 m_1' \equiv 1 \pmod{n_1}$ by the definition of m_1'.

Thus, x is indeed a solution and, just as before, any $x' \equiv x \pmod{N}$ is also a solution, while by uniqueness $\pmod{N}$ this is the only solution. □

Let us return to our system

$$\begin{aligned} x &\equiv 2 \pmod{5}, \\ x &\equiv 8 \pmod{9}, \\ x &\equiv 5 \pmod{7} \end{aligned}$$

with this new method. Then $n_1 = 5$, $n_2 = 9$, and $n_3 = 7$, and $m_1 = 9 \cdot 7 = 63$, $m_2 = 5 \cdot 7 = 35$, and $m_3 = 5 \cdot 9 = 45$. Also, $b_1 = 2$, $b_2 = 8$, and $b_3 = 5$.

Applying Euclid's algorithm to m_1 and n_1 we find (skipping the details) $1 = 63(2) + 5(-25)$, so $m_1' = 2$. Applying Euclid's algorithm to m_2 and n_2 we find $1 = 35(-1) + 9(4)$, so $m_2' = -1$. Applying Euclid's algorithm to m_3 and n_3 we find $1 = 45(-2) + 7(13)$, so $m_3' = -2$. Then

$$x \equiv 63(2)(2) + 35(-1)(8) + 45(-2)(5) = -478 \equiv 152 \pmod{315}$$

agreeing with our previous answer.

B.3 Quadratic Congruences

In this section we wish to study quadratic congruences modulo a prime. We begin by counting the number of solutions.

Lemma B.21. *Let p be an odd prime. For any $a \not\equiv 0 \pmod{p}$, the congruence $x^2 \equiv a \pmod{p}$ either has no solutions or two solutions.*

Proof: If this congruence has no solutions, we are done.

Thus suppose it has a solution x_0, i.e., $x_0^2 \equiv a \pmod{p}$. Then also $(-x_0)^2 = x_0^2 \equiv a \pmod{p}$. Furthermore, we claim that $-x_0 \not\equiv x_0 \pmod{p}$, so we have two solutions. To see this claim, note that if $-x_0 \equiv x_0 \pmod{p}$ then $0 \equiv 2x_0 \pmod{p}$, i.e., p divides $2x_0$. Since p is an odd prime, p does not divide 2, so by Euclid's Lemma p must divide x_0, i.e., $x_0 \equiv 0 \pmod{p}$. But then, on the one hand, $x_0^2 \equiv a \pmod{p}$ by our choice of x_0, and on the other hand, $x_0^2 \equiv 0^2 = 0 \pmod{p}$, so $a \equiv 0 \pmod{p}$, contradicting our choice of a.

Now we have found two solutions, so to complete the proof we must show that there are no other solutions. A moment's reflection reveals we can prove this by showing that if y_0 is any solution (i.e., if $y_0^2 \equiv a \pmod{p}$) then either $y_0 \equiv x_0 \pmod{p}$ or $y_0 \equiv -x_0 \pmod{p}$. To see this, note we have the congruences

$$\begin{aligned} x_0^2 &\equiv a \pmod{p}, \\ y_0^2 &\equiv a \pmod{p}. \end{aligned}$$

Subtracting, we find the congruence

$$x_0^2 - y_0^2 \equiv 0 \pmod{p},$$

i.e., $x_0^2 - y_0^2$ is divisible by p. But $x_0^2 - y_0^2 = (x_0 - y_0)(x_0 + y_0)$, and now we can apply Euclid's Lemma: since p divides this product, it must divide one of the factors, so p divides $x_0 - y_0$, in which case $y_0 \equiv x_0 \pmod{p}$, or p divides $x_0 + y_0$, in which case $y_0 \equiv -x_0 \pmod{p}$. □

We now make an important definition.

Definition B.22. Let p be a prime and let $a \not\equiv 0 \pmod{p}$. Then a is a *quadratic residue* $\pmod{p}$ if $x^2 \equiv a \pmod{p}$ has a solution, and a is a *quadratic nonresidue* $\pmod{p}$ if $x^2 \equiv a \pmod{p}$ does not have a solution.

Remark B.23. We have excluded $a \equiv 0 \pmod{p}$ from our definition. But observe that if p is prime, the congruence $x^2 \equiv 0 \pmod{p}$ has the unique solution $x \equiv 0 \pmod{p}$. To see this, we observe that $x \equiv 0 \pmod{p}$ certainly *is* a solution, and it is the *only* solution, again by Euclid's Lemma: if $x^2 \equiv 0 \pmod{p}$, i.e., if p divides $x^2 = (x)(x)$, then p divides one of the factors, so p divides x, i.e., $x \equiv 0 \pmod{p}$.

p	Quadratic residues $\pmod{p}$	Quadratic nonresidues $\pmod{p}$
3	1	2
5	$1, 4$	$2, 3$
7	$1, 2, 4$	$3, 5, 6$
11	$1, 3, 4, 5, 9$	$2, 6, 7, 8, 10$
13	$1, 3, 4, 9, 10, 12$	$2, 5, 6, 7, 8, 11$
17	$1, 2, 4, 8, 9, 13, 15, 16$	$3, 5, 6, 7, 10, 11, 12, 14$
19	$1, 4, 5, 6, 7, 9, 11, 16, 17$	$2, 3, 8, 10, 12, 13, 14, 15, 18$

Table B.1. Quadratic residues and nonresidues for some small odd primes.

Remark B.24. If $p = 2$ and $a \equiv 1 \pmod{p}$, then $x^2 \equiv 1 \pmod{2}$ certainly has a solution, namely $x = 1$. Thus 1 is a quadratic residue $\pmod{2}$.

Table B.1 is a table of the first few odd primes and their quadratic residues and nonresidues. Note in Definition B.22 that whether a is a quadratic residue $\pmod{p}$ only depends on the congruence class of $a \pmod{p}$, so we only list one representative of each congruence class, and in fact we list the representative a with $1 \leq a \leq p - 1$.

Inspection of this table shows that for these values of p there are the same number of quadratic residues as there are nonresidues, namely $(p - 1)/2$ of each. Let us prove that this is true in general.

Lemma B.25. *Let p be an odd prime. Then among the $p - 1$ integers between 1 and $p - 1$ there are $(p - 1)/2$ quadratic residues $\pmod{p}$, and there are $(p - 1)/2$ quadratic nonresidues $\pmod{p}$.*

Proof: For $i = 1, 2, 3, \ldots, (p - 1)/2$, define a_i by $a_i \equiv i^2 \pmod{p}$, $1 \leq a_i \leq p - 1$. (In other words, a_i is the remainder when i^2 is divided by p.) By definition, each a_i is a quadratic residue (as $x^2 \equiv a_i \pmod{p}$ has the solution $x = i$). We claim first that $\{a_1, \ldots, a_{(p-1)/2}\}$ are distinct, showing that there are *at least* $(p-1)/2$ quadratic residues. We claim next that any quadratic residue must be one of the a_i's, showing that these are *all* the quadratic residues. Putting these two claims together we see that $(p-1)/2$ of the $p - 1$ possible values of a between 1 and $p - 1$ are quadratic residues, so the remaining $p - 1 - (p - 1)/2 = (p - 1)/2$ values of a must be quadratic nonresidues.

Let us first prove the first claim: Suppose $a_{i_1} = a_{i_2}$ for some $i_1 \neq i_2$. Then, by definition, $i_1^2 \equiv a_{i_1} = a_{i_2} \equiv i_2^2 \pmod{p}$, i.e., $i_1^2 \equiv i_2^2 \pmod{p}$. Now we argue as in the proof of Lemma B.21. This congruence gives $i_1^2 - i_2^2 \equiv 0 \pmod{p}$, so p divides $i_1^2 - i_2^2 = (i_1 - i_2)(i_1 + i_2)$, and then by Euclid's Lemma p divides $i_1 - i_2$ or p divides $i_1 + i_2$. Now i_1 and i_2 are both between

1 and $(p-1)/2$, so the smallest $i_1 - i_2$ can be is $1-(p-1)/2 = -(p-3)/2$ and the largest $i_1 - i_2$ can be is $(p-1)/2-1 = (p-3)/2$. Since $(p-3)/2 < p$, the only way p can divide $i_1 - i_2$ is if $i_1 - i_2 = 0$, i.e., if $i_1 = i_2$, which we have ruled out by assumption. On the other hand, the smallest $i_1 + i_2$ can be is $1+1 = 2$, and the largest $i_1 + i_2$ can be is $(p-1)/2+(p-1)/2 = p-1$, and p cannot divide any number in this range. Thus we cannot have $a_{i_1} = a_{i_2}$ for $i_1 \neq i_2$, so $\{a_1, \ldots, a_{(p-1)/2}\}$ are distinct.

Let us now prove the second claim: Suppose a is a quadratic residue $\pmod p$. Then by definition $a \equiv k^2 \pmod p$ for some k. Let $j \equiv k \pmod p$, $1 \le j \le p-1$. There are now two possibilities for j: either $1 \le j \le (p-1)/2$ or $(p+1)/2 \le j \le p-1$ (as we can divide the interval between 1 and $p-1$ into these two halves).

Now if $1 \le j \le (p-1)/2$, then $j = i$ for some value of i in this range, so $a \equiv j^2 = i^2 \equiv a_i \pmod p$ by the definition of a_i. On the other hand, if $(p+1)/2 \le j \le p-1$, then by simple algebra, $1 \le p-j \le (p-1)/2$ so $i = p-j$ is in range. Now if $i = p-j$, then $j = p-i$, so $a \equiv j^2 = (p-i)^2 \equiv (-i)^2 = i^2 \equiv a_i \pmod p$. Thus in either case a is one of the a_i's. □

Remark B.26. We should observe that the proof of Lemma B.25 shows us how to find the quadratic residues $\pmod p$, and hence the entries in the left column of Table B.1: Simply square the integers between 1 and $(p-1)/2$, and find integers between 1 and $p-1$ congruent to these squares. (Those integers will simply be the remainders when these squares are divided by p.) Then the entries in the right column of this table will be the remaining integers between 1 and $p-1$. For example,

$$\begin{array}{llll} p=5: & 1^2 \equiv 1 \pmod 5, & 2^2 \equiv 4 \pmod 5; & \\ p=7: & 1^2 \equiv 1 \pmod 7, & 2^2 \equiv 4 \pmod 7, & 3^2 \equiv 2 \pmod 7; \\ p=11: & 1^2 \equiv 1 \pmod{11}, & 2^2 \equiv 4 \pmod{11}, & 3^2 \equiv 9 \pmod{11}, \\ & 4^2 \equiv 5 \pmod{11}, & 5^2 \equiv 3 \pmod{11}. & \end{array}$$

There is a second, less obvious pattern that can be found in the table.

Lemma B.27. *Let p be an odd prime and let a and b be integers with $a \not\equiv 0 \pmod p$ and $b \not\equiv 0 \pmod p$. Let c be an integer with $c \equiv ab \pmod p$.*

(1) If a and b are both quadratic residues $\pmod p$, then c is a quadratic residue $\pmod p$.

(2) If a is a quadratic residue $\pmod p$, and b is a quadratic nonresidue $\pmod p$, then c is a quadratic nonresidue $\pmod p$. Similarly, if a is a quadratic nonresidue $\pmod p$, and b is a quadratic residue $\pmod p$, then c is a quadratic nonresidue $\pmod p$.

(3) *If a and b are both quadratic nonresidues* $(\bmod p)$, *then c is a quadratic residue* $(\bmod p)$.

Proof:

(1) Since a is a quadratic residue $(\bmod p)$, by definition there is an integer x with $x^2 \equiv a \pmod p$. Similarly, there is an integer y with $y^2 \equiv b \pmod p$. Set $z = xy$. Then

$$z^2 = (xy)^2 = x^2y^2 \equiv ab \equiv c \pmod p,$$

so by definition c is a quadratic residue $(\bmod p)$.

(2) Suppose a is a quadratic residue $(\bmod p)$, and let x be an integer with $x^2 \equiv a \pmod p$. We will prove the lemma in this case by contradiction. So suppose c is a quadratic residue $(\bmod p)$, and let z be an integer with $z^2 \equiv c \pmod p$. Since $a \not\equiv 0 \pmod p$, we see that $x \not\equiv 0 \pmod p$ (as p is a prime), and then we know that there is an integer w with $wx \equiv 1 \pmod p$. Then $(wx)^2 \equiv 1^2 \equiv 1 \pmod p$. But $(wx)^2 = w^2x^2 = w^2a$, so we see that $w^2a \equiv 1 \pmod p$. Now let $y = wz$. Then

$$y^2 = (wz)^2 = w^2z^2 \equiv w^2c \equiv w^2(ab) \equiv (w^2a)b \equiv (1)(b) \equiv b \pmod p,$$

so we see that b is a quadratic residue $(\bmod p)$, contradicting our hypothesis.

(3) As we have seen in Lemma B.25, there are $(p-1)/2$ quadratic residues $(\bmod p)$ between 1 and $p-1$. Call them $d_1, d_2, \ldots, d_{(p-1)/2}$. For each i between 1 and $(p-1)/2$, let e_i be an integer between 1 and $p-1$ with $e_i \equiv ad_i \pmod p$. Then no two e_i's are equal, as if $e_{i_1} = e_{i_2}$, then $ad_{i_1} \equiv ad_{i_2} \pmod p$, and since p is a prime and $a \not\equiv 0 \pmod p$, by Lemma B.10 this implies that $d_{i_1} \equiv d_{i_2} \pmod p$, and hence that $d_{i_1} = d_{i_2}$ (as they are both between 1 and $p-1$), contradicting our choice of the d_i's.

Now by part (2), each e_i is a quadratic nonresidue $(\bmod p)$ (as it is congruent to the product of a, a quadratic nonresidue $(\bmod p)$, and d_i, a quadratic residue $(\bmod p)$). But also, by Lemma B.25, there are $(p-1)/2$ quadratic nonresidues $(\bmod p)$ between 1 and $p-1$, so $e_1, \ldots, e_{(p-1)/2}$ must be all of them. Since b is assumed to be a quadratic nonresidue $(\bmod p)$, it must be congruent to one of the e_i's. So let $b \equiv e_{i_0} \equiv ad_{i_0}^2 \pmod p$. Then, setting $z = ad_{i_0}$, we see that

$$z^2 = (ad_{i_0})^2 = a^2d_{i_0}^2 = a(ad_{i_0}^2) \equiv ae_{i_0} \equiv ab \equiv c \pmod p,$$

so by definition c is a quadratic residue $(\bmod p)$. □

Corollary B.28. *Let p be an odd prime and let a be an integer with $a \not\equiv 0 \pmod{p}$. Let a' be an integer with $aa' \equiv 1 \pmod{p}$. If a is a quadratic residue $\pmod{p}$, then a' is also a quadratic residue $\pmod{p}$; while if a is a quadratic nonresidue $\pmod{p}$, then a' is also a quadratic nonresidue $\pmod{p}$.*

Proof: This follows directly from Lemma B.27, setting $b = a'$ and $c = 1$ (which is certainly a quadratic residue $\pmod{p}$). □

There is a third, far less obvious, pattern that can be found in the table. We ask when $p-1$, or, equivalently, when -1 (as $-1 \equiv p-1 \pmod{p}$), is a quadratic residue $\pmod{p}$. Looking at the table, we see that this is true for $p = 5$, 13, 17, and 29 and false for $p = 3$, 7, 11, 19, and 23. We will be able to generalize this and to prove that it is true for $p \equiv 1 \pmod{4}$ and false for $p \equiv 3 \pmod{4}$. This will take considerable work. Along the way, we will prove two famous theorems, Fermat's Little Theorem (of great importance in itself) and Wilson's theorem (whose main importance is in precisely the use we shall put it to).

Theorem B.29 (Fermat's Little Theorem). *Let p be an odd prime. For any $a \not\equiv 0 \pmod{p}$,*

$$a^{p-1} \equiv 1 \pmod{p}.$$

Proof: Let us first consider $F = (p-1)! = 1 \cdot 2 \cdot 3 \cdots (p-1)$. As p is a prime, and p does not divide any of the factors of F, we conclude (by the contrapositive of Euclid's lemma) that p does not divide F. We then conclude that the gcd of p and F is 1, i.e., that p and F are relatively prime. (The gcd g of p and F divides each of these. Since p is a prime, and g divides p, we see that $g = 1$ or $g = p$. We cannot have $g = p$, as g divides F, and we know that p does not.)

Now consider the set $\{a, 2a, 3a, \ldots, (p-1)a\}$. We claim that no two distinct elements of this set can be congruent $\pmod{p}$. For suppose $i_1 a \equiv i_2 a \pmod{p}$ with $i_1 \neq i_2$. Then, by Lemma B.10, $i_1 \equiv i_2 \pmod{p}$ (as a is relatively prime to p). Since i_1 and i_2 are both between 1 and $p-1$, this can only happen if $i_1 = i_2$, a contradiction.

Clearly no two distinct elements of $\{1, 2, \ldots, (p-1)\}$ are congruent $\pmod{p}$, and we have just shown that no two distinct elements of $\{a, 2a, \ldots, (p-1)a\}$ are congruent $\pmod{p}$. But each element in the second set is congruent to *some* element in the first set (in particular, to the remainder when it is divided by p) and these sets both have the *same* number $p-1$

of elements. Thus, we see that the congruence classes of the elements in the second set are simply a rearrangement of the elements in the first set. Hence the product of the elements in the second set is congruent to the product of the elements in the first set $(\bmod p)$. Now the product of the elements in the first set is simply $1 \cdot 2 \cdot 3 \cdot \ldots (p-1) = F$, while the product of the elements in the second set is

$$a \cdot 2a \cdot 3a \cdot \ldots \cdot (p-1)a = 1 \cdot 2 \cdot 3 \cdot \ldots \cdot (p-1) \cdot a \cdot a \cdot a \ldots \cdot a = Fa^{p-1}.$$

We thus obtain the congruence

$$Fa^{p-1} \equiv F = F \cdot 1 \pmod{p}.$$

We began the proof by showing that F and p are relatively prime, so by Lemma B.10, we conclude that

$$a^{p-1} \equiv 1 \pmod{p}$$

as claimed. □

Using this theorem, we are halfway to our goal.

Corollary B.30. *If p is a prime, $p \equiv 3 \pmod{4}$, then -1 is a quadratic nonresidue* $(\bmod p)$.

Proof: First we observe that $p \equiv 3 \pmod{4}$ means $p = 4k + 3$ for some integer k, so $(p-1)/2 = 2k+1$ is *odd.*

Now suppose -1 is a quadratic residue $(\bmod p)$, i.e., that there is an x with $x^2 \equiv -1 \pmod{p}$. Raising each side to the $(p-1)/2$ power, we have

$$(x^2)^{(p-1)/2} \equiv (-1)^{(p-1)/2} \pmod{p},$$

i.e.,

$$x^{p-1} \equiv -1 \pmod{p},$$

since, as we have just observed, $(p-1)/2$ is odd, and -1 raised to an odd power is -1. But by Fermat's Little Theorem, regardless of the value of x (which is certainly $\not\equiv 0 \pmod{p}$),

$$x^{p-1} \equiv 1 \pmod{p}.$$

Putting these two congruences together, we see that

$$1 \equiv -1 \pmod{p}, \text{ i.e., } 2 \equiv 0 \pmod{p},$$

i.e., p divides 2, which is impossible as p is an *odd* prime. □

Now we turn to the second half of our goal. In the proof of Theorem B.29, we simply needed to know $F \not\equiv 0 \pmod{p}$. Now we will need to find the exact value of $F \pmod{p}$.

Theorem B.31 (Wilson's Theorem). *Let p be a prime. Then*

$$(p-1)! \equiv -1 \pmod{p}.$$

Proof: First, we note that this is true for $p = 2$ as $1! = 1 \equiv -1 \pmod{2}$.

Henceforth we assume p is an odd prime. Then $p-1$ is even, so there are an even number $p-1$ of integers between 1 and $p-1$, i.e., in the set $T = \{1, \ldots, p-1\}$. Let us exclude the two integers 1 and $p-1$, to obtain the set

$$S = \{2, 3, \ldots, p-2\},$$

containing an even number $p-3$ of integers.

We claim that for every element z of S there is an element y of S with $y \neq z$, and $zy \equiv 1 \pmod{p}$.

We already know that for any z in T there is a y in T with $zy \equiv 1 \pmod{p}$ (Theorem B.11). In particular, since S is a subset of T, we know that for any z in S, there is a y in T with $zy \equiv 1 \pmod{p}$. We cannot have $y = 1$, as then $zy \equiv z \not\equiv 1 \pmod{p}$, and we cannot have $y = p-1$, as then $zy \equiv -z \not\equiv 1 \pmod{p}$. Thus y is in S. Finally, we cannot have $y = z$ as then $z^2 \equiv 1 \pmod{p}$. But we know that $x^2 \equiv 1 \pmod{p}$ can have at most two solutions (Lemma B.21), and it certainly has the solutions $x = \pm 1$, so it cannot have the solution $x = z$ as well.

Now let us consider the product $F_0 = 2 \cdot 3 \cdot \ldots \cdot (p-2)$ of the elements of S. Each element of S pairs up with another element of S where the product of the two is congruent to $1 \pmod{p}$, and there are $(p-3)/2$ such pairs of elements, so

$$F_0 \equiv (1)^{(p-3)/2} = 1 \pmod{p}.$$

But now

$$\begin{aligned} F &= (p-1)! \\ &= 1 \cdot 2 \cdot 3 \cdot \ldots \cdot (p-1) \\ &= 1 \cdot (2 \cdot 3 \cdot \ldots \cdot (p-2)) \cdot (p-1) \\ &= 1 \cdot F_0 \cdot (p-1) \\ &\equiv 1 \cdot 1 \cdot (p-1) \pmod{p} \\ &\equiv -1 \pmod{p}, \end{aligned}$$

as claimed. □

Remark B.32. Suppose that n is not a prime. Then n is divisible by some integer a with $1 < a \leq n-1$. Suppose $(n-1)! \equiv -1 \pmod{n}$. Then, by Proposition B.7, $(n-1)! \equiv -1 \pmod{a}$. Thus $(n-1)!$ is *not* divisible by a. On the other hand, by the very definition of $(n-1)! = 1 \cdot 2 \dots (n-1)$, it has a as a factor (as a is one of these integers), a contradiction. We thus conclude that if n is not a prime, then $(n-1)! \not\equiv -1 \pmod{n}$. Putting this together with Theorem B.31, we obtain the statement (often referred to as Wilson's theorem): an integer $n > 1$ is a prime if and only if $(n-1)! \equiv -1 (\bmod\, n)$. (On the face of it, this is a test for primes: to see if n is prime, compute $(n-1)!$, divide it by n, and see what the remainder is. However, because $(n-1)!$ grows so quickly, this is in practice a perfectly useless test.)

Finally, we arrive at the other half of our goal.

Corollary B.33. *If p is a prime, $p \equiv 1 \pmod{4}$, set $E = ((p-1)/2)!$. Then*

$$E^2 \equiv -1 \pmod{p}.$$

In particular, -1 is a quadratic residue $(\bmod\, p)$ *in this case.*

Proof: First we observe that $p \equiv 1 \pmod{4}$ means $p = 4k+1$ for some integer k, so $(p-1)/2 = 2k$ is *even.* As we have just seen, $(p-1)! \equiv -1 \pmod{p}$. Let us rewrite this:

$$\begin{aligned} (p-1)! &\equiv -1 \pmod{p} \\ 1 \cdot 2 \dots \cdot (p-1) &\equiv -1 \pmod{p}. \end{aligned}$$

Let us pair off the integer i with the integer $p-i$ on the left-hand side of this congruence, for $i = 1, 2, \dots, (p-1)/2$, to obtain

$$\begin{aligned} (1 \cdot (p-1))(2 \cdot (p-2))(3 \cdot (p-3)) \cdots (((p-1)/2)(p-(p-1)/2)) &\equiv -1 \pmod{p} \\ (1 \cdot -1)(2 \cdot -2)(3 \cdot -3) \cdots (((p-1)/2)(-(p-1)/2)) &\equiv -1 \pmod{p} \\ (-1^2)(-2^2)(-3^2) \cdots (-((p-1)/2)^2) &\equiv -1 \pmod{p}. \end{aligned}$$

Now we observe that there are $(p-1)/2$ terms on the left-hand side, so pulling out the factors of -1 we see that there are $(p-1)/2$ of them, and $(-1)^{(p-1)/2} = 1$ since $(p-1)/2$ is even, so we see

$$\begin{aligned} (-1)^{(p-1)/2}(1^2)(2^2)(3^2) \cdots (((p-1)/2)^2) &\equiv -1 \pmod{p} \\ (1^2)(2^2)(3^2) \cdots (((p-1)/2)^2) &\equiv -1 \pmod{p} \\ (1 \cdot 2 \cdot 3 \dots (p-1)/2)^2 &\equiv -1 \pmod{p}, \end{aligned}$$

as claimed. □

Corollary B.34. *Let p be an odd prime and let $a \not\equiv 0 \pmod{p}$.*

(1) If $p \equiv 1 \pmod{4}$, then either a and $-a$ are both quadratic residues $\pmod{p}$ or they are both quadratic nonresidues $\pmod{p}$.

(2) If $p \equiv 3 \pmod{4}$, then exactly one of a and $-a$ is a quadratic residue $\pmod{p}$ and the other is a quadratic nonresidue $\pmod{p}$.

Proof: This follows immediately from Lemma B.27, Corollary B.30, and Corollary B.33. □

We have just decided when -1 is a quadratic residue $\pmod{p}$ (and of course 1 always is). We also need to know when 2 and -2 are quadratic residues $\pmod{p}$. We answer that question now.

Lemma B.35. *Let x be any odd integer. Then $x^2 \equiv 1 \pmod{8}$.*

Proof: We simply compute: If $x \equiv 1 \pmod{8}$, then certainly $x^2 \equiv 1 \pmod{8}$. If $x \equiv 3 \pmod{8}$, $x^2 \equiv 9 \equiv 1 \pmod{8}$. If $x \equiv 5 \pmod{8}$, $x^2 \equiv 25 \equiv 1 \pmod{8}$. Finally, if $x \equiv 7 \pmod{8}$, $x^2 \equiv 49 \equiv 1 \pmod{8}$. □

Lemma B.36.

(1) If p is a prime, $p \equiv 3 \pmod{8}$ or $p \equiv 5 \pmod{8}$, then 2 is a quadratic nonresidue $\pmod{p}$.

(2) If p is a prime, $p \equiv 1 \pmod{8}$ or $p \equiv 7 \pmod{8}$, then 2 is a quadratic residue $\pmod{p}$.

Proof:

(1) We prove this by induction on p. If $p = 3$, then this is certainly true. Now suppose p is a prime, $p \equiv 3 \pmod{8}$ or $p \equiv 5 \pmod{8}$ and suppose (1) is true for all primes $p' < p$ with $p' \equiv 3 \pmod{8}$ or $p' \equiv 5 \pmod{8}$.

We proceed by contradiction. Suppose 2 is a quadratic residue $\pmod{p}$, and let x be an integer with $x^2 \equiv 2 \pmod{p}$. We can assume that $1 \leq x \leq p-1$, and, replacing x by $p - x$ if necessary, that x is odd. Hence $x^2 = pq+2$ for some odd integer q. By Lemma B.35, for any odd integer x, $x^2 \equiv 1 \pmod{8}$, so $pq = x^2 - 2 \equiv 1 - 2 \equiv -1 \equiv 7 \pmod{8}$. If $p \equiv 3 \pmod{8}$ this forces $q \equiv 5 \pmod{8}$, and if $p \equiv 5 \pmod{8}$ this forces $q \equiv 3 \pmod{8}$.

So in neither case can we have $q \equiv 1 \pmod{8}$ or $q \equiv 7 \pmod{8}$.

Now q has a prime factorization, of course, and if all of the prime factors of q were congruent to 1 or 7 $(\bmod 8)$, then q itself would be congruent to 1 or 7 $(\bmod 8)$, which is not the case. Hence q must be divisible by some prime p' with $p' \equiv 3$ or $5 \pmod 8$. Since $1 \le x \le p-1$, $x^2 < p$, so $q < p$, and in particular $p' < p$. But

$$x^2 = pq + 2 = p'(pq/p') + 2 \equiv 2 \pmod{p'}$$

and so 2 is a quadratic residue $(\bmod p')$, contradicting our inductive hypothesis.

(2a) First we handle the case $p \equiv 7 \pmod 8$. We proceed as in the proof of part (1) to show the following claim: if $p \equiv 5$ or $7 \pmod 8$, then -2 is a quadratic nonresidue $(\bmod p)$.

For $p = 5$ this claim is certainly true. Now suppose p is prime, $p \equiv 5 \pmod 8$ or $p \equiv 7 \pmod 8$, and suppose the claim is true for all primes $p' < p$ with $p' \equiv 5 \pmod 8$ or $p' \equiv 7 \pmod 8$.

Again we proceed by contradiction. Suppose -2 is a quadratic residue $(\bmod p)$, and let x be an integer with $x^2 \equiv -2 \pmod p$. Again we may assume $1 \le x \le p-1$ and x is odd, so $x^2 = pq - 2$ with q odd and (again using Lemma B.35) $pq = x^2 + 2 \equiv 1 + 2 = 3 \pmod 8$. If $p \equiv 5 \pmod 8$ this forces $q \equiv 7 \pmod 8$, and if $p \equiv 7 \pmod 8$ this forces $q \equiv 5 \pmod 8$. Then the prime factors of q are all less than p, but not all of them can be congruent to 1 or 3 $(\bmod 8)$, as if they were we would have $q \equiv 1$ or $3 \pmod 8$. Hence q has a factor $p' < p$ with $p' \equiv 5$ or $7 \pmod 8$, and $x^2 \equiv -2 \pmod{p'}$, and so -2 is a quadratic residue $(\bmod p')$, contradicting our inductive hypothesis.

Now suppose $p \equiv 5 \pmod 8$. Then $p \equiv 1 \pmod 4$, so by Corollary B.33, -1 is a quadratic residue $(\bmod p)$. Then we can apply Lemma B.27 to conclude that $2 = (-2)(-1)$ is a quadratic nonresidue $(\bmod p)$.

(2b) Now we handle the case $p \equiv 1 \pmod 8$. So as not to interrupt the flow of the argument, we use a result that we shall prove below.

We claim that the congruence $x^4 + 1 \equiv 0 \pmod p$ has a solution. Assuming that claim, let $x = a_0$ be a solution. Now consider $y = a_0^2 + 1$. Then

$$\begin{aligned} y^2 &= (a_0^2+1)^2 \\ &= a_0^4 + 2a_0^2 + 1 \\ &\equiv 2a_0^2 \pmod p, \end{aligned}$$

so $2a_0^2$ is a quadratic residue $(\bmod p)$. Now a_0^2 is certainly a quadratic residue $(\bmod p)$, so, by Lemma B.27, 2 is a quadratic residue $(\bmod p)$ as well. (Letting $y = a_0^2 - 1$, we similarly see that -2 is a quadratic residue $(\bmod p)$.)

Now we must prove the claim. Let $p = 8k + 1$. Then by elementary algebra $x^8 - 1$ divides $x^{8k} - 1 = x^{p-1} - 1$, and $x^4 + 1$ divides $x^8 - 1$, so $x^4 + 1$ divides $x^{p-1} - 1$. Write $x^{p-1} - 1 = (x^4 + 1)f(x)$. Note that $f(x)$ is a polynomial of degree $p - 1 - 4 = p - 5$.

Now for any $a \not\equiv 0 \pmod p$, $a^{p-1} \equiv 1 \pmod p$ by Fermat's Little Theorem, so

$$0 \equiv a^{p-1} \equiv (a^4 + 1)f(a) \pmod p,$$

i.e., p divides $(a^4 + 1)f(a)$. Since p is a prime, p divides $a^4 + 1$ or p divides $f(a)$.

We now proceed by contradiction. Suppose $x^4 + 1 \equiv 0 \pmod p$ does not have a solution. Then p never divides $a^4 + 1$, so for every value of a with $1 \leq a \leq p-1$, p divides $f(a)$. But there are $p - 1$ values of a, and $f(x)$ is a polynomial of degree $p - 5$, and this is impossible by Lemma B.39 below.□

Corollary B.37. *Let p be an odd prime.*

(1) If $p \equiv 1 \pmod 8$, then 2 and -2 are both quadratic residues $(\bmod p)$.

(2) If $p \equiv 3 \pmod 8$, then 2 is a quadratic residue $(\bmod p)$ *and -2 is a quadratic nonresidue* $(\bmod p)$.

(3) If $p \equiv 5 \pmod 8$, then 2 and -2 are both quadratic nonresidues $(\bmod p)$.

(4) If $p \equiv 7 \pmod 8$, then 2 is a quadratic nonresidue $(\bmod p)$ *and -2 is a quadratic residue* $(\bmod p)$.

Proof: This follows directly from Lemma B.36, Corollary B.33, Corollary B.30, and Lemma B.27. □

Up to now we have considered quadratic residues and nonresidues $(\bmod p)$ when p is a prime. But for our applications in Chapter 2, we also have to consider the case when the modulus is composite. The basic definition is the same: Fix an integer n. Let a be an integer relatively prime to n. Then a is a quadratic residue $(\bmod n)$ if the congruence $x^2 \equiv a \pmod n$ has a solution and a is a quadratic nonresidue $(\bmod n)$ if it does not.

Corollary B.38. *Let n be divisible by a prime $p_1 \equiv 3 \pmod 8$ and by a prime $p_2 \equiv 7 \pmod 8$. Then 2 and -2 are both quadratic nonresidues $\pmod n$.*

Proof: On the one hand, suppose 2 is a quadratic residue $\pmod n$. Then $x^2 \equiv 2 \pmod n$ for some x. But then $x^2 \equiv 2 \pmod{p_2}$, i.e., 2 is a quadratic residue $\pmod{p_2}$, which is impossible by Corollary B.37(4).

On the other hand, suppose -2 is a quadratic residue $\pmod n$. Then $x^2 \equiv -2 \pmod n$ for some x. But then $x^2 \equiv -2 \pmod{p_1}$, i.e., 2 is a quadratic residue $\pmod{p_1}$, which is impossible by Corollary B.37(2). □

We now prove the deferred result that we need.

Lemma B.39 (Lagrange). *Let $f(x) = x^n + c_{n-1}x^{n-1} + \ldots + c_1 x + c_0$ be a polynomial with integer coefficients and let p be a prime. Then there are at most n values of a with $0 \leq a \leq p-1$ such that $f(a) \equiv 0 \pmod p$.*

Proof: We proceed by induction on n. In case $n = 1$, $f(x) = x + c_0$ and $f(x) \equiv 0 \pmod p$ only for the single value of a with $a \equiv -c_0 \pmod p$, $0 \leq a \leq p-1$.

Now assume the lemma is true for every such polynomial of degree $n-1$ and let $f(x)$ have degree n. If $f(x) \equiv 0 \pmod p$ has no solutions, we are done. So suppose $f(a_1) \equiv 0 \pmod p$. By the usual division algorithm for polynomials, we can write $f(x) = (x - a_1)g(x) + f(a_1)$ with $g(x)$ a polynomial of degree $n-1$. Then $f(a_1) \equiv 0 \pmod p$.

Now $g(x)$ has degree $n-1$, so by the inductive hypothesis there are at most $n-1$ values of x with $0 \leq x \leq p-1$ and $g(x) \equiv 0 \pmod p$; call these $a_2, \ldots, a_k$ with $k \leq n$. We need to show that if $x = b$, $0 \leq b \leq p-1$, with $f(x) \equiv 0 \pmod p$, then $b = a_1, a_2, \ldots,$ or a_k. So assume $f(b) \equiv 0 \pmod p$. Then

$$0 \equiv f(b) = (b - a_1)g(b) + f(a_1) \equiv (b - a_1)g(b) \pmod p,$$

so p divides $(b - a_1)g(b)$. But p is a prime, so p divides one of the factors. Thus p divides $b - a_1$, in which case $b = a_1$ as each is between 0 and $p-1$, or p divides $g(b)$, in which case $b = a_2, a_3, \ldots,$ or a_k, and we are done. □

Our penultimate result is a famous theorem, the Law of Quadratic Reciprocity, first proved by Gauss. We shall prove this theorem in the next section.

Theorem B.40 (Law of Quadratic Reciprocity (Gauss)). *Let p and q be distinct odd primes.*

(1) If at least one of p and q is congruent to $1 \pmod 4$, *then one of the following is true:*

(a) p is a quadratic residue $(\bmod\, q)$ *and q is a quadratic residue* $(\bmod\, p)$*; or*

(b) p is a quadratic nonresidue $(\bmod\, q)$ *and q is a quadratic nonresidue* $(\bmod\, p)$.

(2) If both p and q are congruent to $3 \pmod 4$, *then one of the following is true:*

(a) p is a quadratic residue $(\bmod\, q)$ *and q is a quadratic nonresidue* $(\bmod\, p)$*; or*

(b) p is a quadratic nonresidue $(\bmod\, q)$ *and q is a quadratic residue* $(\bmod\, p)$.

We need the following corollary of this result.

Corollary B.41. *Let p be a prime congruent to* $3 \pmod 4$ *and let $q \neq p$ be an odd prime. Then q is a quadratic residue* $(\bmod\, p)$ *if and only if $-p$ is a quadratic residue* $(\bmod\, q)$.

Proof: First suppose $q \equiv 1 \pmod 4$. Then by Theorem B.40 (The Law of Quadratic Reciprocity), q is a quadratic residue $(\bmod\, p)$ if and only if p is a quadratic residue $(\bmod\, q)$. But, since $q \equiv 1 \pmod 4$, p is a quadratic residue $(\bmod\, q)$ if and only if $-p$ is a quadratic residue $(\bmod\, q)$, by Corollary B.34, yielding the result.

Next suppose $q \equiv 3 \pmod 4$. Then by Theorem B.40 (The Law of Quadratic Reciprocity), q is a quadratic residue $(\bmod\, p)$ if and only if p is a quadratic nonresidue $(\bmod\, q)$. But, since $q \equiv 3 \pmod 4$, p is a quadratic nonresidue $(\bmod\, q)$ if and only if $-p$ is a quadratic residue $(\bmod\, q)$, by Corollary B.34, again yielding the result. □

B.4 Proof of the Law of Quadratic Reciprocity

In this section we investigate quadratic residues and nonresidues, leading to the proof of the Law of Quadratic Reciprocity (Theorem B.40). It is convenient to introduce the following standard notation.

Definition B.42. Let p be a prime and let a be relatively prime to p. The *Legendre symbol*, or *quadratic residue symbol* (a/p), is defined by

$$(a/p) = 1 \text{ if } a \text{ is a quadratic residue } (\operatorname{mod} p),$$
$$(a/p) = -1 \text{ if } a \text{ is a quadratic nonresidue } (\operatorname{mod} p).$$

Lemma B.43. *Let p be a prime and let a and b be relatively prime to p. Then $(ab/p) = (a/p)(b/p)$.*

Proof: This is merely a restatement of Lemma B.27. □

We have the following criterion, due to Euler, for a to be a quadratic residue $(\operatorname{mod} p)$.

Proposition B.44 (Euler). *Let p be an odd prime and let a be relatively prime to p. Then $a^{(p-1)/2} \equiv \pm 1 \pmod{p}$. In fact, $a^{(p-1)/2} \equiv (a/p) \pmod{p}$, i.e.,*

$$a^{(p-1)/2} \equiv 1 \pmod{p} \text{ if } a \text{ is a quadratic residue } (\operatorname{mod} p),$$
$$a^{(p-1)/2} \equiv -1 \pmod{p} \text{ if } a \text{ is a quadratic nonresidue } (\operatorname{mod} p).$$

Proof: Let $b = a^{(p-1)/2}$. Then $b^2 = a^{p-1} \equiv 1 \pmod{p}$ by Theorem B.29. Thus b is a solution of the congruence $x^2 - 1 \equiv 0 \pmod{p}$. Clearly this congruence has ± 1 as solutions. By Lemma B.39, it cannot have any other solutions, so we must have $b \equiv \pm 1 \pmod{p}$.

Suppose a is a quadratic residue $(\operatorname{mod} p)$. Then $a \equiv c^2 \pmod{p}$ for some c, and so $a^{(p-1)/2} \equiv (c^2)^{(p-1)/2} = c^{p-1} \equiv 1 \pmod{p}$. This now shows that the congruence $x^{(p-1)/2} - 1 \equiv 0 \pmod{p}$ has every quadratic residue as a solution, and there are $(p-1)/2$ of these, so there cannot be any other solutions. Thus if a is a quadratic nonresidue $(\operatorname{mod} p)$, $a^{(p-1)/2} \equiv -1 \pmod{p}$. □

Corollary B.45. *Let p be an odd prime. Then $(-1/p) = 1$ if $p \equiv 1 \pmod{4}$ and $(-1/p) = -1$ if $p \equiv 3 \pmod{4}$.*

Proof: If $p \equiv 1 \pmod{4}$ then $(p-1)/2$ is even, so $(-1)^{(p-1)/2} = 1$. If $p \equiv 3 \pmod{4}$ then $(p-1)/2$ is odd, so $(-1)^{(p-1)/2} = -1$. Now apply Proposition B.44. □

Although the Law of Quadratic Reciprocity had been conjectured earlier, the first person to prove it was Gauss. This was one of Gauss's great achievements, and he returned to this theorem repeatedly, producing several proofs. The proof we present here is a variant, due to Eisenstein, of Gauss's third proof. We begin with a lemma due to Gauss, but it is convenient to establish (some nonstandard) notation first. For a prime p and an integer ℓ relatively prime to p, we let $\tilde{\ell}$ be the remainder when ℓ is divided by p. That is, $\tilde{\ell}$ is the unique integer between 1 and $p-1$ with $\tilde{\ell} \equiv \ell \pmod{p}$.

Lemma B.46 (Gauss's Lemma). *Let p be an odd prime and let a be relatively prime to p. Let $S = \{a, 2a, \ldots, ((p-1)/2)a\}$. Let*

$$n = \#(\{\ell \text{ in } S \mid \tilde{\ell} > p/2\}).$$

Then $(a/p) = (-1)^n$.

Proof: Observe that all of the elements of S are relatively prime to p and that no two of them are congruent $(\bmod\, p)$. Let $m = \#(\{\ell \text{ in } S \mid \tilde{\ell} < p/2\})$. Since, for any ℓ relatively prime to p, either $\tilde{\ell} < p/2$ or $\tilde{\ell} > p/2$, we have $m + n = p$.

Denote by $q_1, \ldots, q_m$ those remainders less than $p/2$ that appear when elements of S are divided by p, and by $r_1, \ldots, r_n$ those remainders greater than $p/2$ that appear when elements of T are divided by p. Clearly the elements of $T = \{q_1, \ldots, q_m, r_1, \ldots, r_n\}$ are all distinct. We claim that in fact the elements of $U = \{q_1, \ldots, q_m, p - r_1, \ldots, p - r_n\}$ are all distinct. To see this, suppose that $q_i = p - r_j$ for some i and j. By definition, $q_i \equiv va \pmod{p}$ and $r_j \equiv wa \pmod{p}$ for some integers v and w between 1 and $(p-1)/2$. Then

$$0 \equiv p = q_i + r_j \equiv va + wa = (v + w)a \pmod{p},$$

which is impossible as $2 \leq v + w \leq p - 1$, and so, in particular, $v + w \not\equiv 0 \pmod{p}$.

Observe that U is a set of $(p-1)/2$ distinct integers, all between 1 and $(p-1)/2$, so in fact we must have $U = \{1, \ldots, (p-1)/2\}$ (where the elements of U appear in some unpredictable but irrelevant order).

Let Π_S be the product of the elements of S, Π_T the product of the elements of T, and Π_U the product of the elements of U. Let us calculate these numbers $(\bmod\, p)$. First, from the definition of S, we see that

$$\begin{aligned}\Pi_S &= a \cdot 2a \cdots ((p-1)/2)a = a \cdots a \cdot 1 \cdot ((p-1)/2) \\ &= a^{(p-1)/2}((p-1)/2)!.\end{aligned}$$

Next, since T simply consists of the remainders when the elements of S are divided by p, we certainly have

$$\Pi_T = q_1 \cdots q_m r_1 \cdots r_n \equiv \Pi_S \pmod{p}.$$

Finally, the punch line: We calculate Π_U two ways. On the one hand, from the definition of U,

$$\begin{aligned}\Pi_U &= q_1 \cdots q_m (p - r_1) \cdots (p - r_n) \equiv q_1 \cdots q_m (-r_1) \cdots (-r_n) \\ &= (-1)^n q_1 \cdots q_m r_1 \cdots r_n \equiv (-1)^n \Pi_T \pmod{p}.\end{aligned}$$

On the other hand, we have observed that $U = \{1, \ldots, (p-1)/2\}$, so

$$\Pi_U = 1 \cdot 2 \cdots ((p-1)/2) = ((p-1)/2)!.$$

Combining these calculations, we see that

$$\begin{aligned}((p-1)/2)! = \Pi_U &\equiv (-1)^n \Pi_T \equiv (-1)^n \Pi_S \\ &\equiv (-1)^n a^{(p-1)/2} ((p-1)/2)! \pmod{p}.\end{aligned}$$

Euler's Criterion (Proposition B.44) tells us that $a^{(p-1)/2} \equiv (a/p) \pmod{p}$. Using that we then obtain

$$1 \equiv (-1)^n a^{(p-1)/2} \equiv (-1)^n (a/p) \pmod{p},$$

and so $(a/p) = (-1)^n$, as claimed. □

In the following lemma, $[\cdot]$ denotes the greatest integer function, as usual.

Lemma B.47. *Let p be an odd prime and let a be an odd integer. Let n be as in the statement of Gauss's Lemma. Then*

$$n \equiv n' = \sum_{k=1}^{(p-1)/2} [ka/p] \pmod{2}$$

and thus $(a/p) = (-1)^{n'}$.

Proof: We keep the notation of the proof of Gauss's Lemma. By definition, for each value of k, with k between 1 and $(p-1)/2$, $ka = [ka/p]p + \widetilde{ka}$. Since a and p are both odd, this gives $\widetilde{ka} \equiv [ka/p] + k \pmod{2}$. Thus

$$\sum_{k=1}^{(p-1)/2} \widetilde{ka} \equiv \sum_{k=1}^{(p-1)/2} [ka/p] + \sum_{k=1}^{(p-1)/2} k \pmod{2}.$$

Now each $\widetilde{ka}$ is either a q_i or an r_j, so

$$\begin{aligned}\sum_{k=1}^{(p-1)/2} [ka/p]\widetilde{ka} &= \sum_{i=1}^{m} q_i + \sum_{j=1}^{n} s_j \\ &\equiv \sum_{i=1}^{m} q_i - \sum_{j=1}^{n} s_j \pmod 2 \\ &= -np + \sum_{i=1}^{m} q_i + \sum_{j=1}^{n}(p - s_j) \\ &\equiv n + \sum_{i=1}^{m} q_i + \sum_{j=1}^{n}(p - s_j) \pmod 2.\end{aligned}$$

But, as we observed in the proof of Gauss's Lemma, $\{q_1, \ldots, q_m, p-s_1, \ldots, p-s_n\} = \{1, \ldots, (p-1)/2\}$. Thus

$$\sum_{k=1}^{(p-1)/2} \widetilde{ka} \equiv n + \sum_{k=1}^{(p-1)/2} k \pmod 2,$$

and comparing the two expressions for $\sum_{k=1}^{(p-1)/2} \widetilde{ka}$ yields the result. □

We now assemble these results to prove the Law of Quadratic Reciprocity, which we restate using the Legendre symbol.

Theorem B.48 (Law of Quadratic Reciprocity (Gauss)). *Let p and q be odd primes. Then*

$$(p/q)(q/p) = (-1)^{\frac{p-1}{2} \cdot \frac{q-1}{2}}.$$

Proof: Let R be the rectangle in the xy-plane whose vertices are $(0,0)$, $(p/2, 0)$, $(0, q/2)$, and $(p/2, q/2)$. Let D be the diagonal of R running from $(0,0)$ to $(p/2, q/2)$. D divides R into two triangles. Let T^+ be the triangle lying above D and T^- be the triangle lying below D. Consider the lattice points (i.e., points with integer coefficients) *strictly* inside R (i.e., in R but not on the boundary of R). Note that there are $r = ((p-1)/2)((q-1)/2)$ such lattice points. Let n^+ be the number of these lattice points that are in T^+ and let n^- be the number of these lattice points that are in T^-. Note that the line D has the equation $y = (q/p)x$, and so none of these lattice points lie on D. Hence we see that $r = n^+ + n^-$, and so

$$(-1)^{\frac{p-1}{2} \cdot \frac{q-1}{2}} = (-1)^r = (-1)^{n^+ + n^-} = (-1)^{n^+}(-1)^{n^-}.$$

We wish to count n^- and n^+. Starting on the x-axis and moving vertically up until we hit D, we see that

$$n^- = \sum_{k=1}^{(p-1)/2} n_k^-$$

where $n_k^- = \#(\{\text{lattice points in } T^- \text{ with } x - \text{coordinate } k\})$, i.e.,

$$n_k^- = \#(\{(k, y) \mid y \text{ is an integer with } 0 < y < (q/p)k\}) = [kq/p],$$

so $n^- = \sum_{k=1}^{(p-1)/2}[kq/p]$. Similarly, starting on the y-axis and moving horizontally to the right until we hit D we obtain that $n^+ = \sum_{k=1}^{(p-1)/2}[kp/q]$. But by Lemma B.47, $(q/p) = (-1)^{n^-}$ and $(p/q) = (-1)^{n^+}$, and we are done. □

Theorem B.49. *Let p be an odd prime. Then*

(1) $(-1/p) = (-1)^{(p-1)/2}$, *and*

(2) $(2/p) = (-1)^{(p^2-1)/8}$.

Proof: This is just a restatement of Corollary B.30, Lemma B.33, and Lemma B.36. □

B.5 Primitive Roots

In this section we introduce, and investigate, the notion of a primitive root. The theory of primitive roots is an important one, but for the most part is tangential to our purposes. However, it provides a very illuminating viewpoint, and an alternate proof of some of our previous results, on quadratic residues, and we carry out our investigation far enough to arrive at these.

We begin with a useful observation. Let n be any integer and suppose that a is relatively prime to n. Then, by Theorem B.11, there is an integer a' with $aa' \equiv 1 \pmod{n}$, and a' is unique $\pmod{n}$. In this situation we write $a^{-1} \equiv a' \pmod{n}$. Observe that with this definition of $a^{-1} \pmod{n}$, all the usual laws of exponents hold $\pmod{n}$.

Definition B.50. Let p be a prime and let a be relatively prime to p. The order $\operatorname{ord}_p a$ of $a \pmod{p}$ is the smallest positive integer k such that $a^k \equiv 1 \pmod{p}$.

Lemma B.51. *Let p be a prime and let a be relatively prime to p. Let ℓ be any integer. Then $a^\ell \equiv 1 \pmod{p}$ if and only if ℓ is divisible by $\operatorname{ord}_p a$.*

Proof: For convenience, let $k = \operatorname{ord}_p a$.

First, suppose that ℓ is divisible by k. Then $a^\ell = (a^k)^{\ell/k} \equiv 1^{\ell/k} = 1 \pmod{p}$. Conversely, suppose that $a^\ell \equiv 1 \pmod{p}$. By the division algorithm, we may write $\ell = kq + r$ where $r = 0$ or r is an integer with $1 \leq r < k$. Then $1 \equiv a^\ell = a^{kq+r} = a^{kq}a^r = (a^k)^q a^r \equiv 1^q a^r = a^r \pmod{p}$. But by the definition of $\operatorname{ord}_p a$, $a^r \not\equiv 1 \pmod{p}$ for $1 \leq r < \operatorname{ord}_p a$, so we must have $r = 0$. □

Corollary B.52. *Let p be a prime and let a be relatively prime to p. Then $\operatorname{ord}_p a$ divides $p - 1$.*

Proof: By Fermat's Little Theorem (Theorem B.29), $a^{p-1} \equiv 1 \pmod{p}$, so this follows immediately from Lemma B.51. □

Lemma B.53. *Let p be a prime and let a be relatively prime to p. The following are equivalent:*

(1) $\operatorname{ord}_p a = p - 1$.

(2) $\{1 = a^0, a, \ldots, a^{p-2}\}$ *are all distinct* $\pmod{p}$.

Proof: We will prove that (1) is false if and only if (2) is false.

Suppose that (1) is false. Then, by Corollary B.52, $a^k \equiv 1 \pmod{p}$ for some k with $1 \leq k < p - 1$. But then a^0 and a^k are not distinct $\pmod{p}$, and (2) is false. On the other hand, suppose that (2) is false. Then $a^i \equiv a^j \pmod{p}$ with $i \neq j$ and i and j both between 0 and $p - 2$. We may assume that $i < j$. Set $k = j - i$. Then $a^k \equiv 1 \pmod{p}$ with $1 \leq k < p - 1$, and (1) is false. □

Definition B.54. Let p be a prime. Then r is a *primitive root* $\pmod{p}$ if $\operatorname{ord}_p r = p - 1$.

Our goal is to show that every prime has a primitive root. We build up to this by proving two results that are useful in themselves.

Lemma B.55. *Let p be a prime and let $\{a_1, \ldots, a_n\}$ be relatively prime to p, and let $a \equiv a_1 \cdots a_n \pmod{p}$. Let $k_i = \operatorname{ord}_p a_i$ for each i, and let $k = k_1 \cdots k_n$. Suppose that $\{k_1, \ldots, k_n\}$ are pairwise relatively prime. Then $\operatorname{ord}_p a = k$.*

Proof: We prove this by induction on n. The case $n = 1$ is trivial. The case $n = 2$ is the crucial case. Let $j = \operatorname{ord}_p a_1a_2$. Certainly

$$(a_1a_2)^{k_1k_2} = (a_1^{k_1})^{k_2}(a_2^{k_2})^{k_1} \equiv 1^{k_2}1^{k_1} \equiv 1 \pmod{p},$$

so j divides k_1k_2.

On the other hand, $a_1^ja_2^j = (a_1a_2)^j \equiv 1 \pmod{p}$ gives $a_1^j \equiv a_2^{-j} \equiv (a_2^j)^{-1} \pmod{p}$. Then

$$1 = 1^j \equiv (a_1^{k_1})^j = (a_1^j)^{k_1} \equiv (a_2^{-j})^{k_1} \equiv (a_2^{jk_1})^{-1} \pmod{p},$$

so k_2 divides jk_1. Since we are assuming that k_1 and k_2 are relatively prime, this implies that k_2 divides j. Similarly, we see that k_1 divides jk_2, and hence that k_1 divides j. But again, since k_1 and k_2 are relatively prime, we obtain that k_1k_2 divides j. Thus $j = k_1k_2$, i.e., $\operatorname{ord}_p a_1a_2 = k_1k_2$.

Now for the general inductive step. Assume that the theorem is true for some value of $n \geq 2$ and consider $\{a_1, \ldots, a_{n+1}\}$. Set $a' = a_1 \cdots a_n$. By the inductive hypothesis, $\operatorname{ord}_p a' = k'$, where $k' = k_1 \cdots k_n$. But then, by the $n = 2$ case, $a = a'a_{n+1} = a_1 \cdots a_{n+1}$ has order $k = k'k_{n+1} = k_1 \cdots k_{n+1}$, as claimed. □

Proposition B.56. *Let p be a prime and let $b_1, \ldots, b_n$ be relatively prime to p. Let $k_i = \operatorname{ord}_p b_i$ for each i, and let $k = \operatorname{lcm}(k_1, \ldots, k_n)$. Then there is an a with $\operatorname{ord}_p a = k$.*

Proof: Factor k as $k = p_1^{e_1} \cdots p_m^{e_m}$ with each p_i a prime. Then, for each i, there is an element b_i with k_i divisible by $p_i^{e_i}$, say $k_i = p_i^{e_i}q_i$ for some q_i. Set $a_i = b_i^{q_i}$. Then $\operatorname{ord}_p a_i = p_i^{e_i}$. Let $a = a_1 \cdots a_m$. Then, by Lemma B.55, $\operatorname{ord}_p a = k$. □

Theorem B.57. *Let p be a prime. Then p has a primitive root.*

Proof: Consider $\{1, 2, \ldots, p-1\}$. Let $\operatorname{ord}_p i = k_i$ for each i. Let $k = \operatorname{lcm}(k_1, \ldots, k_{p-1})$. Then $i^k \equiv 1 \pmod{p}$ for each i. Suppose that $k < p-1$. Then the polynomial congruence $x^k - 1 \equiv 0 \pmod{p}$ has $p-1 > k$ solutions, contradicting Lemma B.39. Hence we must have $k = p-1$. But then, by Proposition B.56, there is an a with $\operatorname{ord}_p a = p-1$, and so, by definition, a is a primitive root $\pmod{p}$. □

Lemma B.58. *Let p be a prime and let r be a primitive root $\pmod{p}$. Let a be relatively prime to p. Then $a \equiv r^k \pmod{p}$ for some k, and k is well defined $(\operatorname{mod}(p-1))$.*

Proof: The $p-1$ powers of r, $\{1 = r^0, r, \ldots, r^{p-2}\}$ are all mutually incongruent $(\bmod p)$. But there are only $p-1$ congruence classes of elements a relatively prime to p (given by $\{1, 2, \ldots, p-1\}$), so any such a must be congruent to $p^k \pmod p$ for some k.

If $k' \equiv k \pmod{(p-1)}$ then $k' = k + m(p-1)$ for some m, and then $r^{k'} = r^{k+m(p-1)} = r^k(r^{p-1})^m \equiv r^k(1)^m = r^k \pmod p$. On the other hand, if $r^{k'} \equiv r^k \pmod p$, then $r^{k'-k} \equiv 1 \pmod p$ so $k' - k$ is divisible by $p-1$, i.e., $k' \equiv k \pmod{(p-1)}$. □

Corollary B.59. *Let p be an odd prime and let a be relatively prime to p. The a is a quadratic residue $(\bmod p)$ if and only if for some, and hence for any primitive root r $(\bmod p)$, $a \equiv r^k \pmod p$ with k even.*

Proof: First let us note that the statement of the corollary makes sense, as if p is an odd prime then $p-1$ is even. By Lemma B.58, k is well defined $(\bmod(p-1))$ and so is certainly well defined $(\bmod 2)$.

Suppose that $a \equiv r^k \pmod p$ with k even. Let $k = 2j$. Then $a \equiv r^k = r^{2j} = (r^j)^2 \pmod p$ and a is a quadratic residue $(\bmod p)$. Conversely, suppose that a is a quadratic residue $(\bmod p)$, so $a \equiv b^2 \pmod p$ for some b. Then $b \equiv r^j \pmod p$ for some j, so $a \equiv b^2 \equiv (r^j)^2 = r^{2j} = r^k \pmod p$ with $k = 2j$ even. □

We now collect a number of the results we have earlier proved about quadratic residues together, and prove them using primitive roots. Note that the only prior results we have used in proving Theorem B.57 are Fermat's Little Theorem (Theorem B.29) and Lagrange's result (Lemma B.39), so our development here is independent of the other results in Section B.3.

Corollary B.60. *Let p be an odd prime.*

(1) Let r be a primitive root $(\bmod p)$. For any a relatively prime to p, let $a \equiv r^k \pmod p$. Then $(a/p) = (-1)^k$.

(2) There are $(p-1)/2$ quadratic residues $(\bmod p)$ and $(p-1)/2$ quadratic nonresidues $(\bmod p)$.

(3) For any a relatively prime to p, $a^{(p-1)/2} \equiv (a/p) \pmod p$.

(4) $(-1/p) = (-1)^{(p-1)/2} = 1$ if $p \equiv 1 \pmod 4$ and $= -1$ if $p \equiv 3 \pmod 4$.

(5) For any a and b relatively prime to p, $(ab/p) = (a/p)(b/p)$.

Proof:

(1) This is just a restatement of Corollary B.59.

(2) By Corollary B.59, $\{1 = r^0, r^2, r^4, \ldots, r^{p-3}\}$ are quadratic residues $(\mathrm{mod}\, p)$ and $\{r^1, r^3, r^5, \ldots, r^{p-2}\}$ are quadratic nonresidues $(\mathrm{mod}\, p)$.

(3) Let $b \equiv a^{(p-1)/2} \pmod{p}$. Then $b^2 \equiv a^{p-1} \equiv 1 \pmod{p}$. The congruence $0 \equiv x^2 - 1 = (x-1)(x+1) \pmod{p}$ has only the two solutions $x \equiv \pm 1 \pmod{p}$, so we see $b \equiv \pm 1 \pmod{p}$. Now $a = r^k$ for some k and then $b \equiv r^{k(p-1)/2} \pmod{p}$. If a is a quadratic residue $(\mathrm{mod}\, p)$, then k is even and $k(p-1)/2$ is divisible by $p-1$ and so $b \equiv 1 \pmod{p}$. If a is a quadratic nonresidue $(\mathrm{mod}\, p)$, then k is odd and $k(p-1)/2$ is not divisible by $p-1$ and so $b \not\equiv 1 \pmod{p}$, in which case we must have $b \equiv -1 \pmod{p}$.

(4) Let $a = -1$ in part (2). Note that $(p-1)/2$ is even if $p \equiv 1 \pmod{4}$ and that $(p-1)/2$ is odd if $p \equiv 3 \pmod{4}$. Note furthermore that, for $p \equiv 1 \pmod{4}$, $-1 \equiv (r^{(p-1)/4})^2 \pmod{4}$.

(5) Let $a \equiv r^k \pmod{p}$ and $b \equiv r^\ell \pmod{p}$. Then $ab \equiv r^{k+\ell} \pmod{p}$. By part (1), $(a/p) = (-1)^k \pmod{p}$, $(b/p) = (-1)^\ell \pmod{p}$, and $(ab/p) = (-1)^{k+\ell} \pmod{p}$. But $(-1)^{k+\ell} = (-1)^k(-1)^\ell$. □

B.6 Exercises

Exercise B.1. Fix a positive integer n. Let k be any integer and consider the set $S = \{k, k+1, k+2, \ldots, k+n-1\}$. Show that, for any integer x, the congruence $x \equiv a \pmod{n}$ is valid for a exactly one of the integers in S. (Compare Corollary B.6.) A set S with this property is called a *complete system of residues* $(\mathrm{mod}\, n)$.

Exercise B.2. Fix a positive integer n. Let k be any integer relatively prime to n and consider the set $S = \{0, k, 2k, \ldots, (n-1)k\}$. Show that S is a complete system of residues $(\mathrm{mod}\, n)$.

Exercise B.3. Let n be any positive integer and let S be any complete system of residues $(\mathrm{mod}\, n)$. Show that S has n elements.

Exercise B.4. Solve each of the following congruences:

(a) $2x \equiv 11 \pmod{17}$;

(b) $7x \equiv -5 \pmod 9$;

(c) $19x \equiv 8 \pmod{14}$;

(d) $5x + 3 \equiv 7 \pmod 9$;

(e) $13x + 4 \equiv 11 \pmod{10}$;

(f) $12x + 8 \equiv 5 \pmod{17}$.

Exercise B.5. Solve each of the following congruences, if possible:

(a) $3x \equiv 6 \pmod 9$;

(b) $12x \equiv 4 \pmod{20}$;

(c) $5x \equiv 8 \pmod{10}$;

(d) $4x \equiv 12 \pmod{20}$;

(e) $10x \equiv 10 \pmod{35}$;

(f) $6x \equiv 11 \pmod{15}$.

Exercise B.6. Use Euclid's Algorithm (from Chapter 2) to solve the following congruences:

(a) $97x \equiv 125 \pmod{127}$;

(b) $323x \equiv 725 \pmod{1001}$;

(c) $12345x \equiv 54321 \pmod{41981}$;

(d) $13579x \equiv 24680 \pmod{99991}$.

Exercise B.7. Solve the following systems of simultaneous congruences:

(a) $x \equiv 4 \pmod 7, x \equiv 4 \pmod 9$;

(b) $x \equiv 3 \pmod 5, x \equiv 6 \pmod 8$;

(c) $x \equiv -4 \pmod{11}, x \equiv 7 \pmod 8$;

(d) $x \equiv 6 \pmod{11}, x \equiv 0 \pmod{13}$.

Exercise B.8. Solve the following systems of simultaneous congruences:

(a) $x \equiv 2 \pmod 3, x \equiv 3 \pmod 5, x \equiv 4 \pmod 7$;

(b) $x \equiv 3 \pmod 4, x \equiv 4 \pmod 7, x \equiv 9 \pmod{11}$.

Exercise B.9. Fix a nonzero integer n. Suppose that b is relatively prime to n. Then, by Theorem B.11, the congruence $bx \equiv a \pmod n$ holds for a unique value of $x \pmod n$. In this case let us write $x \equiv a/b$. Show that with this definition, the "usual" rules of fractions hold:

(a) $b(a/b) \equiv a$;

(b) $(a/b)(b/a) \equiv 1$;

(c) $a(b/c) \equiv (ab)/c$;

(d) $(ab)/(ac) \equiv b/c$;

(e) $(a/b)(c/d) \equiv (ac)/(bd)$;

(f) $(a/c) + (b/c) \equiv (a+b)/c$;

(g) $(a/b) + (c/d) \equiv (ad+bc)/(bd)$;

(h) $(a/b) \equiv (c/d) \Leftrightarrow ad \equiv bc$.

(In all cases we assume that the denominators are relatively prime to n.)

Exercise B.10. Fix a prime p.

(a) Show that for any c, the congruence

$$x^2 + y^2 \equiv c \pmod p$$

has a solution.

(b) More generally, suppose that $a \not\equiv 0 \pmod p$ and $b \not\equiv 0 \pmod p$. Show that for any c, the congruence

$$ax^2 + by^2 \equiv c \pmod p$$

has a solution.

Exercise B.11. Use the properties of the Legendre symbol to find the value of each of the following Legendre symbols with a minimum of hand computation:

(a) $(9767/9931)$;

(b) $(9803/9967)$;

(c) $(-210/991)$;

(d) $(-210/983)$;

(e) $(2747/2897)$;

(f) $(2747/2837)$.

(All of the odd integers above are prime, except for 2747, which is composite.)

Exercise B.12.

(a) Let p be an odd prime and let a be an integer relatively prime to p. Let $n \geq 1$ be an arbitrary integer. Show that the congruence $x^2 \equiv a \pmod{p^n}$ has a solution if and only if the congruence $x^2 \equiv a \pmod{p}$ has a solution, i.e., if and only if a is a quadratic residue $(\operatorname{mod} p)$.

(b) Let $p = 2$ and let a be an odd integer. Observe that the congruence $x^2 \equiv a \pmod{2}$ always has a solution, and that the congruence $x^2 \equiv a \pmod{4}$ has a solution if and only if $a \equiv 1 \pmod{4}$. Let $n \geq 3$ be an arbitrary integer. Show that the congruence $x^2 \equiv a \pmod{2^n}$ has a solution if and only if $a \equiv 1 \pmod{8}$.

Exercise B.13. Let m and n be relatively prime. Show that the congruence $x^2 \equiv a \pmod{mn}$ has a solution if and only if the congruences $x^2 \equiv a \pmod{m}$ and $x^2 \equiv a \pmod{n}$ have solutions.

Exercise B.14. Let $b = 2^{e_0} p_1^{e_1} \cdots p_k^{e_k}$ with $\{p_i\}$ distinct odd primes, $e_0 \geq 0$ and $e_i > 0$ for $i > 0$. Let a be relatively prime to p. Show that the congruence $x^2 \equiv a \pmod{b}$ has a solution if and only if a is a quadratic residue $(\operatorname{mod} p_i)$ for each i and

(a) if $e_0 = 0$ or 1, no further condition,

(b) if $e_0 = 2$, $a \equiv 1 \pmod{4}$, and

(c) if $e_0 \geq 3$, $a \equiv 1 \pmod{8}$.

Exercise B.15. Use Gauss's Lemma (Lemma B.46) to prove Theorem B.49, i.e., that for an odd prime p

(a) $(-1/p) = (-1)^{(p-1)/2}$, and

(b) $(2/p) = (-1)^{(p^2-1)/8}$.

Exercise B.16.

(a) Use Gauss's Lemma (Lemma B.46) directly to prove the Law of Quadratic Reciprocity for $p = 3$. (Hint: consider $q \pmod{12}$. There are four cases.)

(b) Use Gauss's Lemma (Lemma B.46) directly to prove the Law of Quadratic Reciprocity for $p = 5$. (Hint: consider $q \pmod{20}$. There are eight cases.)

Exercise B.17. Let $b > 1$ be an odd integer and let a be relatively prime to b. Let $b = p_1^{e_1} \cdots p_k^{e_k}$ be the prime factorization of b. The *Jacobi symbol* (a/b) is defined by $(a/b) = (a/p_1)^{e_1} \cdots (a/p_k)^{e_k}$ where the symbols on the right are Legendre symbols.

(a) Show that if $(a/b) = -1$, then the congruence $x^2 \equiv a \pmod{b}$ does not have a solution.

(b) Show that if $(a/b) = 1$, then the congruence $x^2 \equiv a \pmod{b}$ may or may not have a solution. (That is, give an example of each possibility.)

(c) Show that $(a_1 a_2/b) = (a_1/b)(a_2/b)$.

(d) Show that $(a/b_1 b_2) = (a/b_1)(a/b_2)$.

(e) Show that $(-1/b) = (-1)^{(b-1)/2}$.

(f) Show that $(2/b) = (-1)^{(b^2-1)/8}$.

(g) Suppose that $a > 1$ and $b > 1$ are relatively prime odd integers. Show that $(a/b)(b/a) = (-1)^{((a-1)/2)((b-1)/2)}$.

Exercise B.18.

(a) For each prime number p between 2 and 19, find all primitive roots $\pmod{p}$.

(b) For each prime number p between 23 and 47, find at least one primitive root $\pmod{p}$.

Exercise B.19. Let p be a prime and let r be a primitive root $(\bmod p)$. Let d be a positive integer.

(a) Suppose that d is relatively prime to $p-1$. Show that $a^d \equiv 1 \pmod{p}$ if and only if $a \equiv 1 \pmod{p}$.

(b) Show that r^d is a primitive root $(\bmod p)$ if and only if d is relatively prime to $p-1$.

(c) Suppose that d divides $p-1$. Show that $a^d \equiv 1 \pmod{p}$ has exactly d solutions, and that these are given by $a \equiv r^{k(p-1)/d} \pmod{p}$ for $k = 0$, $1, \ldots, d-1$.

(d) In general, let $g = \gcd(d, p-1)$. Show that $a^d \equiv 1 \pmod{p}$ has exactly g solutions, and that these are given by $a \equiv r^{k(p-1)/g} \pmod{p}$ for $k = 0$, $1, \ldots, g-1$.

Appendix C

Continuations from Chapter 2

In this appendix we continue some of the items from Chapter 2. In each case, we simply pick up where we left off, with no further ado.

C.1 Continuation of the Proof of Theorem 2.8

We now deal with the cases $D = -11, -7, 5, 6, 7, 11, 13, 17, 21$, and 29.

First, we need to see that in these cases, our previous proof does not work. For example, let us consider $D = -7$ and let us take $\gamma_0 = (2/5)\sqrt{-7}$ corresponding to the point $(0, 2/5)$ of the plane. Now this point is apparently nearest the origin, but $\|(0, 2/5)\|_{-7} = |\ 0^2 + 7(2/5)^2| = 1.12 > 1$, so its actual distance from the origin is greater than 1. As a second example, let us consider $D = 6$ and let us take $\gamma_0 = (4/9)\sqrt{6}$, corresponding to the point $(0, 4/9)$ of the plane. Now this point is apparently nearest the origin but $\|(0, 4/9)\|_6 = 1.185\ldots > 1$, so its actual distance from the origin is greater than 1.

But we can still find an appropriate point γ in each of these cases! This time γ will not be the apparently nearest point. Instead, it will be apparently further away. For example, when $D = -7$ and $\gamma_0 = (2/5)\sqrt{-7}$, we may take $\gamma = (1 + \sqrt{-7})/2$, corresponding to the point $(1/2, 1/2)$ of the plane, and then $\|\gamma_0 - \gamma\|_{-7} = \|(-1/2, -1/10)\|_{-7} = |(-1/2)^2 + 7(-1/10)^2| = 0.32 < 1$, and when $D = 6$ and $\gamma_0 = (4/9)\sqrt{6}$ we may take $\gamma = 1$, corresponding to the point $(1, 0)$, and then $\|\gamma_0 - \gamma\|_6 = \|(-1, 4/9)\|_6 = |(-1)^2 - 6(4/9)^2| = 0.185\ldots < 1$. (In fact, we may take a point apparently even further away from γ_0 that is actually closer to γ_0. For example, if we take $\gamma = 6 - 2\sqrt{6}$, corresponding to the point $(6, -2)$, we see that $\|\gamma_0 - \gamma\|_6 = |(-6)^2 - 6(-22/9)^2| = 0.148\ldots$. But all we need

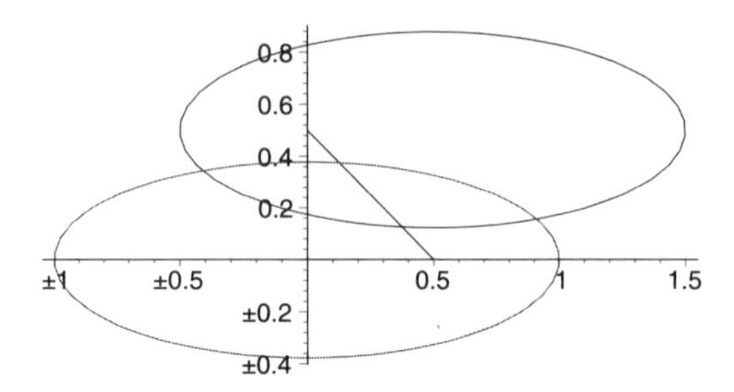

Figure C.1. The case $D = -7$.

is a point γ with $\|\gamma_0 - \gamma\|_6 < 1$, so our first choice $\gamma = 1$ will do, and we do not have to go out apparently this far.)

We have previously noted that we may restrict our attention to $\triangle_0$, the region consisting of points that are apparently closest to the origin. In fact, we will begin by considering $\triangle_0^+$, the portion of this region in the first quadrant. (In case $D \equiv 2$ or $3 \pmod 4$, this is the square with vertices $(0,0), (1/2,0), (1/2,1/2)$, and $(0,1/2)$, and in case $D \equiv 1 \pmod 4$, this is the right triangle with vertices $(0,0), (1/2,0)$, and $(0,1/2)$.)

Again we begin with $D < 0$, where we can consider ellipses.

First consider $D = -7$. Then the points (x,y) with $\|(x,y)-(0,0)\|_{-7} < 1$ are the interior of an ellipse centered at $(0,0)$, and the points (x,y) with $\|(x,y) - (1/2,1/2)\|_{-7} < 1$ are the interior of an ellipse centered at $(1/2,1/2)$, and these two ellipses completely cover $\triangle_0^+$. See Figure C.1. (You should check this. Verify that the ellipse centered at $(0,0)$ crosses the y-axis at the point $(0,\sqrt{1/7})$ and the line $x+y = 1/2$ at the point $(1/8,3/8)$, and the ellipse centered at $(1/2,1/2)$ crosses the y-axis at the point $(0,1/2-(1/2)\sqrt{3/7})$ and the line $x+y = 1/2$ at the point $(3/8,1/8)$, so the "top" of the first ellipse lies above the "bottom" of the second ellipse in $\triangle_0^+$.)

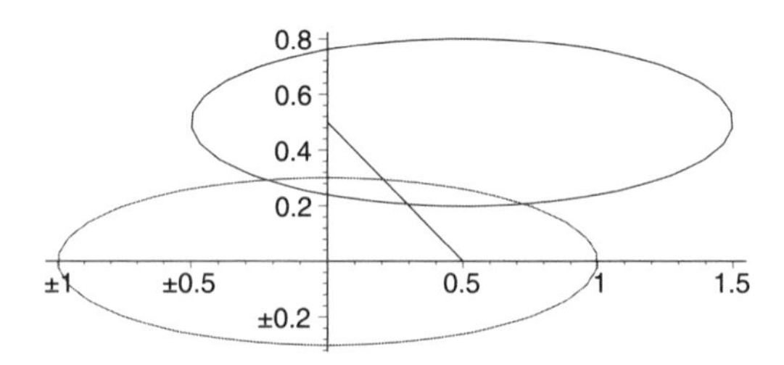

Figure C.2. The case $D = -11$.

Thus all the points in $\triangle_0^+$ are taken care of. Depending on γ_0 in $\triangle_0^+$, we may choose $\gamma = 0$ or $\gamma = 1/2 + (1/2)\sqrt{-7}$ with $\|\gamma_0 - \gamma\|_{-7} < 1$, as required. Then we may take care of all of the points in $\triangle_0$ by using the symmetry of the situation. The points in the second quadrant are taken care of by choosing $\gamma = 0$ or $\gamma = (-1/2)+(1/2)\sqrt{-7}$; the points in the third quadrant are taken care of by choosing $\gamma = 0$ or $\gamma = (-1/2)+(-1/2)\sqrt{-7}$; and the points in the fourth quadrant are taken care of by choosing $\gamma = (1/2) + (-1/2)\sqrt{-7}$.

The situation for $D = -11$ is very similar. Again the ellipses centered at the points $(0, 0)$ and $(1/2, 1/2)$, corresponding to $\gamma = 0$ and $\gamma = (1/2) + (1/2)\sqrt{-11}$, cover $\triangle_0^+$. See Figure C.2. (You should again verify this, but this time we will leave it to you to find the coordinates of the various intersections.) And again, once we have covered $\triangle_0^+$, we may use the symmetry of the situation to reflect the centers of the ellipses and cover $\triangle_0$, as required.

Now we turn to $D > 0$. Our strategy is the same, but it is more complicated to carry out, as we must consider hyperbolas rather than ellipses. The first case to consider is $D = 5$. Then the points (x, y) with $\|(x, y) - (0, 0)\|_5 < 1$ are the interior of a hyperbolic region centered at $(0, 0)$, and the points (x, y) with $\|(x, y) - (1/2, 1/2)\|_5 < 1$ are the interior of a hyperbolic region centered at $(1/2, 1/2)$, and these two regions together cover $\triangle_0^+$. See Figure C.3. (Once again you should verify this. The top curve $x^2 - 5y^2 = -1$ intersects the y-axis at $(0, \sqrt{1/5})$ and the line $x + y = 1/2$ at $((5 - \sqrt{21})/8, (\sqrt{21} - 1)/8)$, and the bottom curve $(x - 1/2)^2 - 5(y - 1/2)^2 = -1$ intersects the y-axis at the origin $(0, 0)$ and

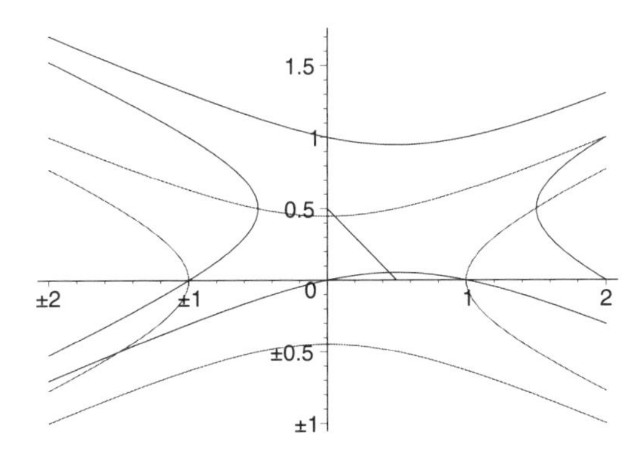

Figure C.3. The case $D = 5$.

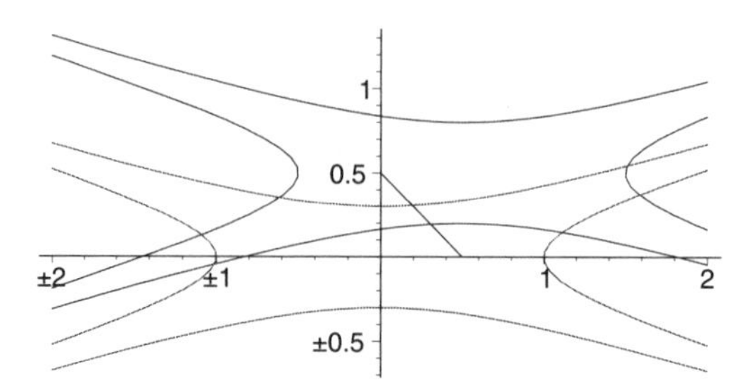

Figure C.4. The case $D = 13$.

the line $x + y = 1/2$ at $((\sqrt{21} - 1)/8, (5 - \sqrt{21})/8)$, so the picture is as shown.) Again, by symmetry, once we have covered $\triangle_0^+$ we can cover $\triangle_0$.

The situation for $D = 13$ is very similar. Again the hyperbolic regions centered at $(0, 0)$ and $(1/2, 1/2)$ cover $\triangle_0^+$. See Figure C.4. (Again you should verify this, figuring out the coordinates of the intersection points for yourself.) Again, by symmetry, once we have covered $\triangle_0^+$ we can cover $\triangle_0$.

Now we turn to $D = 6$. Here the interiors of the hyperbolic regions centered at $(0, 0)$ and $(-1, 0)$ cover all of $\triangle_0^+$. (See Figure C.5.) There is one subtlety here, however, that we need to remark on. The top curve $x^2 - 6y^2 = -1$ and the right-hand curve $(x + 1)^2 - 6y^2 = 1$ intersect at the point $(1/2, \sqrt{5/24})$ on the right-hand border of $\triangle_0^+$. But remember that we are only concerned with $\mathbf{Q}$-points (e, f), i.e., with points (e, f) with both coordinates rational numbers (as these are the points that correspond to elements $e + f\sqrt{D}$ of $\mathbf{Q}(\sqrt{D})$) and the y-coordinate $\sqrt{5/24}$ of this point is not a rational number, so this point is not a $\mathbf{Q}$-point and we have indeed covered $\triangle_0^+$, and again by symmetry we cover $\triangle_0$.

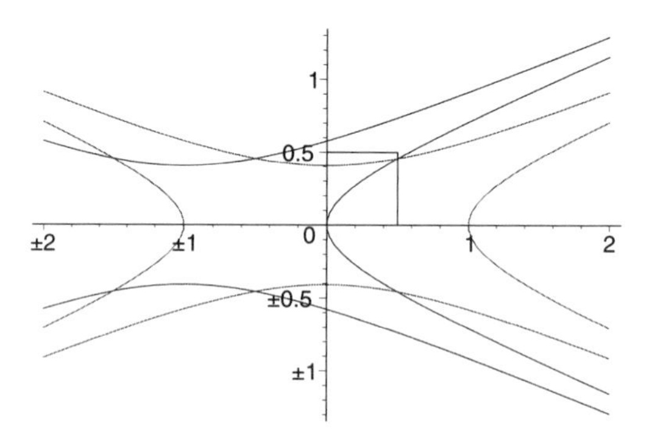

Figure C.5. The case $D = 6$.

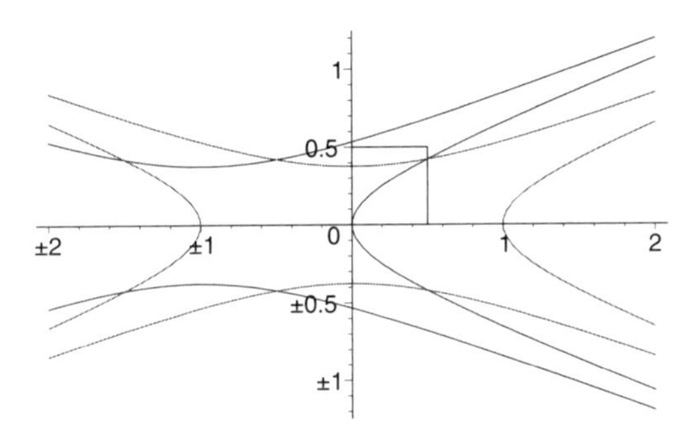

Figure C.6. The case $D = 7$.

The situation for $D = 7$ is very similar. See Figure C.6. Again, using hyperbolas centered at $(0, 0)$ and $(-1, 0)$ we cover all of $\triangle_0^+$. We have the same subtlety. The two hyperbolas intersect at the point $(1/2, \sqrt{5/28})$, but this is not a **Q**-point. Again, by symmetry, once we have covered $\triangle_0^+$ we can cover $\triangle_0$, so we are done.

To recapitulate, in the cases $D = -1$, -2, -3, 2, and 3 we could find a single value of γ so that the associated region covered $\triangle_0^+$, and in the cases $D = -7$, -11, 5, 13, 6, and 7 we could find two values of γ so that the two associated regions together covered $\triangle_0^+$. We can handle other values of D if we use more values of γ for each D. For $D = 17$ we can choose $\gamma = 0$, $(1/2)+(1/2)\sqrt{17}$, or -1. For $D = 21$ we can choose $\gamma = 0$, $(1/2)+(1/2)\sqrt{21}$, or -1. For $D = 29$ we can choose $\gamma = 0$, $(1/2) + (1/2)\sqrt{29}$, -1, or $4 + \sqrt{29}$. pagebreak For $D = 11$ we can choose $\gamma = 0$, -1, $2 + \sqrt{11}$, or $-5 - \sqrt{11}$. We leave the details for the exercises.

C.2 Continuation of Example 2.26

(2) Now we do an example with $R = \mathcal{O}(\sqrt{-7})$. This example is in principle the same as the previous example, where we had $R = \mathcal{O}(\sqrt{-1})$, but actually it is more involved, as it uses, and illustrates, the subtlety in the proof of Theorem 2.8 in the case $D = -7$. To simplify notation, and to bring out the parallel with the previous example, we will set $j = \sqrt{-7}$.

Let $\alpha_1 = 20 + 13j$ and $\alpha_2 = 5 + j$. Then

$$\begin{array}{rcl} 20 + 13j & = & (5+j)((11+3j)/2) + 3 \\ 5 + j & = & 3((3+j)/2) + ((1-j)/2) \\ 3 & = & ((1-j)/2)(1+j) + (-1) \\ 1 + j & = & (-1)(-1-j), \end{array}$$

so the gcd is -1, and then

$$\begin{array}{rcl} -1 & = & 3 + ((1-j)/2)(-1-j) \\ & = & 3 + ((5+j) + 3((-3-j)/2))(-1-j) \\ & = & (5+j)(-1-j) + 3(-1+2j) \\ & = & (5+j)(-1-j) + ((20+13j) + (5+j)((-11-3j)/2))(-1+2j) \\ & = & (20+13j)(-1+2j) + (5+j)((51-21j)/2). \end{array}$$

This certainly needs a lot of explanation. We use the language and the notation of the proof of Theorem 2.8 here.

In the first step, $(20+13j)/(5+j) = (191+45j)/32 = 5.96875+1.40625j$. This corresponds to the point $(5.96875, 1.40625)$ in the plane, and this point is *apparently* nearest the lattice point $(6, 1)$ representing $6+j$. In the usual metric on the plane, its distance from this point is $\sqrt{(-1/32)^2 + (13/32)^2} = \sqrt{170}/32 = 0.407\ldots$. But it is *not* actually nearest this point in the metric $\|\cdot\|_{-7}$ and in fact in this metric the distance between these two points is $\|(6,1) - (191/32, 45/32)\|_{-7} = \|(1/32, -13/32)\|_{-7} = |(1/32)^2 + 7(-13/32)^2| = 37/32 > 1$, so this point will not do. Instead, we choose the lattice point $(11/2, 3/2)$ representing $(11+3j)/2$. This point is apparently further away, as in the usual metric on the plane its distance from the original point is $\sqrt{(15/32)^2 + (3/32)^2} = \sqrt{234}/32 = 0.478\ldots$. But in the metric $\|\cdot\|_{-7}$ this new point *is* actually nearest to the original point. (We check that in this metric the distance between this point and the original point is $\|(11/2, 3/2) - (191/32, 45/32)\|_{-7} = \|(-15/32, 3/32)\|_{-7} = |(-15/32)^2 + 7(3/32)^2| = 288/1024 < 1$, as we expect.)

In the second step, $(5+j)/3 = (5/3)+(1/3)j = 1.66\ldots+0.33\ldots j$. This corresponds to the point $(1.66\ldots, 0.33\ldots)$ in the plane, and this point is apparently nearest the lattice point $(3/2, 1/2)$ representing $(3+j)/2$. But again what matters is distance in the metric $\|\cdot\|_{-7}$. As it happens, in this case this same point is actually nearest to the original point, and we choose it. (Again we check that the actual distance between this point and the original point is $\|(3/2, 1/2) - (5/3, 1/3)\|_{-7} = \|(-1/6, 1/6)\|_{-7} = |(-1/6)^2 + 7(1/6)^2| = 8/36 < 1$, as we expect.)

In the third step, $3/((1-j)/2) = (3/4) + (3/4)j = 0.75 + 0.75j$. This corresponds to the point $(0.75, 0.75)$ in the plane, and this point is apparently, and also actually, equidistant from the points $(0.5, 0.5)$ and $(1, 1)$, which are the apparently, and also the actually, nearest lattice points. We may choose either one of them, and we choose the point $(1, 1)$ representing $1 + j$. (Once again we check that the actual distance between this point and the original point is $\|(1,1) - (3/4,3/4)\|_{-7} = \|(1/4,1/4)\|_{-7} = |(1/4)^2 + 7(1/4)^2| = 8/16 < 1$, as we expect.)

C.3 Exercises

Exercise C.1. Fill in the details of the proof of Theorem 2.8:

(a) in the case $D = -7$;

(b) in the case $D = -11$;

(c) in the case $D = 5$;

(d) in the case $D = 13$;

(e) in the case $D = 17$;

(f) in the case $D = 21$;

(g) in the case $D = 19$;

(h) in the case $D = 11$.

Exercise C.2. Let $R = \mathcal{O}(\sqrt{-7})$. Set $j = \sqrt{-7}$. In each case, find a gcd of the following sets of elements of R, and express that gcd as a linear combination of those elements:

(a) $\{2 + j, 13 + j\}$;

(b) $\{5 + 2j, 4 + j\}$;

(c) $\{14 + j, 3 + j\}$.

Index

algebraic, 144
algebraic integer, 145
algebraic number field, 144, 183
Archimedes cattle problem, 140
ascending chain condition, 150
associates, 51

binary quadratic forms, 178

cakravala, 4, 111
Chinese Remainder Formula, 222
Chinese Remainder Theorem, 69, 218, 219
class number, 80, 157
Complete Induction, 193
complete set of representatives, 207
composition, 93, 113
congruent, 205
conjugate, 13, 101, 112
continued fractions, 177

Dedekind, 155, 180
Dedekind domain, 150, 153
degree, 144
descent, 97
Dirichlet, 185
Dirichlet's class number formula, 177
Dirichlet's Unit Theorem, 182
discriminant, 179

Egyptian fraction decomposition, 203
Eisenstein, 237
equivalence relation, 206
Euclid, 1
Euclid's algorithm, 33, 38, 40, 215, 222
Euclid's Lemma, 49–51, 58, 212
Euclidean domain, 21, 22, 30, 35, 48, 208
Euler, 96, 237

Fermat, 3, 91, 95, 96, 100, 111
Fermat's Little Theorem, 228
Fibonacci numbers, 200
field, 10
fractional ideal, 152
Fundamental Theorem of Arithmetic, 1
fundamental unit, 138

Gauss, 80, 170, 177, 235, 240
Gauss's Lemma, 238
Gaussian integers, 2, 91, 99, 101
GCD-L property, 35, 48
Gelfond-Schneider, 144
greatest common divisor (gcd), 32, 40, 51, 57

Hermite, 144

ideal, 45, 47
ideal class group, 156
ideal element, 180
inert, 105
integral domain, 7
integrally closed, 151
irreducible, 51, 52, 56, 60

Jacobi symbol, 249

Kummer, 180, 181

Lagrange, 111, 235
large period, 127

Law of Quadratic Reciprocity, 75, 235, 240
Legendre symbol, 237
Lindemann, 144

mathematical induction, 93, 192
maximal ideal, 149
minimum polynomial, 144
Minkowski, 157

noetherian, 150
Non-UFD Test, 61–63, 65–67, 71, 73, 76
norm, 13, 20, 156

order, 241
ordinary prime, 60

pairwise relatively prime, 34
Pell's equation, 4, 15, 111, 125
Pigeonhole Principle, 97, 197
prime, 51, 52, 56, 101, 102
prime factorization, 56
prime ideal, 149
primitive root, 242
principal ideal, 45
principal ideal domain (PID), 47, 48, 52, 53, 213
product of ideals, 148

quadratic field, 13, 146
quadratic nonresidue, 224
quadratic residue, 224
quadratic residue symbol, 237

ramification index, 178
ramify, 105
reduced, 112
reduced composition, 114
reduction, 112
relatively prime, 34
representation, 112
residue class field degree, 178
ring of integers, 13, 145

small period, 127
split, 105

Thue, 97, 98
trace, 13
transcendental, 144

unique factorization domain (UFD), 53, 55, 56, 67, 69, 72, 78–80, 99
unit, 11, 31, 51, 138

Well-Ordering Principle, 193
Wilson's Theorem, 230, 231